POLYURETHANE POLYMERS

POLYURETHANE POLYMERS

Blends and Interpenetrating Polymer Networks

Edited by

SABU THOMAS
Mahatma Gandhi University, Kottayam, Kerala, India

JANUSZ DATTA
Gdansk University of Technology, Gdansk, Poland

JÓZEF T. HAPONIUK
Gdansk University of Technology, Gdansk, Poland

ARUNIMA REGHUNADHAN
Mahatma Gandhi University, Kottayam, Kerala, India

ELSEVIER

Elsevier
Radarweg 29, PO Box 211, 1000 AE Amsterdam, Netherlands
The Boulevard, Langford Lane, Kidlington, Oxford OX5 1GB, United Kingdom
50 Hampshire Street, 5th Floor, Cambridge, MA 02139, United States

British Library Cataloguing-in-Publication Data
A catalogue record for this book is available from the British Library

Library of Congress Cataloging-in-Publication Data
A catalog record for this book is available from the Library of Congress

ISBN: 978-0-12-804039-3

For Information on all Elsevier publications visit our website at https://www.elsevier.com/books-and-journals

Publisher: Matthew Deans
Acquisition Editor: David Jackson
Editorial Project Manager: Sarah Jane Watson
Production Project Manager: Mohanapriyan Rajendran
Designer: Mark Rogers

Typeset by MPS Limited, Chennai, India

CONTENTS

LIST OF CONTRIBUTORS

Vayyaprontavida K. Abitha
Mahatma Gandhi University, Kottayam, Kerala, India

Jiji Abraham
Mahatma Gandhi University, Kottayam, Kerala, India

ukasz Byczyński
Rzeszow University of Technology, Rzeszow, Poland

Pawe Chmielarz
Rzeszów University of Technology, Rzeszów, Poland

Janusz Datta
Gdańsk University of Technology, Gdańsk, Poland

Omdeo Gohatre
North Maharashtra University, Jalgaon, Maharashtra, India

Sainath Ashok Jadhav
Institute of Chemical Technology, Mumbai, India

Karingamanna Jayanarayanan
Amrita University, Coimbatore, Tamil Nadu, India

Kuruvilla Joseph
Indian Institute of Space Science and Technology, Thiruvananthapuram, Kerala, India

Nandakumar Kalarikkal
Mahatma Gandhi University, Kottayam, Kerala, India

Vinayak Kamble
Institute of Chemical Technology, Mumbai, India

Krishnan Kanny
Durban University of Technology, Durban, South Africa

Cheol S. Kim
Chonbuk National University, Jeonju, Republic of Korea

Bozena Król
Rzeszów University of Technology, Rzeszów, Poland

Piotr Król
Rzeszów University of Technology, Rzeszów, Poland

Mehar A. Minnath
DGM MES College Mampad, Malappuram, Kerala, India

Raghvendra K. Mishra
Mahatma Gandhi University, Kottayam, Kerala, India; Indian Institute of Space Science and Technology, Thiruvananthapuram, Kerala, India

Saket S. Mulge
Institute of Chemical Technology, Mumbai, India

Thomas E. Müller
RWTH Aachen University, Aachen, Germany

Chan H. Park
Chonbuk National University, Jeonju, Republic of Korea

Barbara Pilch-Pitera
Rzeszow University of Technology, Rzeszów, Poland

Etathil Purushothaman
University of Calicut, Malappuram, Kerala, India

Ajay V. Rane
Durban University of Technology, Durban, South Africa

Arunima Reghunadhan
Mahatma Gandhi University, Kottayam, Kerala, India

Sarawut Rimdusit
Department of Chemical Engineering, Faculty of Engineering, Chulalongkorn University, Bangkok, Thailand

Nikesh Samarth
Institute of Chemical Technology, Mumbai, India

Arathyram R.K. Sasikala
Chonbuk National University, Jeonju, Republic of Korea

Mo Song
Loughborough University, Loughborough, United Kingdom

Jan Subocz
West Pomeranian University of Technology, Szczecin, Poland

Sabu Thomas
Mahatma Gandhi University, Kottayam, Kerala, India; Indian Institute of Space Science and Technology, Thiruvananthapuram, Kerala, India

Sunan Tiptipakorn
Department of Chemistry, Faculty of Liberal Arts and Science, Kasetsart University, Nakhon Pathom, Thailand

Yao Tong
Loughborough University, Loughborough, United Kingdom

Afeesh R. Unnithan
Chonbuk National University, Jeonju, Republic of Korea

Alpesh U. Vala
Institute of Chemical Technology, Mumbai, India

Marcin W och
Gdańsk University of Technology, Gdańsk, Poland

Marek Zenker
West Pomeranian University of Technology, Szczecin, Poland

LIST OF FIGURES

LIST OF TABLES

LIST OF SCHEMES

CHAPTER 1

Polyurethanes: Structure, Properties, Synthesis, Characterization, and Applications

Arunima Reghunadhan and Sabu Thomas
Mahatma Gandhi University, Kottayam, Kerala, India

Contents

1.1 INTRODUCTION

Polyurethanes (PUs) are a versatile class of polymers in both synthesis and applications. The superior properties make them suitable for a wide range of applications. They have very good mechanical properties, chemical resistance, and resilience. The presence of reactive functionalities in the main chain of polyurethane increases its compatibility with most polymers and so is suitable for the formation of networked structures. Polyurethane differs from normal polymers in that there are no urethane monomers. The chemical linkage existing in the polymer is denoted as urethane (−NHCOO−) and is the most reactive moiety in the polyurethane. Dependent on the starting materials, the end product differs in its nature, properties, and applications. From its first synthesis in 1937 by the genius German chemist Prof. Otto Bayer through a normal polyaddition reaction, it has been the most demanded plastic all over the world. Otto Bayer is recognized as the father of polyurethane. The polyurethane

Polyurethane Polymers: Blends and Interpenetrating Polymer Networks
DOI: http://dx.doi.org/10.1016/B978-0-12-804039-3.00001-4

synthesis chemistry is easy in its basics, but becomes complex in the laboratory setting. The variety in the starting material and the additives and catalysts makes the synthetic route complex.

Polyurethanes are synthesized by the addition of alcohols and isocyanates. Diols or polyols are the alcoholic part and the diisocyanates are reacted with them. As a good variety of both these components are available easily, a large variety of polyurethanes can be obtained. PU can be soft, as in the case of foams, or hard, as in the case of automotive parts. Polyurethane exists in may forms in our lives, from clothes to footwear, from your bed to the house's roofing material, from cars through to the construction field [1]. They are in every part of our daily life.

1.2 TYPES OF POLYURETHANES, THEIR PROPERTIES AND APPLICATIONS

Different types of polyurethanes are available according to their synthetic route and applications. Commonly available polyurethanes are PU foams. Polyurethane foams may be low density flexible foams or low density rigid foams [2]. Low density flexible foams are used for furniture, truck seating, and cushions. The properties required for the usage in such industries are the flexibility, resilience, high mechanical strength, and durability. Rigid foams have high mechanical strength, low heat conduction, low moisture absorption, and low density. They are used as core materials in the manufacturing of thermal insulations in refrigerators and building insulation panels.

Another type is PU elastomers. They have low cost, resistance to high loads, high compression strength, and they can be easily formulated and colored. Polyurethane elastomers are considered as substitutes for plastics and rubbers. They have high abrasion resistance, resistance to solvents and chemicals, high impact strength, and low moisture uptake. These superior mechanical properties make them suitable for the manufacturing of packaging materials, in the health product industry, and the printing industry. They are also used in wheels, rollers, pulleys, shock absorbers, and bushings. Low density elastomers are used in the footwear industry for making shoe soles.

The most important class of polyurethanes are thermoplastic polyurethane. This type of material varies in its starting material and properties. Thermoplastic polyurethanes can be synthesized according to the required properties. They are similar to plastics and they have good elasticity and transparency. Thermoplastic polyurethanes (TPUs) are segmented polyurethanes with alternating hard and soft blocks. The length of the soft and hard blocks can be controlled during the synthesis. They are widely used in the medical equipment, castor wheels, power tools, sports, and the footwear industry. TPUs have high abrasion strength, high chemical and solvent resistance, and resistance to oil and grease.

1.3 SYNTHESIS AND CHEMISTRY OF POLYURETHANES

Polyurethanes are reactive polymers similar to epoxies and polyesters. Polyurethanes are obtained from the reaction between diols and diisocyanates. Diols are systems with two −OH groups and an isocyanate contains −NCO group as the end point. PUs can be synthesized in a single step (the one shot method) by direct mixing of the components or in the prepolymer method. The prepolymer method involves the addition of diol and isocyanate to form the polyurethane prepolymer with −NCO as end groups. This prepolymer is then extended with chain extenders to form polyurethanes. The most commonly used isocyanates for the synthesis of polyurethane are diphenylmethane diisocyanate (MDI) or toluene diisocyanate (TDI), hexamethylene diisocyanate (HDI) or isophorone diisocyantate (IPDI). Aromatic isocyanates are more reactive than the aliphatic varieties. Polyols used for the synthesis of PUs can be polyether polyols and polyester polyols. The polyether polyols are ethylene glycol, propylene glycol, poly-ethylene propylene oxide and the polyester polyols are condensed products of adipic acid or phthalic acid with 1,4-butane diol, or 1,6-hexane diol. The formation involves a rearrangement step in which the hydrogen from the alcohol to the nitrogen of isocyanate and the oxygen links to the carbon. The simple reaction between −OH and −NCO is represented in the following step

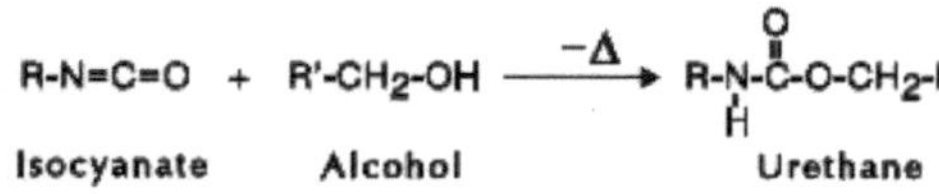

Many side products like urea, allophanate, uric acid, and biuret are possible from the above reaction. A simple representation of the side reactions occurring during the formation of urethane are given in Fig. 1.1.

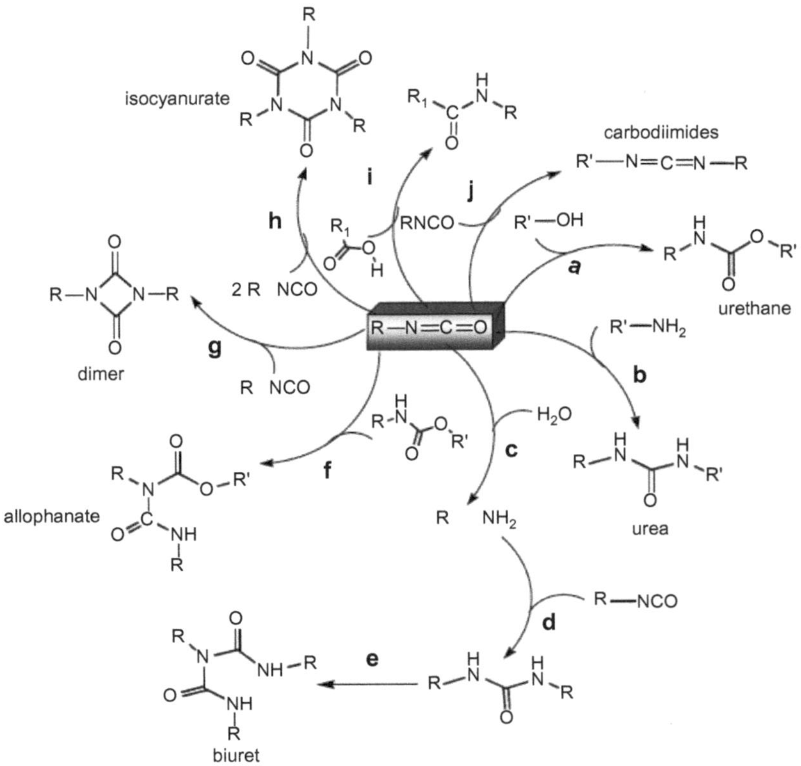

Figure 1.1 Schematic representation of the side reactions during the reaction between diols and isocyanates.

Polyurethanes normally have a segmented structure. A hard segment derived from the isocyanates and a soft segment from the diols are normally formed. The length and molecular weight of the monomer molecule has a direct influence on the nature of the polyurethane formed. The low molecular weight monomers can produce highly concentrated urethanes and so they will be hard and stiff. On the other hand, long and high weight monomers when combined to form PUs result in a high amount of soft segments. As the concentration of the soft segment increases, the resultant material becomes flexible and elastomeric. The main factor determining the product nature is the NCO–OH ratio. Side reaction inhibitors and the catalyst are unavoidable in the PU synthesis.

A schematic representation of the PU synthesis is given below in Scheme 1.1.

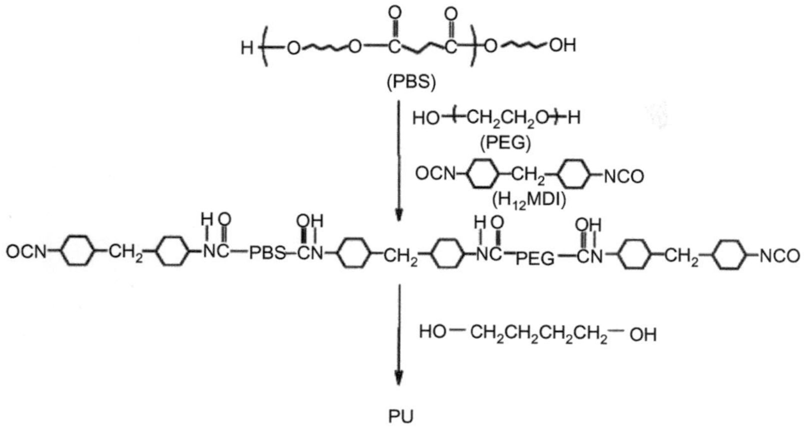

Scheme 1.1 General synthetic route of polyurethane.

One of the special features of polyurethane is that they can be derived from natural sources such as vegetable oils. The synthesis of PU from renewable resources is a method adopted by many researchers. This is a greener approach for the synthesis to be applied for PUs [3,4]. Different types of vegetable oils will give different polyols that can be used for the synthesis of PUs, examples include caster oil, soybean oil, and sunflower oil as source materials. Yagneh and Talemi synthesized and characterized the polyurethanes from caster oil. They found that the synthesized materials are suitable for using as biodegradable implants. They followed the prepolymer method and derived polyol from the caster oil was reacted with isocyanate. The synthetic route is given in Scheme 1.2 [5].

Lignin can be used as a raw material for the synthesis of polyurethanes. Rigid foams, elastomers and sealants can be obtained from the lignin based synthesis. Bio based PU synthesis is economic and promotes green approach of chemistry [6].

Kessler and coworkers [36] compared the −OH values of different vegetable oils and synthesized polyurethane using the polyols derived accordingly. They utilized olive, canola, grape seed, linseed and castor oils. The method was proposed to be solvent- and catalyst free. First they oxidized the triglyceride oils into epoxidized vegetable oil with formic acid and peroxide. Then ring opening polymerization was undertaken with caster oil (Scheme 1.3).

Polyurethanes involve toxic isocyanates and phosgene. Avoiding the toxicity in PU synthesis is an important concern for the synthetic

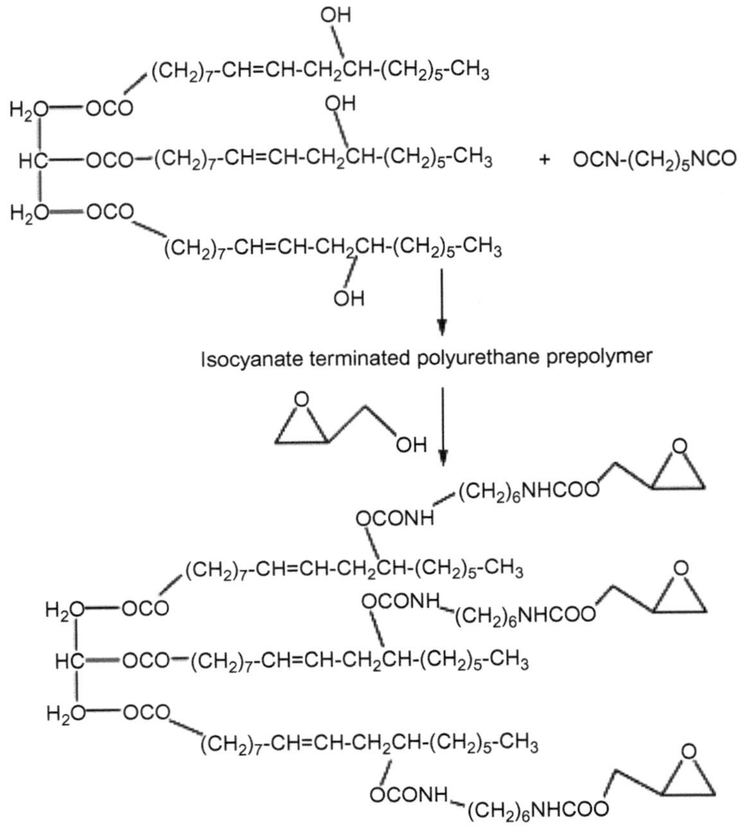

Scheme 1.2 Synthesis of polyurethane from caster oil.

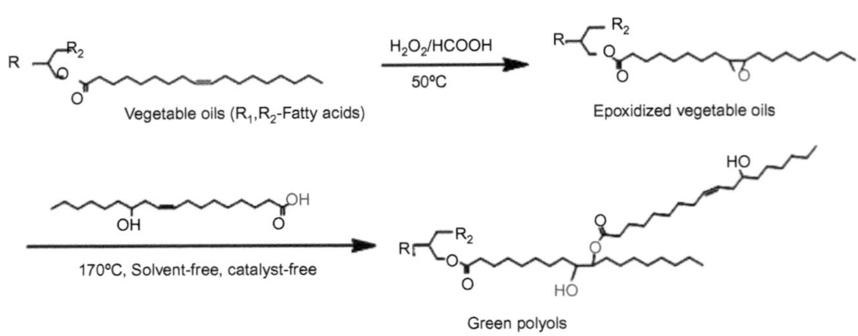

Scheme 1.3 Preparation of vegetable oil based polyols.

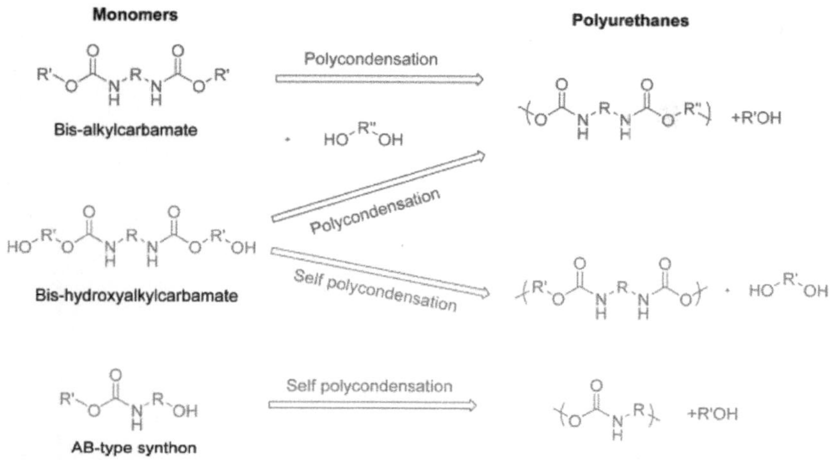

Scheme 1.4 Three main steps of PU synthesis by a trans urethane polymerization.

chemistry. This is the role of nonisocyante routes which is applied nowadays to make polyurethanes. Nonisocyante PUs can be synthesized from amines and diols or using cyclo carbonates [7]. In their review work, Cramail et al. explained the normal synthesis of PU and the nonisocyanate method known as transurethanization. Here the alcohol is mixed with carbamate and results in polyurethanes with similar properties that are obtained via the isocyanate method [8].

The nonisocyante polyurethanes possess higher chemical resistance and lower permeability than the conventional polyurethanes [9,10] (Scheme 1.4).

1.4 STRUCTURAL ANALYSIS OF POLYURETHANES

Thermoplastic polyurethanes have a segmented structure. Segmented polyurethanes are also considered as block copolymers of type (AB)n with alternating soft (A) and hard (B) segments. The specific segments have specific applications and properties. The detrimental factor in the PU applications are the length of each segment. The hard segments (HS) appear as dispersed in a matrix of soft segments (SS). A schematic representation of the segmented polyurethane is given in Scheme 1.5 [11].

The hard segments are normally aligned and induce crystallinity to the segmented PUs. Hence, some PUs are semicrystalline and have improved properties. The crystalline structure is observed at lower temperatures. At

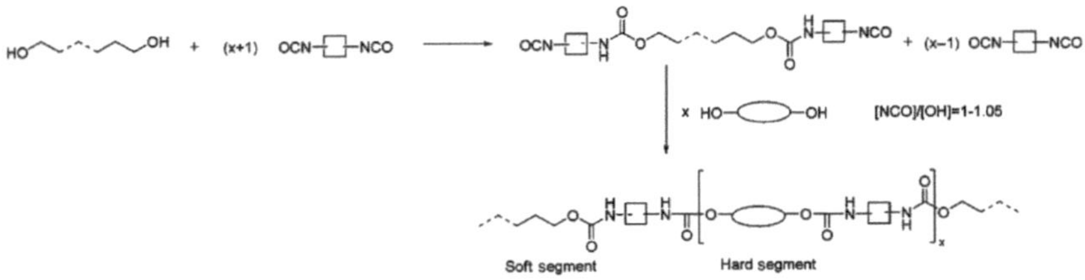

Scheme 1.5 Segmented polyurethane.

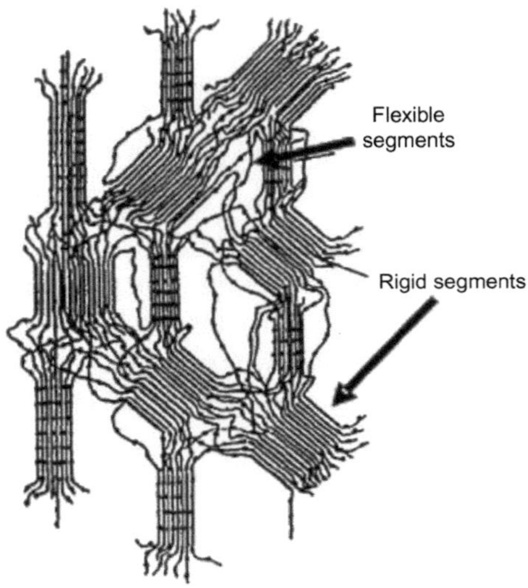

Figure 1.2 The segmented structure of polyurethane.

high temperatures, the crystallites melt and thus the structural changes with the application of temperature can be studied by Differential Scanning Calorimetry (DSC) and X-Ray diffraction (XRD) [12,13] (see Fig. 1.2).

1.5 PHASE SEPARATION AND MISCIBILITY

The segmented polyurethanes are of importance due their special behavior of phase separation. The two segments, hard and soft, have high interphase adhesion, but they do not fail at the interphase. Polyurethanes will tend to phase separate because of the incompatibility between the soft and hard segments [14,15]. The hard segments normally act as crosslinks in the matrix [16]. The enhanced properties of PUs are considered as a result of this phase separation. Petrovic and Ferguson [17] explained the phenomena of phase separation in polyurethanes and the thermodynamics of miscibility and phase separation. They considered the PU system to be similar to a polymer blend system, where the resultant blend is an average of the properties of the two polymers. In the case of PUs, the property of the material is the added sum of the hard segments and the soft segments [17]. Phase separation phenomena in segmented PUs depends mainly on the length of the hard and soft segments. If the soft segments are too long then they will phase separate and if they are too short, then the possibility of phase separation is decreased [18]. Phase separation process in PUs are affected by the hydrogen bonding interactions between urethane linkages, the reaction conditions during synthesis, the length of hard domains, and the temperature. The phase separation process is determined by the Differential Scanning Calorimentry [19], small angles X-ray scattering (SAXS) [20] and morphological analysis. Li and coworkers considerd the phase separation in segmented polyurethanes with the help of the SAXS technique. They concluded that the kinetics factor could control whether it was phase mixed or phase separated [21,22]. The DSC measurements are based on the fact that the two segments have different T_g and the variation in the glass transition temperature is useful in the study of phase separation in PUs. The hydrogen bonding interactions inside the polyurethane chains also plays a role in the phase separation process [23]. The variation in the temperature is a governing factor in the phase separation process. The separated structure at room temperature is extremely different from that at high temperature. A schematic representation of the separated structure at different temperatures is given in Fig. 1.3.

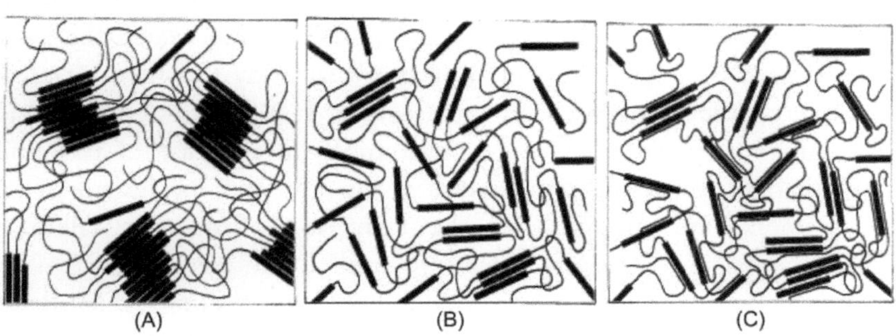

Figure 1.3 Schematic representations of polyurethane structures at various temperatures: (A) room temperature phase-separated morphology; (B) most disassociated structures at high temperature; (C) dispersed phase "trapped" at extremely low temperature.

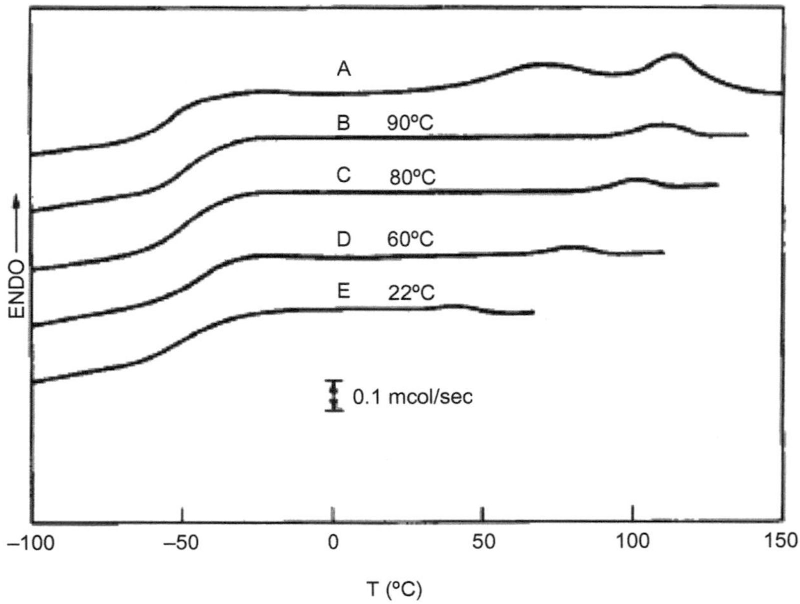

Figure 1.4 DSC thermograms of PU (A) as received and (B–E) annealed at 190°C for different time periods.

Hsu and coworkers studied the microstructure of polyurethane using FTIR technique [24]. They analyzed the variation in the absorbance peak of different functionalities and the extent of hydrogen bonding was studied quantitatively. From their studies they proved that the spectroscopy data can be correlated to macroscopic structural properties, mainly the phase separation kinetics and the mixing process [25]. In another study, Kwei proposed that the glass transition temperature of the soft segment has a logarithmic variation during phase separation. He heated the polyurethane to 190°C at different time intervals to study the effect of phase separation on the T_g of soft segments (see Fig. 1.4) [26]. The microphase separation or segregation is more pronounced in the case of polyether polyurethanes. When the temperature is too low and the length of the hard segments are enough then the formation of microcrystallites is possible.

Morphological analysis such as Atomic Force Microscopy (AFM) determines the type of phase separation occurring in the polyurethane chains. Petrovic and coworkers suggested that the variation in soft and hard segments will vary the morphology of the end result. They chose thermoplastic PUs with 50% soft segment and 70% soft segment concentration. The

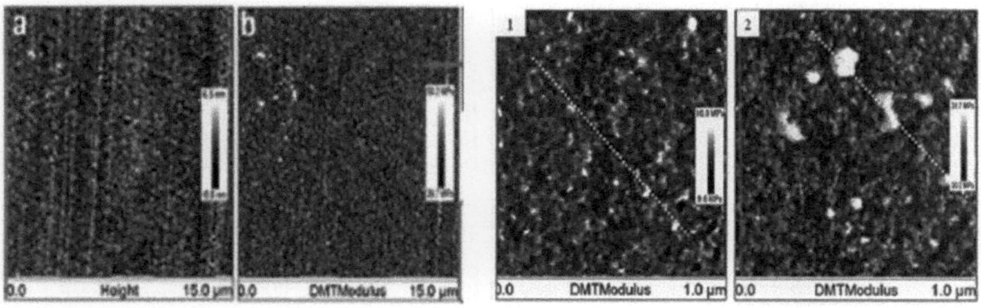

Figure 1.5 AFM images of phase separated structure of TPU 50 (left) and TPU 70 (right).

TPU with 50% soft segments showed a cocontinuous structure whereas that with 70% SS showed globules of micro-sized domains inside the matrix [27] (see Fig. 1.5).

1.6 CHARACTERIZATION OF POLYURETHANES

The polyurethane synthesis involves a change in functional groups. The −OH groups and the −NCO group react to form a −NHCOO− group. After the formation of the urethane prepolymer, polymerization takes place in the presence of a catalyst and chain extender. Spectroscopic techniques are useful in the detection of the products. The side reactions and the byproducts can be identified using infrared and magnetic resonance spectroscopy. A clear pathway of the reaction is explained with the help of these techniques. The extent of hydrogen bonding between different groups in polyurethanes are normally studied quantitatively using Fourier-transform infrared (FTIR). Possible formation of the hydrogen bonding in between the polyurethane chains is given in Scheme 1.6.

In a polyurethane chain, both proton-donating groups like N−H and proton-accepting functionalities like C = O, C−O−C are present. Upon the formation of hydrogen bonds, these functionalities show a change in the fundamental frequency. This change is related to the strength of the hydrogen bond formed [28].

The most common functional frequencies in polyurethanes are given in Table 1.1.

Polyurethanes contain an aromatic ring usually when derived from aromatic isocyanates such as MDI, and the important functionalities are $-CH_2$, C = O (1700 cm^{-1}), −C−O−C (1329 cm^{-1}), NH

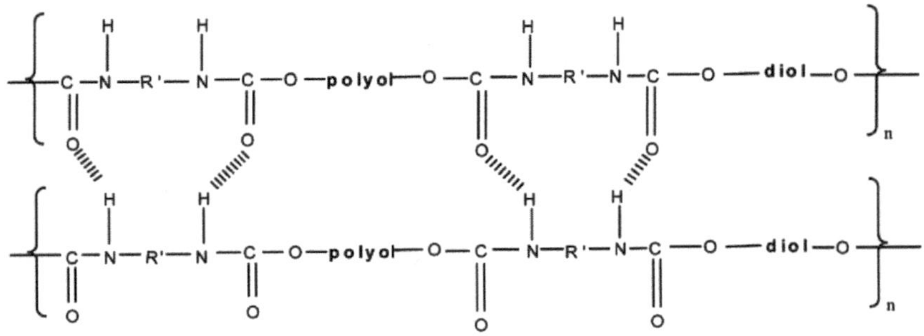

Scheme 1.6 Formation of hydrogen bond in polyurethane.

Table 1.1 Common functional frequencies in polyurethanes

Group	Mode	Frequency (cm^{-1})
N–H	Free	3445–3450
N–H	N–H...N–H	3315–3340
N–H	N–H...O (ether)	3260–3290
C = O (Urethane)	Free	1730–1740
C = O (Urethane)	C = O...H–N	1703–1710

(3328 cm^{-1}), –OH, –CH (2939 cm^{-1}), –C = C. The FTIR spectra of polyurethane [29] is represented in Fig. 1.6.

Nuclear magnetic resonance (NMR) spectroscopy is also an important tool to identify reactions during PU synthesis and the quantification of the hydrogen bonding interactions. Both ^{1}H NMR and ^{13}C NMR techniques are well known for PU analysis [30]. An example for the ^{1}H NMR spectra is given in Fig. 1.7. Galbiati and coworkers synthesized polyurethane prepolymer from toluene diisocyanate and polypropylene glycol. They studied the important peaks and their variations. They found that at room temperature the chemical shifts of the groups are 7.78 ppm (–NH), 7.1–7.2 ppm (aromatic), 3.2–3.9 ppm (–CH), 1.1 ppm (CH_3). The chemical shift values change according to the precursor materials [31].

X-ray diffraction analysis is useful in the case of samples that involve crystalline PUs [32]. The d-spacing is calculated from the XRD patterns using Bragg's equation. Trovati and coworkers analyzed the polyurethanes from MDI using FTIR, DSC, and XRD techniques. They considered three different types of PUs: rigid, semi-rigid, and soft PUs. The change in the crystalline structure with the type of PU is clear from the XRD pattern given in Fig. 1.8. Rigid PUs are more crystalline when

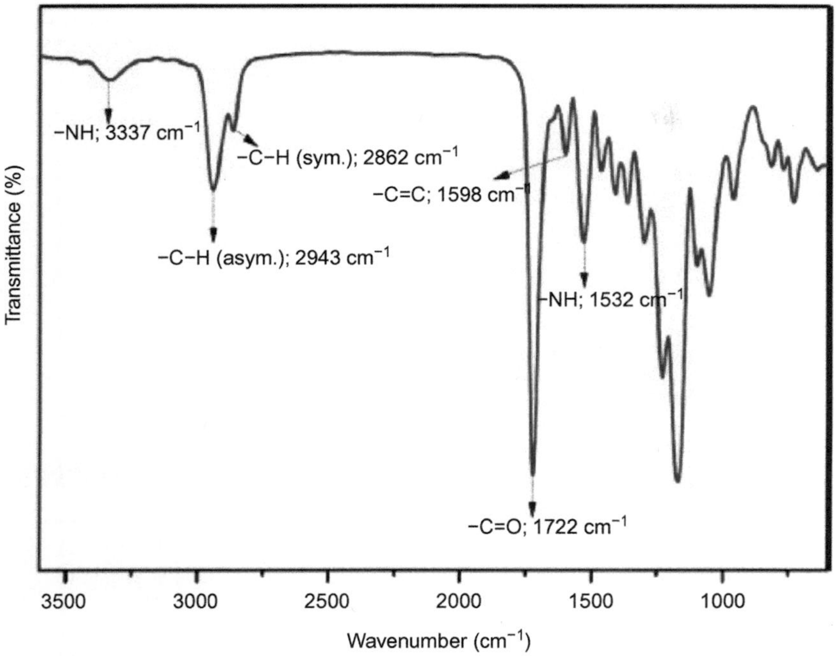

Figure 1.6 FTIR spectra of a polyurethane model compound.

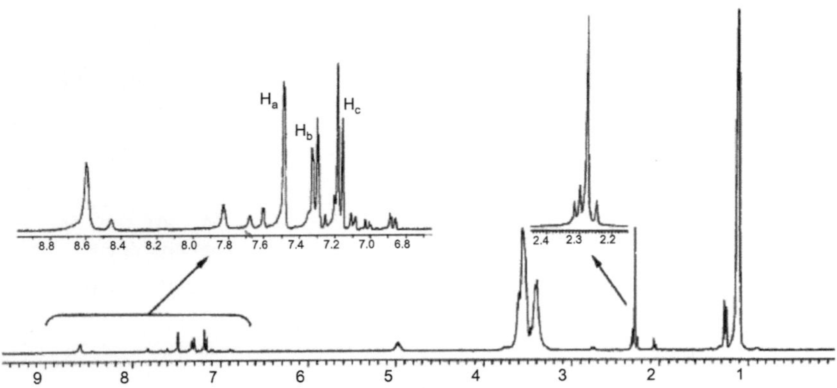

Figure 1.7 ^{1}H NMR spectra of polyurethane prepolymer derived from TDI and PPG.

compared to soft and semi-rigid. The percentage crystallinity for rigid PUs was 37% while that of semi-rigid and soft were 29% and 24% respectively [33].

Thermal properties of polyurethanes are measured using thermogravimetric analysis and differential scanning calorimetry [34,35]. The

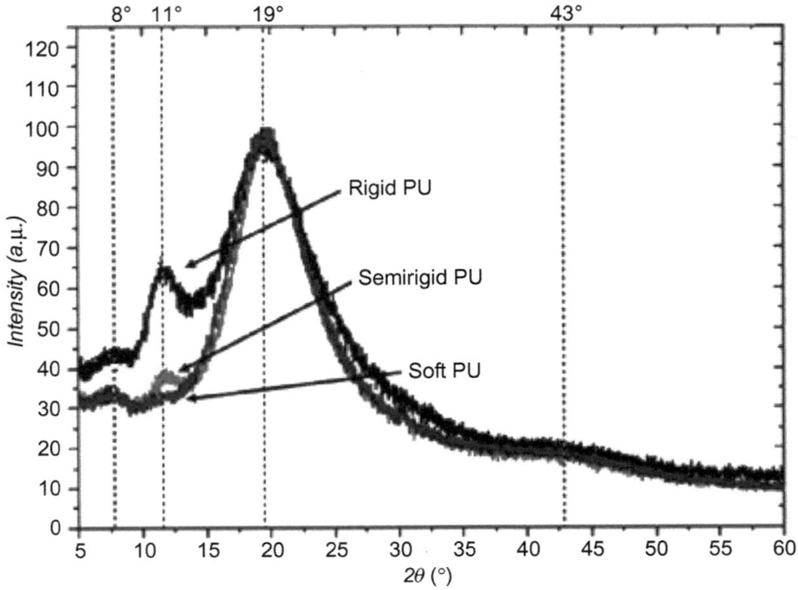

Figure 1.8 XRD pattern of three types of polyurethanes.

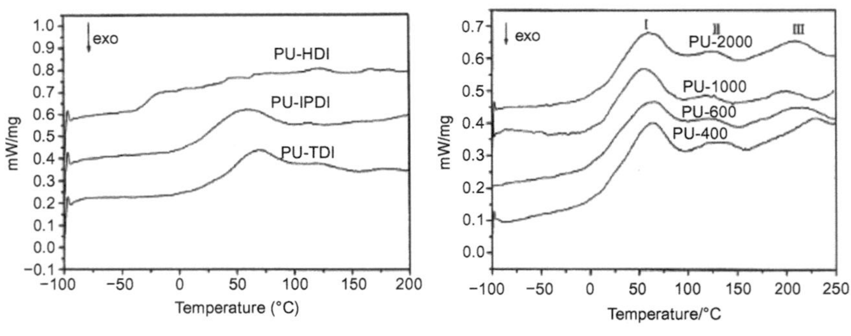

Figure 1.9 DSC curves of PU derived from different isocyanates (left) and with different weight fractions of PEG (right).

segmented polyurethanes have two glass transition temperatures. The T_g values are affected by the type of isocyanate and the amount used in the synthesis. Cheng et al investigated the thermal properties and the phase separation of polyurethane synthesized from different isocyantes and with different molecular weights. The nonisothermal DSC curves of PUs synthesized from TDI, IPDI, and HDI under air atmosphere are given in Fig. 1.9.

1.7 CONCLUSIONS

Polyurethanes are the most demanding type of polymers in the present era because of the versatility in properties and in the synthesis. The one shot method of mixing and prepolymer method are used in the production of polyurethane. As a large number of ingredients are available for the synthesis of PU, a good variety with tunable properties can be obtained. They have been the unsubstituted components in the furniture industry. They have been used as the implants and biodegradable membranes in the medical field. Polyurethanes are available as soft foams and hard tooling materials. Thermoplastic elastomers of polyurethanes are the most wanted materials in the industry. The generation from natural sources and recyclability are the attractive advantages of polyurethanes. Synthesized polyurethanes can be characterized via spectroscopy, calorimetry, and scattering techniques. They maintain the versatility in the results. Phase separated structure of TPUs and the induced crystallinity are interesting to study. Phase separation and crystallinity are temperature dependent features of PUs.

REFERENCES

[1] M. Rogulska, A. Kultys, W. Podkos, 43 (2007) 1402–1414.

[2] M. Szycher, Szycher's Handbook of Polyurethanes, Second Edition, CRC Press, United States, 2013.

[3] E. Yilg, I. Yilg, G.L. Wilkes, Polymer 87 (2014) 1–36. Available from: http://dx.doi.org/10.1016/j.polymer.2014.12.014.

[4] P. Schön, K. Bagdi, K. Molnár, P. Markus, B. Pukánszky, G.J. Vancso, 47 (2011) 692–698.

[5] H. Yeganeh, P. Hojati-Talemi, Polym. Degrad. Stab. 92 (2007) 480–489.

[6] E.A.B. da Silva, M. Zabkova, J.D. Araújo, C.A. Cateto, M.F. Barreiro, M.N. Belgacem, et al., Chem. Eng. Res. Des. 87 (2009) 1276–1292.

[7] M.S. Kathalewar, P.B. Joshi, A.S. Sabnis, V.C. Malshe, R.S.C. Adv. 3 (2013) 4110.

[8] L. Maisonneuve, O. Lamarzelle, E. Rix, E. Grau, H. Cramail, Chem. Rev. 115 (2015) 12407–12439.

[9] O.L. Figovsky, L. Shapovalov, O. Axenov, Surf. Coat. Int., Part B 87 (2004) 83–90.

[10] M. Tryznowski, A. Widerska, Z. Zołek-Tryznowska, T. Gołofit, P.G. Parzuchowski, Polym. (United Kingdom) 80 (2015) 228–236.

[11] E. Delebecq, J.-P. Pascault, B. Boutevin, F. Ganachaud, Chem. Rev. 113 (2013) 80–118.

[12] B. Pukánszky, K. Bagdi, Z. Tővölgyi, J. Varga, L. Botz, 44 (2008) 2431–2438.

[13] K. Bagdi, K. Molnár, A. Wacha, A. Bóta, B. Pukánszky, Polym. Int. 60 (2011) 529–536.

[14] D.K. Chattopadhyay, B. Sreedhar, K.V.S.N. Raju, Polymer (Guildf) 47 (2006) 3814–3825.

[15] K. Hong, S. Park, React. Funct. Polym. 42 (1999) 193–200.

[16] J. Blackwell, K.H. Gardner, Polymer (Guildf) 20 (1979) 13–17.

[17] Z.S. Petrovic, J. Ferguson Prog. Polym. Sci. 16 (1991) 695–836.
[18] M. Rutkowska, A. Eisenberg, Macromolecules 17 (1984) 821–824.
[19] Y. Camberlin, J.P. Pascault, J. Polym. Sci. Polym. Chem. Ed. 21 (1983) 415–423.
[20] Y.J. Li, T. Gao, B. Chu, Macromolecules 25 (1992) 1737–1742.
[21] L.S.T.J. Korley, B.D. Pate, E.L. Thomas, P.T. Hammond, Polymer (Guildf) 47 (2006) 3073–3082.
[22] P.C. Caracciolo, F. Buffa, G.A. Abraham, J. Mater. Sci. Mater. Med. 20 (2009) 145–155.
[23] Z.S. Petrovic, J. Budinski-Simendic, Rubber Chem. Technol. 58 (1985) 701–712.
[24] H.S. Lee, Y.K. Wang, S.L. Hsu, Macromolecules 20 (1987) 2089–2095.
[25] C.W. Meuse, X. Yang, D. Yang, S.L. Hsu, Macromolecules 25 (1992) 925–932.
[26] T.K. Kwei, J. Appl. Polym. Sci. 27 (1982) 2891–2899.
[27] Z.S. Petrovi, D. Hong, I. Javni, N. Erina, F. Zhang, J. Ilavský, Polymer 54 (2013) 372–380.
[28] E. Ylgor, I. Ylgor, E. Yurtsever, Polymer (Guildf) 43 (2002) 6551–6559.
[29] S.S. Mahapatra, S.K. Yadav, H.J. Yoo, M.S. Ramasamy, J.W. Cho, Sensors Actuators B Chem. 193 (2014) 384–390.
[30] J. Spěváček, W. Hiller, W. Hettrich, D. Joel, Eur. Polym. J 25 (1989) 1239–1243.
[31] M. Pegoraro, A. Galbiati, G. Ricca, J. Appl. Polym. Sci. 87 (2003) 347–357.
[32] R. Bonart, E.H. Müller, J. Macromol. Sci. Part B 10 (1974) 177–189.
[33] G. Trovati, E.A. Sanches, S.C. Neto, Y.P. Mascarenhas, G.O. Chierice, J. Appl. Polym. Sci. 115 (2010) 263–268.
[34] R.W. Seymour, S.L. Cooper, Macromolecules 6 (1973) 48–53.
[35] J.T. Haponiuk, A. Balas, T. Kawka, J. Therm. Anal. 36 (1990) 2249–2252.
[36] C. Zhang, S.A. Madbouly, M.R. Kessler, ACS Applied Materials & Interfaces 7 (2) (2015) 1226–1233.

CHAPTER 2

Blends and Interpenetrating Polymer Networks Based on Polyurethane Polymers With Natural and Synthetic Rubbers

Mehar A. Minnath[1] and Etathil Purushothaman[2]
[1]DGM MES College Mampad, Malappuram, Kerala, India
[2]University of Calicut, Malappuram, Kerala, India

Contents

2.1 INTRODUCTION

Remarkable developments in the field of polymer science in the recent past have resulted in novel materials that have immense commercial potential. The demand for high functional, high-performance, and smart materials of high technological properties has led to the production of exciting new products at a very rapid pace. Polymers have replaced many of the conventional engineering materials like steel, rock, and wood in various applications because of the advantages they offer over conventional materials. The important advantages of using polymers are the ease

Polyurethane Polymers: Blends and Interpenetrating Polymer Networks
DOI: http://dx.doi.org/10.1016/B978-0-12-804039-3.00002-6

of processing and productivity, and the reduction of cost. Both natural and synthetic polymers are important for mankind and, hence, the demand for polymers is increasing day by day. Polymer science and technology is one of the exciting and fast-growing areas in the field of material sciences. As with all branches of science, polymer chemistry is continually advancing and recent topics in polymer chemistry involve novel concepts. Research on developing new polymers for more innovative use by replacing existing materials has been undertaken in research laboratories around the world [1]. The resultant numerous research papers and industrial patents in the scientific world are sufficient enough to understand the importance of this field in this era of science and technology.

The properties of polymers can be modified in different ways. Being advantageous in both the economic point of view and the ease of preparation, the widely followed techniques for preparing new materials are through the creation of polymer blends and interpenetrating polymer networks (IPNs). The development of polymer blends and IPNs is an easy and economical pathway for the production of new polymeric materials with excellent properties. These are not only methods of polymer modification, but important route for developing novel materials with unique properties. Academic and industrial research in these areas is flourishing exponentially.

2.1.1 Polymer Blends

All polymers have shortcomings in at least one property, if not more. The blending of two or more polymers often yields a polymer with novel properties that cannot be achieved from any of the individual components. Blending is the way of the present and future plastics [2]. Polymer blends are being used extensively in numerous applications. The wide use and industrially acceptable properties of these materials in modern life can be evidenced by noting the fact that polymer blends constitute 36 wt% of the total polymer consumption, and this continues to increase [3]. A blend can offer a set of properties that may give it the potential of entering application areas not possible with either of the polymers constituting the blend. A significant number of commercial polymer blends have become available, and continued effort to create new materials with enhanced chemical or mechanical performance is expected [4]. In addition, some modification in terms of processing characteristics, durability, and cost can be achieved via polymer blending. The modification in the behavior of polymer systems has been the subject of many studies. Blending also creates materials with excellent combinations of mechanical, thermal, chemical, and morphological properties [5,6].

Blending provides an easy and relatively inexpensive method to develop polymeric materials with desirable properties. During the past few decades, the importance of polymer blends has increased as it is possible to tailor desirable properties by the simple blending of polymers. Polymer blending has already been established as an effective means for constructively altering the properties of polymeric materials. Blending of polymers also provides an easy and useful means of reuse and recycling of polymer wastes. The benefits of polymer blending can be summarized as follows:

- Increased use and maximum diversification of the existing polymers.
- Very low developmental cost.
- Easy processability and extended service range.
- Improved properties than those of individual homopolymers.
- Means of waste recycling and scrap reduction.

2.1.1.1 Types of Polymer Blends

Polymer blends can be classified into different classes on the basis of aspects such as the number of homopolymers used for blending, biodegradability, nature of constituent polymers, and miscibility of the constituent polymers. Fig. 2.1 shows a schematic representation of the classification of polymer blends.

2.1.2 Interpenetrating Polymer Networks

Since the first synthesis by Miller in 1960, the term IPNs has been used to describe the combination of two polymers in network form, at least one of which is synthesized and/or cross-linked in the immediate presence of the other [7,8].

IPNs are a special class of polymers in which the cross-linked polymers are held together by permanent entanglement. The networks are at least partially interlaced on a polymer scale but not covalently bonded to each other. The network cannot be separated unless chemical bonds are broken. Two or more networks can be envisioned to be entangled in such a way that they are linked together and cannot be pulled apart. The conditions required for the synthesis of IPNs are: (1) the component polymers are synthesized and/or cross-linked in the presence of the other; (2) the component polymers have similar kinetics; and (3) the component

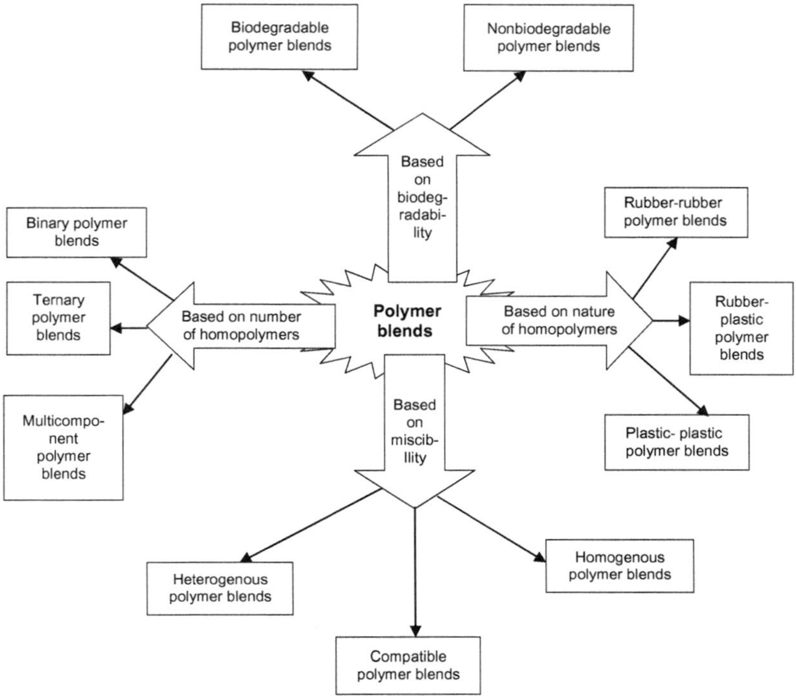

Figure 2.1 Schematic representation of classification of polymer blends.

polymers are not dramatically phase-separated [9]. These are also termed "intelligent polymers" and are of great importance in scientific research due to their unique properties and potential technological applications. They can be considered the next generation of smart materials. IPNs offer the possibility of combining polymers in a network form which are otherwise non-compatible polymers with opposite properties [10]. They are shape memory polymers with increased elastic and mechanical properties, phase stability, thermal stability, dielectric properties and optical clarity. They exhibit excellent characteristics of a multipolymeric system associated to favorable cooperative properties [11].

Some of the significant applications of IPNs include artificial implants, dialysis membranes, drug-delivery systems, and burn dressings. They can also be used in other fields of science like aerospace, textiles, civil engineering, bionics engineering, energy, electronic engineering, household products, fuel cells, solar cells, and food packaging materials. The advancement and future applications of IPNs will be of immense help to mankind.

The morphology of IPNs can be considered as continuous because the molecular disposition of one of the components "reduces" and "restricts" the domain size of the other, so that both phases span over the entire sample. Thermoplastic IPNs (t-IPNs) are a different version of IPNs, being multicomponent polymeric systems whose structure is constructed by physical cross-links rather than chemical cross-links. The mutual penetration of their components (semi-crystalline polymers, block-copolymers, and/or polymers with ionic groups) gives rise to regions mainly built up by segments of one component which are matched to regions based on the second component, with substantial network interpenetration occurring at the interface. As a result, the mechanical properties and the molecular mobility of the whole network are strongly affected by the domain and interface sizes.

2.1.2.1 Classification of IPNs

Many different kinds of IPNs exist. IPNs can be classified into different classes on the basis of arrangement patterns. The following definitions identify some of the more important types:

Sequential IPN: One monomer is polymerized followed by the polymerization of the other monomer to produce a network in a sequential manner.

Novel IPN: Polymer comprising two or more polymer networks which are at least partially interlocked on a molecular scale but not covalently bonded to each other and cannot be separated unless chemical bonds are broken.

Simultaneous interpenetrating networks (SINs): The monomers and/or pre-polymers plus the cross-linkers and initiators of both networks are mixed, followed by simultaneous polymerization.

Semi-IPNs: These are sequential IPNs in which one of the constituent polymers is cross-linked while the other is linear.

Fig. 2.2 shows a schematic representation of the classification of IPNs.

Even though polymer blends and IPNs have similarity in their properties and applications, they do differ from each other. A polymer blend is a

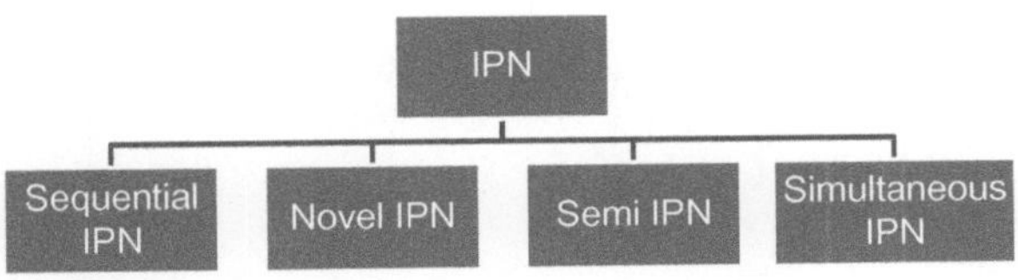

Figure 2.2 Schematic representation of classification of IPNs.

mixture of two or more polymers which are not cross-linked. An IPN is a blend of two or more polymers in a network form, at least one of which is synthesized and/or cross-linked in the immediate presence of the others. A polymer blend melts and, in principle, there are solvents that can dissolve the components but not necessarily one solvent for both components. An IPN is insoluble in almost all solvents, they only swell in solvents [12] and has no specific melting point, only a decomposition point. In fact, the components of a blend can be separated by using suitable solvents. However, the components of an IPN cannot be separated in a similar manner [13].

2.2 POLYURETHANE

Polyurethanes (PU) represent an important class of polymeric material. They exhibit thermoplastic, elastomeric, and thermoset behaviors. A PU has versatile properties and is suitable for foams, thermoplastic elastomers, adhesives, coatings, sealants, and fibers [14]. The properties of PU depend on its chemical and morphological composition. All industrial PUs are made by the polyaddition of polyisocyanates and polyols, whereby polyurethane groups are formed in the polymer backbone, as shown in Scheme 2.1 [15].

Polyurethanes contain alternating rigid and flexible blocks. Generally, the two segments are incompatible, resulting in microphase separation, which is primarily responsible for their excellent mechanical properties [11]. The incompatibility is mainly due to the difference in polarity and chemical nature of the two blocks. The urethane linkage (−NHCO−O−) in the polymer is formed due to the reaction between isocyanates and hydroxyl groups. Most hard segments have urethane and/or urea functional groups that contain strong proton donor NH group and a proton acceptor, CO group, that can form hydrogen bonds to each other. This type of self association would be an additional driving force for phase separation, as the soft segments usually contain functional groups that do not self-associate but have hydrogen bond acceptor groups,

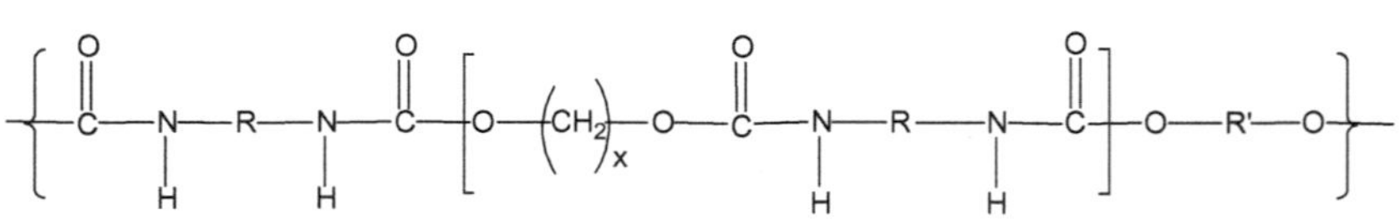

Scheme 2.1 Structure of PU.

such as ether oxygen or ester carbonyl functionalities [16]. These polymers are very versatile because they present structures ranging from rigid to flexible and are commonly used in various forms such as foams, elastomers, and adhesives. Because of this structural peculiarity, they have a wide range of applications including automotive seating, furniture, packaging, and medical devices. Polyurethanes are also widely used as thermal insulators in many industries such as footwear, construction, and electronics. Flexible PU foams are blocks of copolymers whose elastic properties depend on the phase separation between hard and soft domains. Hard blocks are rigid structures that are physically cross-linked and give the polymer its firmness; soft blocks are stretchable chains that give the polymer its elasticity. Therefore, polyurethane foams can be customized by adapting the composition and the ratio of these blocks [17]. Because it is resistant to chemicals, soft, waterproof, and weatherproof, it can be used for various other applications [18].

PU exhibits excellent abrasion resistance and high resilience. As a result of these properties, they have been widely used in the manufacturing of oil and water pipes [19]. Rigid and flexible PU foams are widely used in building construction and in the automotive, bedding, packaging, and medical device industries, because they offer numerous advantages, including being lightweight, have excellent strength, low thermal conductivity, and high weight-carrying capacity [20–22]. On account of the excellent physical and chemical properties and good biological compatibility, PU has found many applications in the field ranging from automotives, rollers systems, shape memory materials, and films to biomedical products [15, 23–25]. However, under extreme conditions, PU materials are prone to cracking and damage, which can potentially affect its end use product properties. To overcome such problems, modification is required. One method for modification is by blending PU with another polymer. Polymers that have been blended with PUs and thermoplastic polyurethanes (TPUs) include polycarbonate [26], polyoxymethylene [27,28], polystyrene [29], butadiene rubber (BR) [30], poly(vinyl chloride) [31,32], styrene acrytonitrile plastic (SAN) [33], and polyamide [34,35].

2.3 BLENDS AND IPNs OF POLYURETHANE

2.3.1 PU Blends and IPNs with Natural Rubber

Natural rubber (NR) is the most important polymer basis of thermoset rubber, and it covers a wide variety of applications. Furthermore, NR can be considered one of the most important natural polymers; it is renewable and ecofriendly. NR is a versatile raw material on account of its unique chemical and physical properties. NR is characterized by good elastic properties, good resilience, and damping behavior but it has poor chemical resistance and processability. Thermoplastic elastomers based on the blending of NR and thermoplastic are classified as thermoplastic NRs (TPNRs), in which NR is the elastomeric phase dispersed in a thermoplastic polymer matrix. NR has an inherent affinity with some olefinic thermoplastics, which permits the formation of TPNRs that are characterized by submicron scale morphologies and excellent physical properties [36]. In some research, NR has been blended with many polymers such as PP, Polyethene (PE), Polyvinyl alcohol (PVA), Nitrile butadiene rubber (NBR), Ethylene propylene rubber (EPR), Ethylene propylene diene terpolymer (EPDM) [37–39]. Epoxidized NR (ENR) has also been used to prepare TPNRs. The use of renewable and ecofriendly polymers such as NR, as the main component of these blends helps to preserve resources and, thus, contributes to the sustainable development of new products. TPNRs are environmentally friendly materials and have high potential for making new impacts in the polymer industry.

Polyurethane is a class of polymers that covers a great variety of different semi-crystalline copolymers with excellent mechanical properties. By varying hard and soft phases, we can easily adapt the properties of PU to the respective applications. A blend or an IPN of PU and NR will result in the development of TPNRs with outstanding properties. Much research is being undertaken in the development and modification of TPNRs based on PU and NR. A detailed study of TPU/NR blends were carried out by Mehar et al. [40,41]. The blends show better mechanical and transport properties than the component polymers. The stress strain curve of TPU/NR blends are shown in Fig. 2.3. T and N stand for TPU and NR respectively and the subscripts denote the weight percentage of each component.

The DSC studies of these blends gave a single T_g value for all compositions, indicating the miscibility of TPU and NR. The investigation was extended by blending ENR with TPU [42]. It was found that ENR-based blends had better properties than NR-based blends. This improvement was attributed to the better compatibility of ENR and TPU due to the formation of H^- bond between the epoxy group in ENR and the isocyanate group in TPU as shown in Scheme 2.2.

ENR and TPU blends were prepared by simple blend and dynamic vulcanization by Pichaiyuta et al. [43] with an intention to develop a low

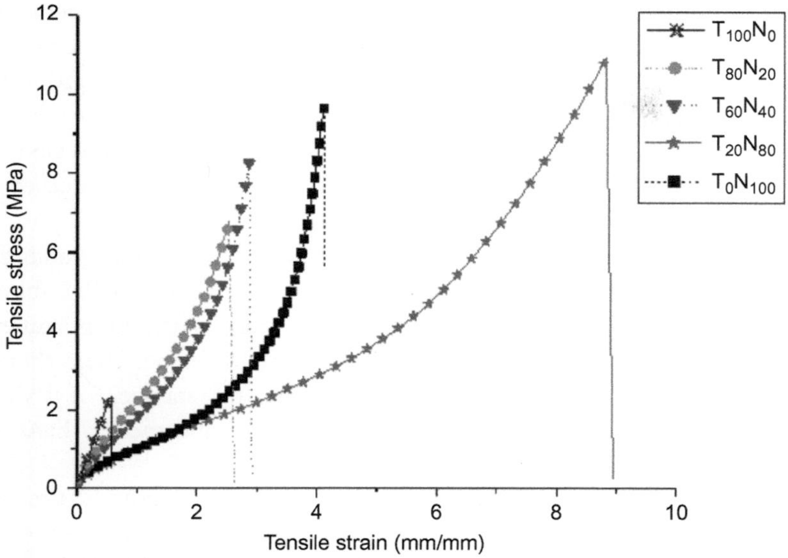

Figure 2.3 The stress strain curve of TPU/NR blends.

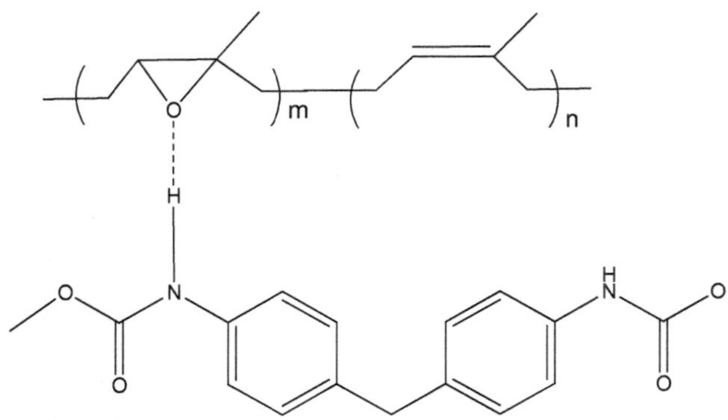

Scheme 2.2 Hydrogen bond between epoxy group of ENR and isocyanate group of TPU.

hardness TPU material with good damping, elastic, and mechanical properties. It was found that the incorporation of ENR into the blend shows a reduction in Young's modulus, hardness, damping properties, and tension set compared to the pure TPU. This indicates the formation of softer TPU materials with superior damping and elastomeric properties. However, incorporation of ENR sacrificed mechanical properties in terms of tensile strength and elongation at break, but these still remain in the range of applicability for industrial uses. It was also found that dynamic vulcanization caused enhancement of mechanical properties, relaxation, damping, rheological properties, and elasticity of the blends. Temperature scanning stress relaxation measurements revealed an improvement in stress relaxation properties and thermal resistance of the dynamically cured ENR/TPU blend.

Radhakrishnan and Gopinathan [44] studied the thermal stability of solution-cast blends of poly(vinyl chloride) and NR-b-PU block-copolymers of three different chain extender diols by thermogravimetry. As the block copolymer content in the blends increased, their thermal stability was found to increase. Enhanced thermal stability of PVC is believed due to the favorable interaction with PVC and the PU hard segments of the block copolymer.

The influence of mechanochemical pretreatment of ground tire rubber (GTR) on the acoustic absorption properties of polyurethane (PU)/GTR foamed composites were studied by Zhang et al. [45]. The mechanochemical pretreatment also enhanced foamability of the composites. The results indicated that PU/GTR foamed composites have improved damping properties as well as acoustic absorption.

In another study, PU was used as a compatibilizer for blending polyvinyl chloride (PVC) with NR [46]. A block copolymer of NR and polyurethane (PU) based on propylene glycol (PG) and toluene diisocyanate (TDI) was used as the compatibilizer. The NR/PU block copolymer was an efficient compatibilizer for the PVC/NR (90/10) blend system. Solution viscometry and optical microscopy studies revealed that optimum compatibilization of the PVC/NR (90/10) blend occurs at 1.5% of the block copolymer concentration. The tensile and impact properties of the blend significantly improved in the presence of the compatibilizer and reached optimum values at the compatibilizer concentration of 1.5 wt%. The addition of NR to PVC causes deterioration in tensile properties of the blend which is regained with the incorporation of 1.5 wt% of the compatibilizer.

Kalkornsurapranee et al. [47] prepared TPVs based on blends of NR and TPU with a dynamic vulcanization process to develop and characterize high-performance TPVs of low Shore A hardness based on NR. The effects of the sulfur curing systems, accelerator types, optimization of mixing process, and NR types on the properties of the TPVs were investigated. They found that the EV curing system was the best among these alternatives for the extreme processing conditions of TPV. TPVs prepared with DCBS as an accelerator exhibited the highest tensile strength, and the action of the accelerators were ranked as follows: DCBS>TBBS>CBS >Mixed TBBS/TMTD and >TMTD. The tensile strength of the TPVs was related to the ts1 values of the rubber compounds. Mechanistically, a slower ts1 provides a longer effective time for melt mixing and, thus, a reduction in the particle size of the rubber phase in the TPU matrix. ENR performed better than unmodified NR in the dynamically cured blends with TPU. Increasing the epoxide content in ENR improved the properties of the blends; this may have resulted from stronger polar interactions and possibly even the chemical bonding of ENR and TPU at the interface of the phases.

To improve the resistance to solvents (especially toward hydrocarbons), abrasion, and UV irradiation, NR is often blended with polyurethane rubber (PUR). Varghese et al. [48] developed NR, polyurethane rubber (PUR), and NR/PUR-based nanocomposites from the related lattices by adding a pristine synthetic layered silicate (LS; sodium fluorohectorite) in 10 parts per hundred parts rubber (phr). It was found that LS is more compatible and thus better intercalated by PUR than by NR. Further, LS was preferably located in the PUR phase in the blends, which exhibited excellent mechanical properties despite the incompatibility between NR and PUR. Nano-reinforcement was best reflected in stiffness- and strength-related properties of this rubber blend nanocomposites.

2.4 PU BLENDS AND IPNs WITH SYNTHETIC RUBBER

2.4.1 PU/EPDM Blends

Ethylene propylene diene monomer rubber (EPDM) is a material with low density and is highly capable as a filling material. It has excellent heat resistance, oxidation resistance, ozone resistance, and weatherproofing properties. It also has good age resistance and has good compatibility with many polymers. Moreover, the temperature and processing method of EPDM are suited to the temperature range and extrusion processing of TPUs [49].

Dhamodharan et al. developed blends of EPDM rubber and TPU [50,51] to understand the effect of the maleation level of EPDM on the compatibility, morphology, and mechanical strength of the blends. It was found that maleation level of more than 0.5 wt% was required for the

EPDM blends with TPU. However, best mechanical properties were obtained for 0.5% maleated EPDM and TPU blends.

The heat shrinkability of the blends of grafted ethylene propylene polymer and polyurethane elastomer were studied by Mishra et al. [52]. They found that the interchain crosslinking between grafted ethylene propylene polymer and polyurethane elastomer improves the heat shrinkability. The high temperature performance of the sample is dependent on interchain crosslinking. From torque rheometry it was observed that the maximum amount of interchain crosslinking occurred in the sample containing glycidyl acrylate as grafting material.

The influence of processing conditions on the morphology, mechanical properties, and rheology of the blends of TPU and EPDM rubber were investigated by Lei Wu et al. [53]. They found that the dual-phase continuous morphology of the blends was strongly dependant on the EPDM composition, processing temperature, and the shear rates. The network structure of the EPDM domain in TPU matrix became the finest and most regular for the blends containing 7 wt% EPDM. It was also found that high shear rate favored the formation of the perfect network structure. Also, the blends prepared at 180°C present a finer and more perfect network structure than those at the other processing temperatures. The tensile properties improved significantly with addition of EPDM, and obtained the optimum value for the blends containing 7 wt% EPDM. In another study, they employed EPDM as second component to blend with TPU, and found an interesting cocontinuous morphological structure for these blends, in which a network-type EPDM domain existed in the TPU matrix. The results also indicated that EPDM was thermodynamically miscible with the soft segments of TPU and incompatible with the hard segments; therefore, the formation of the network was resulted from the competition of compatible and incompatible segments of TPU with EPDM. This network-type morphological structure enhanced the tensile properties [54].

The effect of EPDM as an additive on the abrasion resistance of TPU resin was investigated by Jing Tan et al. [55]. They reported that the addition of EPDM greatly improved both the mechanical properties and abrasion resistance of the TPU resin. A TPU/EPDM composite with 8 wt% EPDM demonstrated the highest tensile strength, the largest elongation at break, and the best overall performance.

In an attempt to obtain a TPU with low hardness without sacrificing abrasion resistance and tensile properties, TPU was blended with EPDM [56].

A TPU/EPDM (90/10) blend exhibited higher NBS abrasion resistance, tensile strength and tensile modulus, than TPU.

2.4.2 PU/Acrylonitrile Butadiene Rubber Blends

Acrylonitrile butadiene rubber (NBR) is used in gaskets, for lining valves, in expandable shafts, and has many other engineering applications. It has excellent oil and abrasion resistance, but poor ozone resistance and processibility. To overcome certain limitations of NBR and to attain certain desired properties, blending of NBR with another polymer like PU is carried out. PUs are important engineering materials used as barrier materials, sealants, foams, and shoe soles.

NBR and TPU are both polar in character and hence, owing to orientational and dispersional interactions [57,58] in mutual blends, they form associations of NBR–NBR, NBR–TPU and TPU–TPU type. These associations act as multifunctional junctions, or intermolecular linkages in the blends, forming a secondary network that represents their morphological structure when in amorphous state [59]. The blend of these two can exhibit a variety of properties. Attempts in this direction have been undertaken by various researcher teams [60–64].

The studies on PU/NBR blends by Desai et al. [65] showed that the two polymers behave synergistically where the incorporation of NBR increases the elongation and incorporation of PU increases the tensile strength. The blends exhibited better mechanical properties than the individual polymers. NBR also improved the thermal stability and solvent resistance of the PU network. The T_gs of the two components also shifted inwards indicating increasing miscibility.

The blends of PU with hydrogenated nitrile rubbers (HNBR) were prepared by different blending techniques and different blend ratios by Maity et al. [66]. They reported that the processability and weather resistant properties of polyurethane elastomer was improved by blending with HNBR. The thermal stability of the blends was also improved.

A blend composite of PU/NBR blend with dodecyl sulfate intercalated Mg–Al-based layered double hydroxide (DS-LDH) nanofillers were prepared by Kotal et al. [67]. In comparison to the neat PU/NBR blend, the tensile strength and elongation at break showed maximum improvement for 1 wt% filler loading. The storage and loss moduli, thermal stability and limiting oxygen index of the nanocomposites were higher compared to the neat PU/NBR blend. Glass transition temperature and swelling measurements increase up to 3 wt% DS-LDH loading in PU/NBR compared to either neat PU/NBR or its other corresponding nanocomposites.

A blend of thermoplastic polyurethane urea (TPUU) and carboxylated acrylonitrile butadiene rubber (XNBR) were prepared and characterized by Mahmood et al. [68]. They identified a new peak in the FTIR-ATR spectra indicating the existence of chemical reaction between the carboxylic functional group of XNBR and the amide group of the TPUU. This was supported by the detection of an increase in the network crosslink density of the blend using 1H-NMR spectroscopy. The scanning and transmission electron micrographs of the blend morphology showed a uniform dispersion of the minor TPUU phase in the XNBR. A single glass transition peak confirmed the enhancement in the interfacial miscibility.

Another group blended TPUU with NBR [69]. The TPUU was synthesized in situ from the polyaddition of 4, 40-diphenylmethane diisocyanate-terminated prepolymer with Bisaniline-M during blending with NBR in an internal mixer. The blend showed improved tensile characteristics, tear and abrasion resistance. However a temperature dependent softening of amorphous TPUU phase diminishes the dynamic-mechanical behavior of blend vulcanizates. Improvement in physical properties of blends is attributed to good interfacial adhesion between dispersed TPUU domains and continuous NBR matrix, which was confirmed by TEM.

Manoj et al. [70] prepared ternary blends of TPU and a PVC/NBR blend. The blends with ratios of 20/80 and 80/20 were miscible, whereas the 40/60 and 60/40 blends were partially miscible. IR spectroscopy studies showed shifts in the peaks due to specific interactions in the blends. The miscible blends were found to have good fracture toughness, whereas the other blends had lower toughness. Similar behavior was observed for the tensile properties too.

A peroxide-curable hydrogenated nitrile rubber (HNBR) was modified by in situ cross-linked polyurethane (PU) in a 1:1 weight ratio by Sebenik et al. [71]. They incorporated carbon black (CB) of different types (N234, N330, and N550) into the prepared HNBR/PU blends. The DMA results suggested that a quasi-interpenetrating network structure was developed in the HNBR/PU blend. This was confirmed by SEM, TEM, and AFM inspections. It was established that CB was exclusively located in the HNBR phase. CB N234 proved to be more efficient reinforcement than N330 and N550.

2.4.3 PU/PDMS Blends

Polydimethylsiloxane (PDMS), a border line polymer between organic and inorganic polymers, has found wide applications in industries due to its unique features, such as excellent low temperature flexibility, lower surface energy and very high thermal stability. In addition, it has excellent bio-inertness [72]. TPU, when blended with PDMS rubber, develops properties like excellent mechanical strength, and fatigue resistance, combined ease in processability and higher thermal stability, and is hence considered to be an ideal biomaterial for applications where higher strength and flexibility is desired [73—75].

Organosilicone, which exhibits huge potential for improving hydrophobic behavior, corrosion resistance, adhesion, and mechanical properties as well as resistance to heat, has been widely used to modify waterborne polyurethane (WPU) by structural manipulation at the molecular level [76]. Generally, alkoxysilane, a kind of organosilicone, has alkoxy groups at the end of the molecular chains. In the presence of water, alkoxy groups hydrolyze to generate silanol groups and subsequently condense with other silanol groups to form a cross-linked structure of Si-O-Si groups [77—80]. This organic—inorganic hybrid special structure in polymers donates the inherent features of inorganic materials to waterborne polyurethane and results in enhancing the performance of WPU. Pathak et al. [76] prepared a WPU coating on an aluminum alloy with methyltrimethoxysilane and 3-glycidoxypropyltrimethoxysilane. The study indicated that silane modification boosted the corrosion resistance, thermal stability, and mechanical properties of WPUs. In addition, one kind of antifouling WPU coating with higher thermal and mechanical strength was obtained concurrently by Rahman et al. [81] by adding N-[3-(trimethoxysilyl) propyl]-ethylenediamine.

In another report, a hydroxyl-component WPU with a cross-link structure of Si-O-Si groups was prepared by means of 3-aminopropyltriethoxysilane (APTES) and trimethylolpropane (TMP) end-capping concurrently [82]. They reported that, with the addition of hydrophilically modified polyisocyanates, a series of novel solvent-resistant 2K-WPUs bearing internal and external crosslinking structure could be obtained. Two-component waterborne polyurethane mini-emulsions with 3-aminopropyltriethoxysilane were successfully synthesized. They reported that as the amount of APTES content increase, the mean particle size of the hydroxyl-component WPU mini-emulsion increased and the appearance of the mini-emulsion gradually changed from semi-transparent with light blue to slightly white with faint light blue. The viscosity of the hydroxyl-component WPU mini emulsion increased slightly. The 2K-WPUs formed a cross-linked structure of Si-O-Si groups. The thermal stability of 2K-WPU with APTES end-capping was improved in comparison with pure 2K-WPU. The solvent resistance of 2K-WPU films also was reported to be increased with an increase of APTES content. In their study, 2K-WPU films containing 4 wt% APTES showed the strongest resistance to ethanol, acetone and, to a certain extent, DCM. With addition of APTES, 2K-WPU films exhibited excellent tensile strength and breaking elongation.

Microphase separation was investigated by Rebeca et al. [83] for segmented polyurethane copolymers synthesized from 4,4¢-methylenediphenyl diisocyanate and 1,4-butanediol as the hard segments and poly (hexamethylene oxide diol) and hydroxyl-terminated polydimethylsiloxane as soft segments. The neat PDMS-based diol exhibited two segmental relaxations corresponding to the principal repeating unit, $Si(CH_3)_2O$, and to the hydroxyl end-group segments. When incorporated in the polyurethanes, the siloxane units formed their own phase, while the end-group segments were mixed with the second macrodiol and some short hard segment sequences.

The effect of processing parameters on the morphology of TPU/PDMS blends has been investigated by Jineesh and Golok [84]. They quantified the effect of processing parameters and ranked them in the order temperature>time of mixing>rotor speed of the Brabender. The optimum processing conditions for TPU-PDMS were found to be with a processing temperature of 190°C, a mixing time of 2 min and a rotor speed of 80 rpm. They observed a considerable change in the distribution of PDMS domain sizes and the area of each domain with change in the processing conditions. The lowest viscosity ratio was found for the blends which were prepared at the best processing conditions.

Siriporn and coworkers [85] developed a melt blend of TPU and PDMS and studied their tensile properties and morphology. They observed that the maximum amount of PDMS that can be mixed with TPU is only 1%. Such a blend showed an increase in the elongation at break and Young's modulus. Morphological studies showed that the domain size of PDMS dispersed in TPU matrix decreased with increase in the amount of PDMS.

2.4.3.1 PU/EVM Blends

EVM is the accepted abbreviation for ethylene vinyl acetate copolymers with between 40% and 90% vinyl acetate (VA) and having elastomeric properties. EVM is a specialty elastomer with good performance at high temperatures and in oil environments with excellent aging resistance. It also possesses inherent flame retardant properties and are capable of being highly filled with inactive fillers. It is widely used in cables and rubber rollers, as well as for appliances and components in the automotive industry as an irreplaceable rubber material [86].

The preparation of EVM/TPU blends with various ratios and their compatibility were investigated by Shi and coworkers [87,88]. The test results showed that in preparation of the blend, fillers should be added to the blend to improve the processability and that among the fillers used, silica showed the best reinforcing effect on the blends. The best parameters for blending EVM and TPU in a HAAKE rheometer was: mixing temperature 160°C, rotor speed 45 rpm and mixing time 15 min. The test results also showed that the compatibility of EVM 700 (VA = 70%) with TPU was better than that of EVM400 (VA = 40%). They reported that the addition of a compatibilizer EVM-g-MAH and EVM-g-FME improved the processability of the blends. While the addition of EVM-g-MAH could also improve the compatibility of EVM 700/TPU blend along with an improvement in the mechanical properties and hot air aging properties, the addition of EVM-g-FME did not improve the compatibility of EVM/TPU blends. For the reinforcement of EVM/TPU blends they reported silica to be a more active filler than carbon black. The silica-TPU interaction was found to be thermally unstable at 190°C and mechanically unstable at a shear frequency of 1 Hz.

In another investigation, Xinyan et al. [89] studied the effect of two peroxide curing agents, DCP (dicumyl peroxide) and BIPB (bis(tert-butyl peroxy isopropyl)benzene) on the mechanical properties, hot air aging, and oil resistance of the TPU/EVM blends. The results showed that the optimum parameters for dynamically vulcanized TPU/EVM by peroxide-DCP or BIPB in a HAAKE rheometer were: mixing temperature 140–150°C and rotor speed 30 rpm. The mechanical properties and oil resistance of these blends were improved by dynamic vulcanization. It was found that BIPB is a better curing agent than DCP for the dynamic vulcanization of TPU/EVM and its optimum content was 0.8 phr in the blend. FTIR spectra showed that EVM and TPU could both be cured by peroxide in the blend and the curing reaction occurred at $-CH_2-$

groups that were linked with −C− instead of −O− and $-CH_3$ groups in the blend. PCM photographs showed that dynamically vulcanized TPU/EVM blends had "sea-island" phase structure when the curing agent content was low and it had "interlocked/cocontinuous" phase structure as the curing agent content was increased. The spin-lattice relaxation constant, T_{21}, measured with MR-CDS proved that the cross-link density of the cured blends increased with increasing curing agent content.

2.5 PU BLENDS AND IPNs WITH OTHER SYNTHETIC POLYMERS

Blending PU with other polymers overcame the problems resulting from the specific structures of PU macromolecules, interphase interactions, and microphase transformations. The literature survey reveals that most of the reported works were carried out through interpenetrating polymer network/solution process using a thermosetting type of PU and different acrylic materials [90−92]. However, TPU may be mixed with poly(ethylene-co-methyl methacrylate) [93], polypropylene (PP) [94], styrene-acrylonitrile (SAN) [95], poly (butyleneterephalate) [96], polycarbonate (PC) [97], acrylonitrile-butadiene-styrene (ABS) [98], to improve the mechanical properties.

Blends of PMMA and TPU in different compositions viz., 95/5, 90/10, 85/15, and 80/20 (by wt/wt, % of PMMA/TPU) were prepared by melt mixing using a twin-screw extruder [2]. The impact strength of the PMMA/TPU blends was found to increase significantly with an increase in the percentage of TPU up to 20%, by retaining the tensile strength of PMMA. In this article, the authors have reported on the melt blending of PMM A and TPU, with the objective of improving the toughness of PMMA by retaining its optical clarity. A series of PMMA/TPU blends have been prepared by a melt blending method. A marginal reduction in tensile strength and tensile modulus was noticed with an increase in TPU content from 5% to 20% by weight ratio in the blends. A significant improvement in percentage elongation at break and impact strength was noticed with an increase in the TPU content in the blends. From the chemical aging behavior, it can be concluded that the percent elongation behavior is sensitive towards chemical aging. A drastic reduction in percentage elongation at break and tensile modulus of the blends were observed after exposure in different chemical reagents, which they reported to be due to the plasticization effect of chemical reagents in the blends.

Shadi et al. [99] prepared blends with different ratios of TPU/PP by melt mixing using an internal Haake mixer. Properties of the blends were investigated and were compared with those of PVC blood bags. The effect of the addition of the ethylene vinyl acetate (EVA) copolymer on the TPU/PP blend properties was also investigated by them. The results indicated that a TPU/PP/EVA 80/20/5 blend could be used as a new blood bag material. They reported that the blend is homogeneous with higher mechanical strength than that of the commercial PVC blood bag. This blend also showed a compatible cell response in contact with L929 fibroblast cells and fewer tendencies to interaction with platelets compared to the PVC blood bag. Although the blends were immiscible and no chemical reaction at the interface could be found, the blood compatibility of the blends was reported to be improved.

Phase behavior of the binary polymer system consisting of TPU and SAN copolymers of various compositions (100/0, 90/10, 70/30, 50/50, 30/70, 0/100) has been investigated by Jaisankar and coworkers [100]. The phase separation on heating, i.e., lower critical solution temperature (LCST) behavior, was found to occur when these blends were heated to elevated temperatures. The cloud-point temperature or phase separation was found to vary with TPU/SAN blend compositions. They came to the conclusion that the extent of partial miscibility of the two components is very limited.

The flame retardancy, and mechanical properties of a pure PP/TPU blend, and PP/TPU blends containing nanoclay as flame retardants, were investigated by Kannan and coworkers [101]. The limiting oxygen index (LOI) value of the neat PP/TPU blend was 18% and those of the halogen based bromine-containing (flame retardant)-based PP/TPU blend and PP/TPU(5% nanoclay-filled)/MA-g-PP (maleic anhydride-grafted-PP) blends were increased by 32% and 28.5% compared to the pure PP/TPU blend, which showed that the above flame retardants and nanoclay could greatly increase the LOI value of the system. However, they could observe a deterioration of mechanical properties for bromine-containing (flame retardant) compositions. The introduction of the nanoclay to the TPU/PP blend system could also increase the mechanical properties along with flame retardant properties. The tensile stress required to make 200% strain in 5% nanoclay added TPU/PP/MA-g-PP nanocomposite increased by 120%, and the flexural modulus increased by 80–90%.

An attempt was made to prepare PVC/PU blends via direct formation of the thermosetting minor phase from its reactants during blending into PVC by Hamed and Navid [102]. The mechanical properties of PVC/PU blends prepared by reactive blending showed that incorporation of PU elastomer into PVC resulted in an increase in the elongation at break and toughness of the brittle PVC.

The sorption–desorption–resorption–redesorption experiments have been conducted to evaluate the transport behavior of polyethylene glycol (PEG) based polyurethane/polymethyl methacrylate (PU/PMMA, 50/50) semi interpenetrating polymer network (SIPN) with aromatic probe molecules [103]. Molecular migration depends on the nature of the organic solvent, membrane–solvent interaction, temperature, solubility parameter, molecular volume and free volume available within the polymer matrix. The transport parameters like sorption (S), diffusion (D), and permeation (P) coefficients have been calculated for IPN-aromatic probe molecule systems. Sorption and diffusion results measured at different temperatures viz. 20, 40, and 60°C are found to follow Fickian mechanism.

A blend of PU and PMMA was prepared by Patrıcio et al. [104] for the gas separation. They observed that the amorphous system changed from fully elastomeric to partially glassy as PMMA is added to PU. They reported that the blend materials presented partial miscibility associated with hydrogen bond interaction between urethane and PMMA. The permeation of the gases H_2, N_2, O_2, CH_4, and CO_2 were studied. The gas can be predicted to pass through the loose amorphous region of the membrane. Therefore, it can be expected a decrease in permeability with an increase in the gas transport selectivity of these blends. The blends with 90, 80, 70, and 60 wt% PU showed lower permeabilities than those of the PU membrane. The decrease in permeability is well correlated with the decrease in average free volume size. In the case of the H_2/N_2 gas pair, selectivity increases with PMMA content in the blend, and no marked changes were observed for the other gas pairs. The improvement in H_2 selectivity is associated with the increase in rigid amorphous phases, which induces a complex phase-separated morphology in the blends.

Polyether flexible segment glass transition was reported to be at −51°C and no significant changes with PMMA content were observed in PU continuous phase in the blends. On the other hand, a PMMA rich phase presented glass transitions between 74°C and 90°C for blends with 80, 60, 40, and 20 wt% PU. Pure PMMA shows T_g of 117°C. In addition, they observed that the relative free volume content of the samples was not changed significantly.

2.6 APPLICATIONS OF BLENDS AND IPNs OF POLYURETHANE

These blends and IPNs find application in a wide range of areas such as manufacturing of automotive seating, furniture, packaging, footwear, construction, electronics, and biomedical devices. They are used for these applications as they offer numerous advantages, including being lightweight, having excellent strength, low thermal conductivity, high weight-carrying capacity, and haemocompatibility.

Polyurethanes are combined with nylon to make more lightweight, stretchable garments, such as sports clothes. With its excellent strength-to-weight ratio, insulation properties, durability, versatility and corrosion resistance, polyurethane blends and IPNs are frequently used in automotive, building and construction applications too. They are suitable materials for using as foam to make cushions, carpets, interior lining and ceilings. PUs are materials with good biocompatibility, haemocompatibility and other performance characteristics, which make them suitable for a wide range of medical application such as for blood bags, surgical gloves, catheters, synthetic veins, and, more recently, wound dressings. Some common articles made of PU are shown in Fig. 2.4.

2.7 FUTURE SCOPE OF PU BLENDS AND IPNs

The excellent properties of PU-based blends and IPNs open up new avenues and opportunities in this field. The end use of a product depends upon the properties of the material from which it is developed. These blends can be used for the fabrication of many household articles, blood bags, shoe soles, seals, gaskets, insulation materials, cables, molded articles, and toys. There is enough scope for further investigations on these blends. The effect of different fillers of various dimensions on the properties of these blends can be explored. The use of polymeric compatibilization additives to polymer to improve mechanical compatibility in phase-separated blends can be a subject of future research programs. The applications of these blends in membrane technology, however, needs further exploration.

Figure 2.4 Examples of articles made from PU material.

2.8 CONCLUSION

The shortcomings of individual polymers can be overcome by blending them with other polymers. The blends and IPNs developed are have properties better than the individual polymers. By blending, we can arrive at new polymeric material of desired and improved properties. Polyurethane can be blended with both NR and synthetic rubber such as NBR, PDMS, EVM, and EPDM. They also form compatible blends with other polymers like PP, PMMA, SAN, and PVC. The blends and IPNs of PU with other polymers have improved flame retardency, mechanical properties, transport properties, thermal stability, corrosion resistance biocompatibility rather than that seen with individual polymers.

REFERENCES

[1] W.N. John, Special topics in polymer chemistry, in: The Chemistry of Polymers, fourth edition, Royal Society of Chemistry, United Kingdom, 2012.
[2] P. Poomalai, Siddaramaiah, Studies on poly(methyl methacrylate) (PMMA) and thermoplastic polyurethane (TPU) blends, J. Macromol. Sci. A 42 (10) (2005) 1399–1407.
[3] L.A. Utracki (Ed.), Polymer Blends Handbook 1, Kluwer Academic Publishers, Dordrecht, 2002.
[4] T.K. Kim, B.K. Kim, S.Y. Lee, Y.L. Cho, M.S. Kim, H.M. Jeong, Thermoplastic polyurethane elastomer/thermoplastic polyolefin elastomer blends compatibilized with a polyolefinic segment in TPU, Macromol. Res. 18 (2010) 177–184.
[5] D.R. Paul, S. Newman, Polymer Blends 2, Academic Press, New York, 1978, pp. 35–89.
[6] J.G. Boner, P.S. Hope, Polymer Blends and Alloys, Chapman & Hall, Norwell, 1993.
[7] L.H. Sperling, An Introduction to polymer networks and IPNs, in: Interpenetrating Polymer Networks and Related Materials, Plenum Press, New York, 1981, pp. 1–10.
[8] V.D. Athawale, S.L. Kolekar, S.S. Raut, Recent developments in polyurethanes and poly(acrylates) interpenetrating polymer networks, J. Macromol. Sci. C 43 (1) (2003) 1–26.
[9] M.A.G. Jimenez, J.L.R. Armenta, A.M.M. Martinez, J.G.R. Muniz, N.A.R. Vazquez, E.T. Rojas, Interpenetrating polymer networks based on castor oil polyurethane/cellulose derivatives and polyacrylic acid, Lat. Am. Appl. Res. 39 (2009) 131–136.
[10] P. Pissis, G. Georgoussis, V.A. Bershtein, E. Neagu, A.A. Fainleb, Dielectric studies in homogeneous and heterogeneous polyurethane/polycyanurate interpenetrating polymer networks, J. Non-Cryst. Solids 305 (2002) 150–158.
[11] O. Grigoryeva, O. Slisenko, E. Lebedev, A. Bartolotta, Giovanni Carini, Giuseppe Carini, et al., Effect of blending on the structure of thermoplastic interpenetrating polymer networks, J. Macromol. Sci. B 46 (1) (2007) 55–69.
[12] F.O.V. da Cunha, D.H.R. Melo, V.B. Veronese, M.M.C. Forte, Study of castor oil polyurethane – poly(methyl methacrylate) semi-interpenetrating polymer network (SIPN) reaction parameters using a 23 factorial experimental design, Mater. Res. 7 (2004) 539–543.
[13] L.H. Sperling, R. Hu, Interpenetrating Polymer Networks, in: Polymer Blends Handbook, Springer, Dordrecht, Netherlands, 2003, pp. 417–447.
[14] M. Szycher, Szycher's Handbook of Polyurethanes, CRC Press, Boca Raton, USA, 1999.
[15] Z.S. Petrovic, J. Ferguson, Polyurethane elastomers, Prog. Polym. Sci. 16 (1991) 695–836.
[16] J. Mattia, P. Painter, A comparison of hydrogen bonding and order in a polyurethane and poly(urethane – urea) and their blends with poly(ethylene glycol), Macromolecules 40 (2007) 1546–1554.
[17] J. Bernardini, P. Cinelli, I. Anguillesi, M.-B. Coltelli, A. Lazzeri, Flexible polyurethane foams green production employing lignin or oxypropylated lignin, Eur. Polym. J 64 (2015) 147–156.
[18] H.-T. Chiu, J.-H. Wu, A study on the curing behavior of silicone/polyurethane/epoxy blends by rigid-body pendulum rheometer, Polym. Plast. Technol. Eng. 45 (2006) 1081–1085.

[19] M. Fu, H.J. Sun, Polyurethane Elastomers and Their Applications, Second ed., China Chemical Industry Press, Beijing, 1999.
[20] G. Woods, Flexible Polyurethane Foams: Chemistry and Technology, Applied Science Publishers, London, 1982.
[21] M. Szycher, Szycher's Handbook of Polyurethanes, Second ed., Taylor & Francis, Boca Raton, FL, 2013.
[22] C. Zhang, M.R. Kessler, Bio-based polyurethane foam made from compatible blends of vegetable-oil-based polyol and petroleum-based polyol, Sustain. Chem. Eng. 3 (2015) 743–749.
[23] H.M. Jeong, J.H. Song, S.Y. Lee, B.K. Kim, Miscibility and shape memory property of poly(vinyl chloride)/thermoplastic polyurethane blends, J. Mater. Sci. 36 (2001) 5436–5457.
[24] M. Raja, S.H. Ryu, A.M. Shanmugharaj, Thermal, mechanical and electroactive shape memory properties of polyurethane (PU)/poly (lactic acid) (PLA)/CNT nanocomposites, Eur. Polym. J 49 (2013) 3492–3500.
[25] Y. Sun, L. Li, D. Wen, X. Bai, Bistable electrical switching characteristics and memory effect by mixing of oxadiazole in polyurethane layer, J. Phys. Chem. C 119 (2015) 19520–19525.
[26] A.T. Oan, J. Soonwk, Influence of interchange reactions on the miscibility of polyesterurethanes/polycarbonate binary blends, J. Appl. Polym. Sci. 64 (1997) 2363.
[27] G. Kumar, M.R. Arindam, N.R. Neelakanian, N. Subramanian, Stress relaxation behavior of polyacetal–thermoplastic polyurethane elastomer blends, J. Appl. Polym. Sci. 50 (1993) 2209–2216.
[28] W.Y. Chiang, M.S. Lo, Properties of copolymer-type polyacetal/polyurethane blends, J. Appl. Polym. Sci. 36 (1988) 1685–1700.
[29] Y.Q. Liu, Y.F. Zhu, J.R. Xu, J. Sun Yat-Sen Univ. 41 (2002) 49.
[30] X.Q. Xu, X.H. Xu, X.M. Tian, Y.N. Zhao, Chin. Plast. 9 (1999) 46.
[31] C.B. Ye, J. Zhang, Chin. Plast. 10 (2003) 1.
[32] C.B. Ye, J. Zhang, S.Z. Zhou, Chin. Plast. 8 (2004) 48.
[33] S.N. Jaisankar, G. Radhakrishnan, Effect of compatibilizer on morphology and mechanical properties of TPU/SAN blends, Polym. Eng. Sci. 40 (2000) 621–626.
[34] S. Zhang, G. Wang, Z.H. Jiang, W.C. Wu, R.T. Ma, Z.W. Wu, Melting and crystallization behavior of PA1010/TPU blends, J. Appl. Polym. Sci. 94 (2004) 839–844.
[35] S. Zhang, G. Wang, Z.H. Jiang, W.C. Wu, R.T. Ma, Z.W. Wu, Impact properties, phase structure, compatibility, and fracture morphology of polyamide-1010/thermoplastic poly(ester urethane) elastomer blends, J. Polym. Sci. B Polym. Phys. 43 (2005) 1177–1185.
[36] C. Nakason, K. Nuansomsri, A. Kaesaman, S. Kiatkamjornwong, Dynamic vulcanization of natural rubber/high-density polyethylene blends: effect of compatibilization, blend ratio and curing system, Polym. Test 25 (6) (2006) 782–796.
[37] S. Ramarad, M. Khalid, C.T. Ratnam, A. Luqman Chuah, W. Rashmi, Waste tire rubber in polymer blends: a review on the evolution, properties and future, Prog. Mater. Sci. 72 (2015) 100–140.
[38] J. Jobish, C. Nakason, P. Praveen, Dielectric properties and AC conductivity studies of novel NR/PVA full-interpenetrating polymer networks, J. Non-Cryst. Solids 358 (8) (2012) 1113–1119.
[39] N. Hayeemasae, H. Ismail, Improving the tensile properties of natural rubber compounds containing ground ethylene propylene diene rubber waste by two-stage processing, Proc. Chem. 19 (2016) 810–815.
[40] M.A. Minnath, G. Unnikrishnan, E. Purushothaman, Transport studies of thermoplastic polyurethane/natural rubber (TPU/NR) blends, J. Membr. Sci. 379 (2011) 361–369.

[41] M.A. Minnath, E. Purushothaman, Natural rubber/engineering thermoplastic elastomer blends, in: S. Thomas, K.R. Rajisha, H.J. Maria, C.H. Chan, L.A. Pothen (Eds.), Natural Rubber Materials, Volume 1: Blends and IPNs.(RSC Polymer Chemistry Series No. 7)., Royal Society of Chemistry, London, 2014.
[42] M.A. Minnath, PhD Thesis, Development and characterisation of thermoplastic poyurethane – natural rubber / epoxidised natural rubber (TPU-NR/ENR) blends and their chitin reinforced composites. University of Calicut, 2013.
[43] S. Pichaiyuta, C. Nakasona, C. Kummerlöweb, N. Vennemann, Thermoplastic elastomer based on epoxidized natural rubber/thermoplastic polyurethane blends: influence of blending technique, Polym. Adv. Technol. 23 (2012) 1011–1019.
[44] M.N. Radhakrishnan Nair, M.R. Gopinathan Nair, Thermogravimetric analysis of PVC/NR-b-PU blends, J. Therm. Anal. Calorim. 103 (2011) 863–872.
[45] X. Zhang, Z. Lu, D. Tian, H. Li, C. Lu, Mechanochemical devulcanization of ground tire rubber and its application in acoustic absorbent polyurethane foamed composites, J. Appl. Polym. Sci. 127 (2013) 4006–4014.
[46] M.N. Radhakrishnan Nair, M.R. Gopinathan Nair, Compatibility studies and characterisation of a PVC/NR blend system using NR/PU block copolymer, Polym. Bull 56 (2006) 619–631.
[47] E. Kalkornsurapranee, C. Nakason, C. Kummerlowe, N. Vennemann, Development and preparation of high-performance thermoplastic vulcanizates based on blends of natural rubber and thermoplastic polyurethanes, J. Appl. Polym. Sci. 128 (2013) 2358.
[48] S. Varghese, K.G. Gatos, A.A. Apostolov, J. Karger-Kocsis, Morphology and mechanical properties of layered silicate reinforced natural and polyurethane rubber blends produced by latex compounding, J. Appl. Polym. Sci. 92 (2004) 543–551.
[49] Y.F. Su, Master's Thesis, Beijing University of Chemical Technology, (2003).
[50] R. Dhamodharan, P. Maiti, G. Radhakrishnan, Study on blends of ethylene propylene diene rubber and thermoplastic polyurethanes, Polym. Plast. Technol. Eng. 46 (2007) 163.
[51] R. Dhamodharan, P. Maiti, G. Radhakrishnan, Studies on ethylene propylene diene rubber and thermoplastic polyurethane blends: the effect of maleation, Polym. Plast. Technol. Eng. 47 (11) (2008) 1081–1089.
[52] J.K. Mishra, S. Gupta, S. Raychowdhury, C.K. Das, On the heat shrinkability and interchain crosslinking of blends consisting of grafted ethylene-propylene polymer and polyurethane elastomer, Macromol. Mater. Eng. 286 (2001) 516–523.
[53] L. Wu, X. Luo, X. Wang, Influence of processing conditions on dual-phase continuous blend system of thermoplastic polyurethane with ethylene-propylene-diene monomer elastomer, J. Appl. Polym. Sci. 102 (2006) 5472–5482.
[54] X. Wang, X. Luo, A polymer network based on thermoplastic polyurethane and ethylene–propylene–diene elastomer via melt blending: morphology, mechanical properties, and rheology, Eur. Polym. J 40 (2004) 2391.
[55] J. Tan, Y.M. Ding, X.T. He, Y. Liu, Y. An, W.M. Yang, et al., Abrasion resistance of thermoplastic polyurethane materials blended with ethylene–propylene–diene monomer rubber, Polym. Sci. 110 (2008) 1851–1857.
[56] J.-H. Kim, G.-H. Kim, Effect of rubber content on abrasion resistance and tensile properties of thermoplastic polyurethane (TPU)/rubber blends, Macromol. Res. 22 (5) (2014) 523–527.
[57] P.C. Painter, M.M. Coleman, Fundamentals of Polymer Science, Technomic Publ. Co, Lancaster, 1997, pp. 212–217.
[58] D. Frenkel, in: M.E. Cates, M.R. Evans (Eds.), Soft and Fragile Matter, SUSSP Publications and Institute of Physics Publishing, Edinburg, London, 2000, pp. 115–119.

[59] Z. Susteric, I. Dimitrievski, Morphology and finite strain rheology of NBR/TPU blends. Int, J. Polym. Mater. Polym. Biomater. 52 (6) (2003) 527–548.
[60] K.I. Suresh, E.T. Tachil, Study on blending of millable polyurethane derived from castor oil with natural rubber, Angew. Macromol. Chem. 224 (1995) 55–60.
[61] T. Tang, C.P. Hu, S.K. Ying, Y.X. Zhang, Adv. Polym. Blends Alloys Technol. 4 (1993) 1.
[62] S.K. Roy, A. Singh, C.K. Das, Polym. Compos. 3 (6) (1995) 403.
[63] I. Dimitrievski, T. Malavasic, Kovine Zlitine Tehnol. 28 (1–2) (1994) 311.
[64] T. Saito, T. Kawasaki, Y. Higashide, Jpn. Kokai Tokyo Koho JP 09087972 A2, (1997).
[65] S. Desai, I.M. Thakore, A. Brennan, S. Devi, Polyurethane-nitrile rubber blends, J. Macromol. Sci. A 38 (7) (2001) 711–729.
[66] M. Maity, B.B. Khatua, C.K. Das, Polyblend systems of polyurethane and HNBR elastomers (peroxide cure). Int, J. Polym. Mater. Polym. Biomater. 49 (4) (2001) 407–417.
[67] M. Kotal, S.K. Srivastava, A.K. Bhowmick, Thermoplastic polyurethane and nitrile butadiene rubber blends with layered double hydroxide nanocomposites by solution blending, Polym. Int. 59 (2010) 2–10.
[68] N. Mahmood, A.U. Khan, Z. Ali, M. Soh, Khan, A. - Ul Haq, et al., Preparation and characterization of thermoplastic polyurethane–urea and carboxylated acrylonitrile butadiene rubber blend nanocomposites, J. Appl. Polym. Sci. 123 (2012) 3635–3643.
[69] M. Tahir, K.W. Stockelhuber, N. Mahmood, H. Komber, G. Heinrich, Reactive blending of nitrile butadiene rubber and in situ synthesized thermoplastic polyurethane-urea: novel preparation method and characterization, Macromol. Mater. Eng. 300 (2015) 242–250.
[70] N.R. Manoj, P. Sivaraman, S. Praveen, R.D. Raut, L. Chandrasekhar, B.C. Chakraborty, Miscibility and fracture toughness of thermoplastic polyurethane/poly (vinyl chloride)/nitrile rubber ternary blends with special reference to the blend composition, J. Appl. Polym. Sci. 97 (2005) 1763–1770.
[71] U. Sebenik, J. Karger-Kocsis, M. Krajnc, R. Thomann, Dynamic mechanical properties and structure of in situ cured polyurethane/hydrogenated nitrile rubber, J. Appl. Polym. Sci. 125 (2012) 41–48.
[72] F. Abbasi, H. Mirzadeh, A.-A. Katbab, Bulk and surface modification of silicone rubber for biomedical applications, Polym. Int. 51 (2002) 882–888.
[73] J.A. Gopi, G.B. Nando, Optimization of the processing parameters in melt blending of thermoplastic polyurethane and poly dimethyl siloxane rubber, J. Elastom. Plast. 44 (2012) 189–204.
[74] K. Rajan, A. Al-Ghamdi, P. Ramesh, G.B. Nando, Blends of thermoplastic polyurethane (TPU) and polydimethyl siloxane rubber (PDMS), part-I: assessment of compatibility from torque rheometry and mechanical properties, J. Polym. Res. 19 (2012) 1–13.
[75] K. Rajan, A. Al-Ghamdi, P. Ramesh, G.B. Nando, Blends of thermoplastic polyurethane and polydimethylsiloxane rubber: assessment of biocompatibility and suture holding strength of membranes, Int. J. Biomater. (2013) 58–65.
[76] S.S. Pathak, A. Sharma, A.S. Khanna, Value addition to waterborne polyurethane resin by silicone modification for developing high performance coating on aluminum alloy, Prog. Org. Coat 65 (2009) 206–216.
[77] H. Sardon, L. Irusta, M.J.F. Berridi, M. Lansalot, E. Bourgeat-Lami, Synthesis of room temperature self-curable waterborne hybrid polyurethanes functionalized with (3-aminopropyl)triethoxysilane (APTES), Polymer 51 (2010) 5051–5057.

[78] L.L. Zhai, R.W. Liu, F. Peng, Y.H. Zhang, K. Zhong, J.X. Yuan, et al., Synthesis and characterization of nanosilica/waterborne polyurethane end-capped by alkoxysilane via a sol-gel process, J. Appl. Polym. Sci. 128 (2013) 1715–1724.
[79] S.G. Zhang, M. Guo, Z.D. Chen, Q.H. Liu, X. Liu, Grafting photosensitive polyurethane onto colloidal silica for use in UV-curing polyurethane nanocomposites. Colloids Surf, A Physicochem. Eng. Aspects 443 (2014) 525–534.
[80] H. Sardon, L. Irusta, P. Santamaria, M.J.F. Berridi, Thermal and mechanical behaviour of self-curable waterborne hybrid polyurethanes functionalized with (3-aminopropyl)triethoxysilane (APTES), J. Polym. Res. 19 (2012) 9956.
[81] M.M. Rahman, H.H. Chun, H. Park, Preparation and properties of waterborne polyurethane-silane: A promising antifouling coating, Macromol. Res. 19 (2011) 8–13.
[82] S. Yue, Z. Zhang, X. Fan, P. Liu, C. Xiao, Effect of 3-aminopropyltriethoxysilane on solvent resistance, thermal stability, and mechanical properties of two-component waterborne polyurethane, Intern. J. Polym. Anal. Character. 20 (4) (2015) 285–297.
[83] R. Hernandez, J. Weksler, A. Padsalgikar, Microstructural organization of three-phase polydimethylsiloxane-based segmented polyurethanes, J. Runt Macromol. 40 (2007) 5441–5449.
[84] J.A. Gopi, G.B. Nando, Effect of processing parameters on phase morphology and mechanical properties of thermoplastic polyurethane and polydimethylsiloxane rubber blends, Int. J. Plast. Technol. (2015) 54–63.
[85] S. Damrongsakkul, R. Sinweeruthai, J.S. Higgins, Processability and chemical resistance of the polymer blend of thermoplastic polyurethane and polydimethylsiloxane, Macromol. Symp. 198 (2003) 411–419.
[86] Q.H. Zhang, Spec. Rub. Prod. 22 (2) (2001) 18–23.
[87] X. Shi, P. Zhang, Y. Liang, C. Ren, S. Zhao, Preparation and compatibility of EVM/TPU blend, J. Macromol. Sci. B. Phys. 46 (2007) 807–824.
[88] X. Shi, F. Zhao, J. Zhai, S. Zhao, SiO_2 reinforced EVM/TPU blends: bound rubber and rheology behavior, J. Macromol. Sci. B. Phys. 47 (2008) 1211–1227.
[89] X. Shi, L. Jiwen, S. Zhao, Properties and structure of dynamically vulcanized TPU/EVM blends, J. Macromol. Sci. B 48 (6) (2009) 1263–1274.
[90] L. Wu, H. Yu, J. Yan, B. You, Structure and composition of the surface of urethane/acrylic composite latex films, Polym. Int. 50 (12) (2001) 1288–1293.
[91] D. Kukanja, J. Golob, A. Zupancic-valant, M. Krajnc, The structure and properties of acrylic-polyurethane hybrid emulsions and comparison with physical blends, J. Appl. Polym. Sci. 78 (1) (2000) 67–80.
[92] A.S. Vatalis, C.G. Delides, O.P. Grigoryeva, L.O.M. Sergeeva, A.A. Brovko, N. Zimich, et al., Thermoplastic apparent interpenetrating polymer networks of polyurethane and styrene/acrylic acid copolymer obtained by melt mixing. Structure-property relationships, Polym. Eng. Sci. 40 (9) (2000) 2072–2085.
[93] N.S. Robin, K.C. Tapan, R. Sanjay, G.B. Nando, Studies on miscibility of blends of thermoplastic polyurethane and poly(ethylene-co-methyl acrylate), Die Angew. Makromol. Chem. 213 (1993) 7–13.
[94] W.K. Poetschke, H. Stutz, Influence of compatibilizer addition on particle size and coalescence in TPU/PP blends, J. Appl. Polym. Sci. 65 (11) (1997) 2217–2226.
[95] S.N. Jaisankar, G. Radhakrishnan, Effect of compatibilizer on morphology and mechanical properties of TPU/SAN blends, Polym. Eng. Sci. 40 (3) (2000) 621–626.
[96] K. Palanivelu, P. Sivaraman, M. Dasaratha, Studies on thermoplastic polyurethane toughened poly(butylene terephthalate) blends, Reddy Polym. Test 21 (3) (2002) 345–351.
[97] K. Palanivelu, P. Sivaraman, S.K. Sharma, Intl. J. Plast. Technol. 5 (1) (2002) 36–42.

[98] K. Palanivelu, P. Sivaraman, S.K. Sharma, S.K. Verma, Int. J. Plast. Technol. 7 (2) (2003) 133–140.
[99] H.A. Shadi, N.G. Ebrahimi, M.T. Khorasani, Study on thermoplastic polyurethane/polypropylene (TPU/PP) blend as a blood bag material, J. Appl. Polym. Sci. 89 (2003) 2496–2501.
[100] S.N. Jaisankar, Y. Lakshminarayana, G. Radhakrishnan, Miscibility and phase behavior in melt blends of thermoplastic polyurethane and styrene acrylonitrile copolymers, J. Macromol. Sci. B 43 (4) (2004) 755–762.
[101] M. Kannan, S. Thomas, K. Joseph, Flame-retardant properties of nanoclay-filled thermoplastic polyurethane/polypropylene nanocomposites, J. Vinyl Addit. Technol. (2015) 1–9.
[102] H. Haddadi, N.M. Famili, A reactive blending method for preparation of polyvinyl chloride/thermosetting polyurethane blends, Polym. Plast. Technol. Eng. 47 (2008) 255–263.
[103] H. Kumara, Siddaramaiah, A study of sorption/desorption and diffusion of substituted aromatic probe molecules into semi interpenetrating polymer network of polyurethane/polymethyl methacrylate, Polymer 46 (2005) 7140–7155.
[104] Effect of blend composition on microstructure, morphology, and gas permeability in PU/PMMA blends.

CHAPTER 3

Blends and IPNs of Polyurethane Polymers with Thermosetting Polymers

Sunan Tiptipakorn[1] and Sarawut Rimdusit[2]
[1]Department of Chemistry, Faculty of Liberal Arts and Science, Kasetsart University, Nakhon Pathom, Thailand
[2]Department of Chemical Engineering, Faculty of Engineering, Chulalongkorn University, Bangkok, Thailand

Contents

3.1 INTRODUCTION

Polyurethane (PU), first discovered in 1937, is a type of engineering polymer with various physical, chemical, and mechanical characteristics [1]. This kind of polymer possesses many beneficial aspects including a wide range of flexibility, great adhesion for various surfaces, great abrasion resistance, a transparent and glossy appearance, ability to resist environment severity, biocompatibility, durability, toughness, and processability [2,3]. The polymer can be synthesized by the reaction of a polyol and a diisocyanate. Due to a variety of polyol and diisocyanate types, a large spectrum of the polymeric materials can be formed to meet the requirements of specific uses [4]. This versatile polymeric materials has been utilized in fields that include building construction, automobile, coatings, adhesives, foams, elastomers, and medical devices [4–7]. Although PU possesses many beneficial characteristics as previously mentioned, some drawbacks of the polymer limit its applications, such as a slight yellowness, being relatively expensive and a drying requirement before thermal fabrication [1]. In

Polyurethane Polymers: Blends and Interpenetrating Polymer Networks
DOI: http://dx.doi.org/10.1016/B978-0-12-804039-3.00003-8

order to extend the usage range of applications, blending with other polymers and creating an interpenetrating network (IPN) has been proposed.

IPN is defined as a combination of at least two crosslinked polymer networks which possess full or partial physical interlocking between them [8]. IPNs can be typically classified into two main types: semi-interpenetrating network (semi-IPN) or full interpenetrating network (full-IPN) [9]. In the semi-IPN system, only one polymeric component is crosslinked and another component is linear; for the full-IPN systems, the networks of both polymeric components are formed. In addition, an IPN can be classified into two types based on the preparation technique: sequential IPN (SIPN) and simultaneous interpenetrating network (SIN) [10–11]. SIPN is prepared by swelling a polymeric network with a monomer, which is polymerized in situ; SIN is prepared by blending all reactants (monomers or prepolymers, initiators and curing agents) before either component is polymerized. When comparing IPNs obtained from the two different synthesis techniques, the SIN usually provides a higher degree of intermixing because the compatibility of the monomer mixture is greater than that of polymer mixture [12].

There have been many studies on the blends and IPNs of polyurethane. In some systems, the obtained IPN provides outstanding thermal properties, high damping properties, and excellent mechanical properties. For example, Pandit et al. [13] prepared SIPNs of poly(ester-urethane) and polystyrene by curing polyurethane at room temperature and further heating to complete the polymerization of styrene. The results from dynamic mechanical analysis revealed that only one glass transition temperature (T_g) was found in the IPN with PS content up to 30 wt%. However, at the PS content of 50 wt%, the IPN exhibited two glass transition temperatures. Yang et al. [14] synthesized three IPN types of PU-polyacrylate (two SIPNs and one SIN). In these IPN systems, the urethane formed with thermal curing and the acrylate formed with photo-curing. The values of T_g of the resulting IPNs were close to the those of pure components. Recently, the SINs of UV-curable acrylate and epoxy resins were studied. [15] The results presented that the network of urethane acrylate resin was strengthened by an epoxy network, even though the phase separation behavior was found. Mahesh et al. [16] studied the mechanical and thermal properties of bismaleimide modified polyurethane-epoxy IPNs. They reported that the presence of polyurethane in the IPN system could increase the tensile and flexural strengths; however, the decreases of T_g and thermal stability were found.

In this chapter, some of the blends and IPNs of PU polymers with thermosetting polymers will be discussed.

3.2 BENZOXAZINE–URETHANE ALLOYS

Polybenzoxazine, a novel type of phenolic resin, has drawn attention as a high-performance engineering material due to its great mechanical properties and high thermal properties. In addition, many other beneficial aspects of this thermosets have been reported such as no need to use a catalyst upon curing, near zero shrinkage during polymerization, and low water absorption [17–18]. The poly(benzoxazine–urethane) alloy was first studied by Takeichi et al. [19]. The urethane prepolymer was prepared by the reaction of 2,4-butylene polyethylene adipate polyol (molecular weight of approximately 10,000 Da) and 2,4-tolylene diisocyanate (TDI); the benzoxazine monomer was synthesized from bis(3-phenyl-3,4-dihydro-2H-1,3-benzoxazinyl)-isopropane. The chemical bonds between hydroxyl group (HO–) of benzoxazine monomer and isocyanate group (–OCN) of urethane prepolymer were reported. Additionally, the rendered films of cured poly(benzoxazine–urethane) alloys were transparent, exhibiting no phase segregation between two components. These results could be confirmed by the single glass transition temperature determined by differential scanning calorimetry (DSC) [19] and corresponded to the work of Rimdusit et al. [17,20] who studied urethane prepolymer (PU). Fig. 3.1 illustrates the proposed reaction between the urethane prepolymer and benzoxazine monomer.

Rimdusit et al. [21] prepared the uniform copolymer films of benzoxazine resin (BA-a) and urethane prepolymer (PU) at various BA-a/PU weight ratios (100/0, 80/20, 60/40, 20/80, and 0/100) via sequential cure method; two steps were conducted: (1) moisture cured, and then (2) thermal cured. The spectra from Fourier Transform Infrared (FT-IR) from the thermal cure of the BA-a/PU blends at 20/80 ratio is exhibited in Fig. 3.2. The FT-IR spectra revealed the decrease of absorption peak at 2264 cm^{-1} which indicated that NCO-terminated group reacted with atmospheric moisture to produce crosslink networks. In addition, the moisture cure reaction produced a crosslink network which is further reacted to produce urea bonds at characteristic absorption bands at 1639 and 3291 cm^{-1} in the alloy network.

In Fig. 3.2A and B, the FT-IR spectra of the mixture (at BA-a/PU = 20/80) during the step of moisture cure presented the disappearance of NCO-terminated group. That means the progress of intermolecular or intramolecular reaction to form the urea linkage. However, the NCO-terminated group could be observed again beyond 150°C as seen

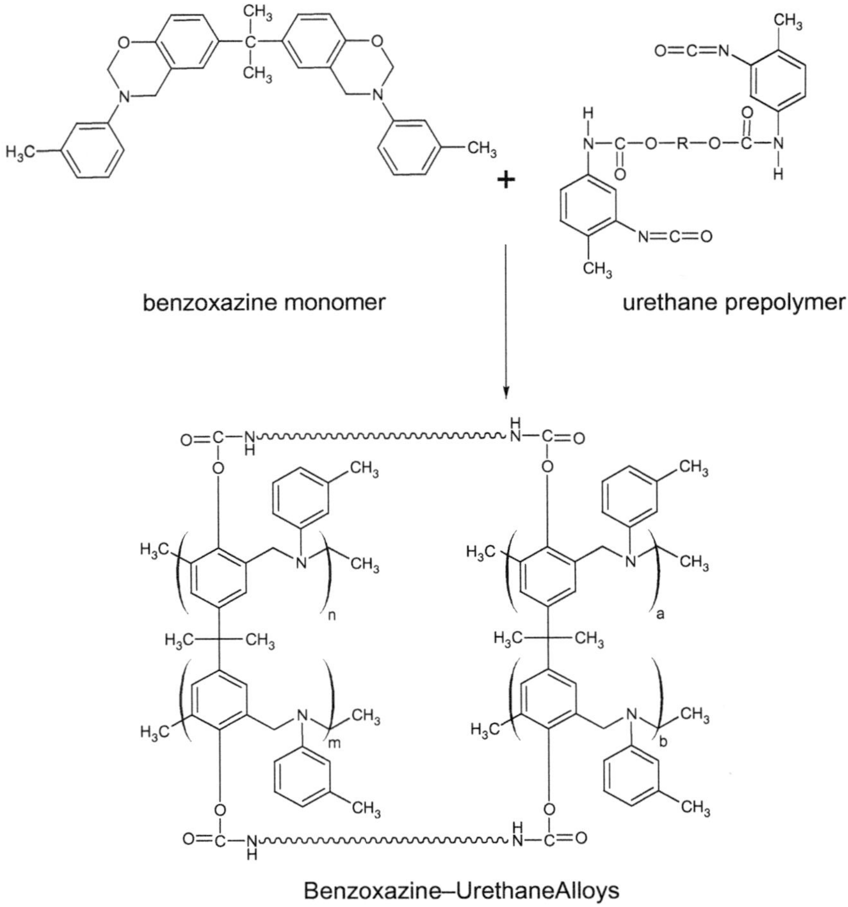

Figure 3.1 Proposed reaction of benzoxazine monomer and urethane prepolymer [13].

in Fig. 3.2a′–c′ because the R-NHCO- bond of biuret was destroyed and transformed to be a NCO-terminated group. The allophanate and biuret in this system was generated from the excess isocyanate in side reactions [22,23]. The allophanate was generated by the reaction of urethane group with isocyanate, while a biuret is formed by the reaction of –N-H groups of urea and isocyanate. The mechanism of revisable of isocyanate at high temperature was presented in the work of Narayan et al. [24]

The processing windows of BA-a/PU resin mixtures at various compositions are exhibited in Fig. 3.3. With the heating rate of 2°C/min,

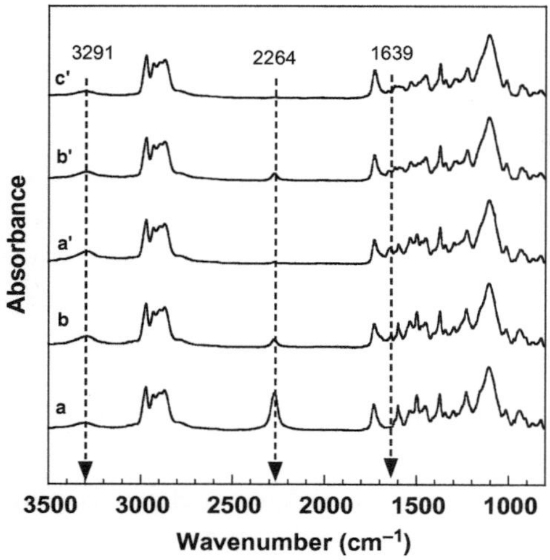

Figure 3.2 FT-IR spectra of the BA-a/PU blends at 20/80 ratio by moisture exposure: (A) 0 h; (B) 8 h; and thermal treated after moisture exposure: (a′) 150°C/1 h; (b′) 190°C/1 h, and (c′) 200°C/4 h.

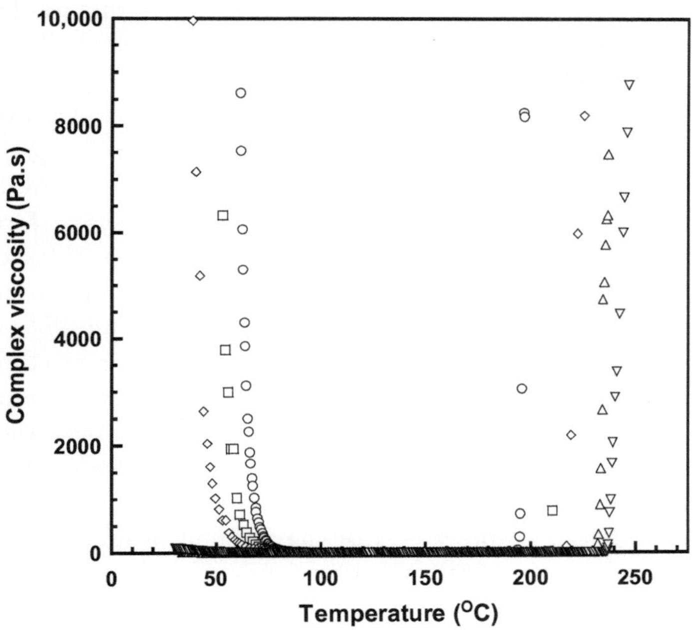

Figure 3.3 Processing windows of BA-a/PU resin mixtures at various compositions: (○) BA-a/PU 100/0, (□) BA-a/PU 80/20, (◇) BA-a/PU 60/40, (△) BA-a/PU 40/60, and (▽) BA-a/PU 20/80.

the temperature of all samples was ramped from 30°C to the temperature beyond the gel point; then dynamic viscosity was measured. On the left hand side of Fig. 3.3, the increase of urethane prepolymer mass fraction could lower their liquefying temperatures. This is attributed to the fact that the used urethane prepolymer was in liquid phase while the used BA-a resin was in solid phase at room temperature. When the mass fraction of PU was higher than mass fraction of benzoxazine resins, no liquefying temperature was observed. Lowering the liquefying temperature was obviously able to lower the processing temperature of a compounding process, which is desirable in many uses.

Furthermore, at the end of the A-stage viscosity, the resin mixtures underwent crosslinking reactions past their gel points, related to a sharp increase in their viscosities. In this case, the maximum temperature at which the viscosity was rapidly raised above 1000 Pa was used as gel temperature for each resin [25—27]. On the right hand side of Fig. 3.3, it can be seen that the gel point was found to increase with increasing the mass fraction of urethane prepolymer. This could be due to the dilution effect of the urethane on the polymerization of benzoxazine resin. The phenomenon of widening processing window is similar to that was found in the system of benzoxazine resin alloyed with epoxy diluent [28].

The thermomechanical properties of the BA-a/PU alloys obtained by traditional cure method (thermal cure method without prior moisture exposure) at various PU compositions are exhibited in Figs. 3.4 and 3.5. From Fig. 3.4, the storage modulus at glassy state tends to decrease with increasing the PU mass fraction in the polymer alloys. Moreover, the storage modulus at room temperature of the BA-a/PU alloys was reduced from 3.12 GPa to 0.1 GPa with the addition of the urethane prepolymer mass fraction from 0 to 40 wt%. It is because PU possesses higher flexibility than poly(BA-a). Additionally, the increase of urethane prepolymer content was able to increase the rubbery plateau. That means the crosslink density of the polymer network was increased with PU mass fraction [29]. Fig. 3.5 shows the loss modulus curves of the BA-a/PU alloys obtained by the traditional cure method as a function of temperature. It can be observed that the addition of urethane prepolymer into the benzoxazine can substantially increase the T_g of the polymer alloys [30].

The storage modulus and the loss modulus of the alloys obtained from moisture cure method reveal the same results as those obtained by traditional cure method. The effects of processing method on crosslink

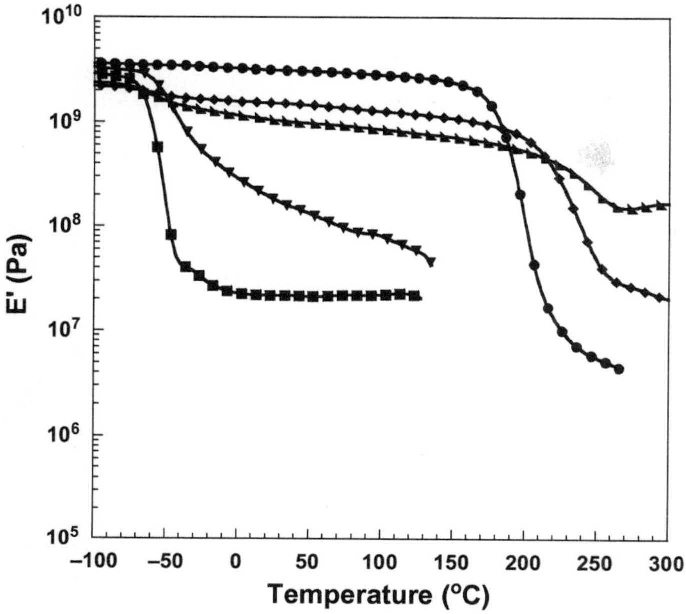

Figure 3.4 Storage modulus of BA-a/PU alloys obtained by traditional cure method at various PU compositions: (●) BA-a/PU 100/0, (♦) BA-a/PU 80/20, (◣) BA-a/PU 60/40, (▲) BA-a/PU 40/60 and (■) BA-a/PU 0/100.

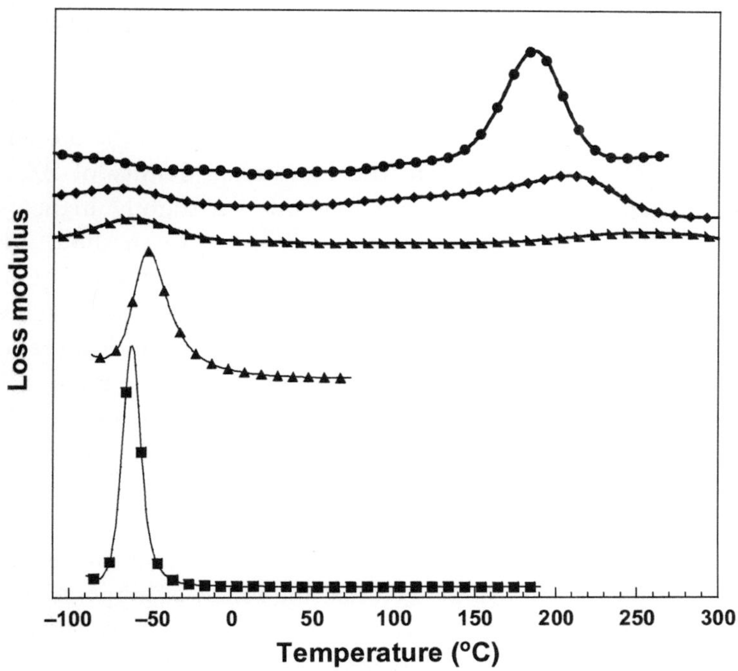

Figure 3.5 Loss modulus of BA-a/PU alloys synthesized by traditional cure method at various PU compositions: (●) BA-a/PU 100/0, (♦) BA-a/PU 80/20, (◣) BA-a/PU 60/40, (▲) BA-a/PU 40/60 and (■) BA-a/PU 0/100.

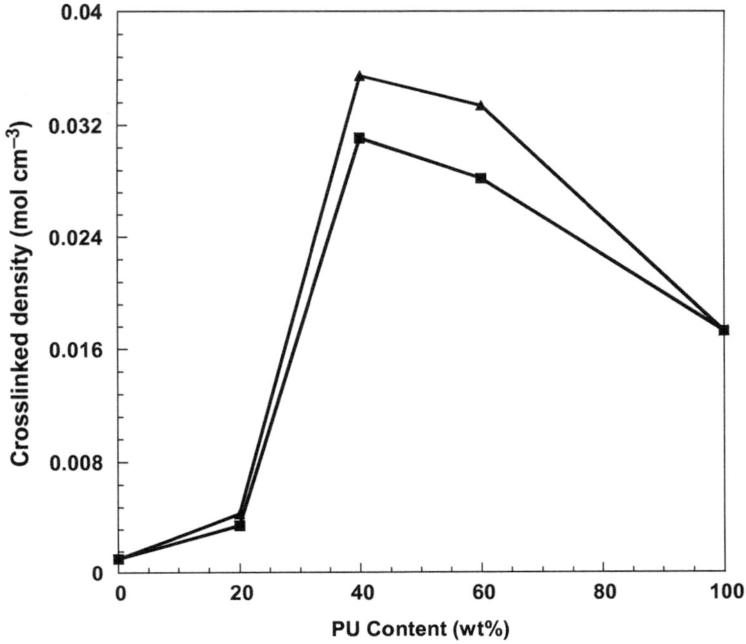

Figure 3.6 Crosslinked density of BA-a/PU alloys at various PU compositions: (▲) Traditional cure method, (■) Sequential cure method.

density and glass transition temperature are shown in Figs. 3.6 and 3.7 respectively. As seen in Fig. 3.6, the crosslinked density of BA-a/PU alloys obtained by traditional cure method was slightly higher value than BA-a/PU alloys obtained by sequential cure method. This is implied that the molecular motion in the BA-a/PU alloys obtained by traditional cure method is higher restricted than the alloys obtained by the sequential cure method. From Fig. 3.7, the glass transition temperature of BA-a/PU alloys obtained by the sequential cure method was a slightly higher value than the alloys obtained by the traditional cure method. It was reported that there was no significant difference of glass transition temperature between the alloys treated with different curing methods.

The initial decomposition temperatures of different processing are compared in Fig. 3.8. It can be seen that the values of initial decomposition temperature for the traditional cure method were higher than those of the sequential cure method. The maximum initial decomposition temperatures of all alloys were observed at the BA-a/PU ratio of 60/40 of both processing methods. The char yields of different processing are

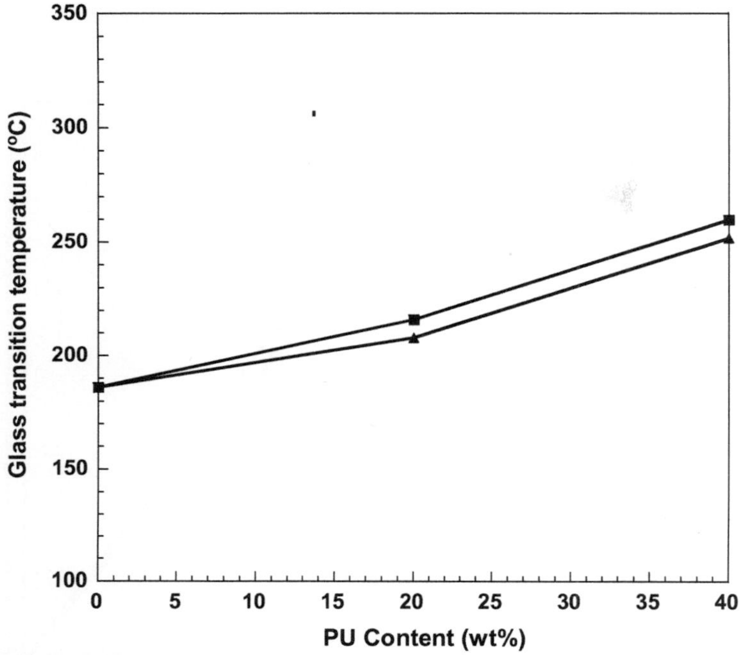

Figure 3.7 Glass transition temperature of BA-a/PU alloys at various PU compositions: (▲) Traditional cure method, (■) Sequential cure method. Note: the glass transition temperatures of the alloys at 60–100 wt% are in the minus range; therefore, the values are not shown in the figure.

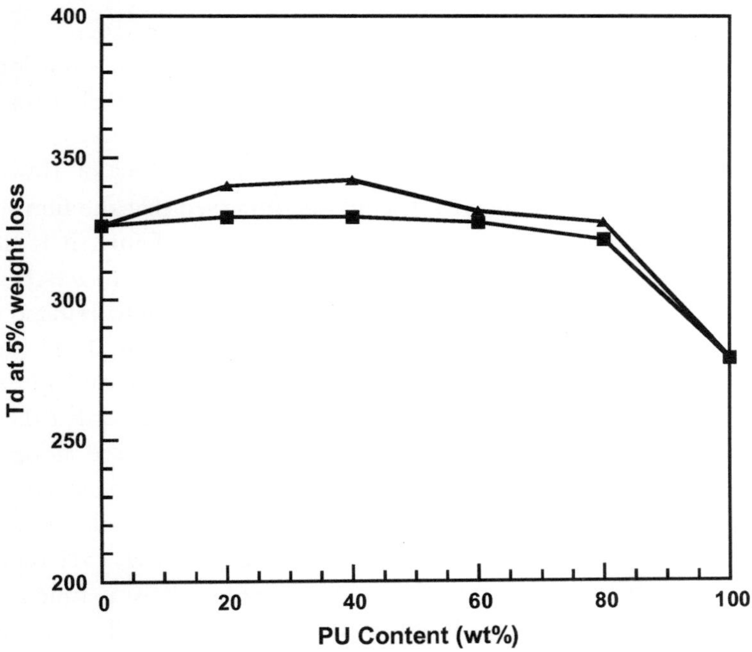

Figure 3.8 Initial degradation temperature at 5% weight loss of BA-a/PU alloys at various PU compositions: (▲) Traditional cure method, (■) Sequential cure method.

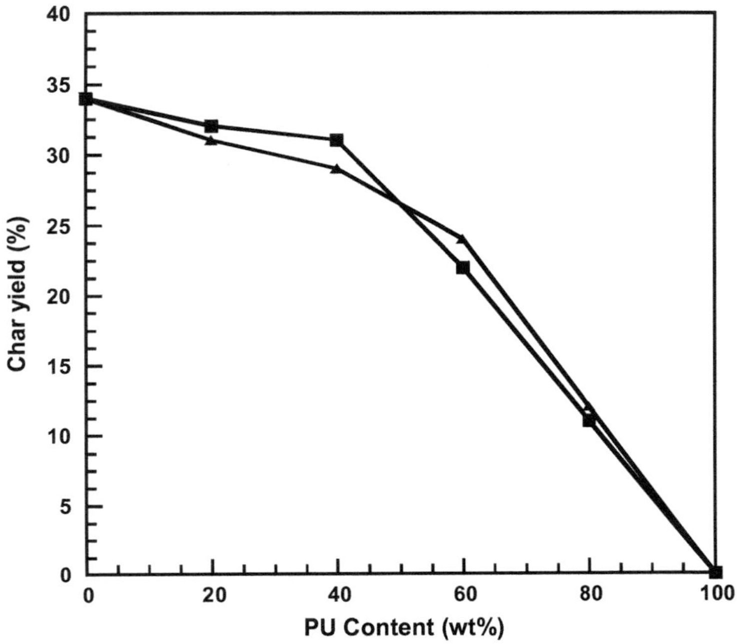

Figure 3.9 Char yield of BA-a/PU alloys at various PU compositions: (▲) Traditional cure method, (■) Sequential cure method.

compared in Fig. 3.9. Note that the addition of benzoxazine portion led to the increase of char yield. Furthermore, the results revealed that the both processing conditions have no effect on char yield.

The effects of PU content on the degree of swelling of BA-a/PU alloys are illustrated in Fig. 3.10. It could be observed that the film specimens swell at different degrees in chloroform. In this figure, it is clearly observed that the volumetric swelling degree of the BA-a/PU alloys under both processing conditions increased along with polyurethane content. The same behavior was reported by Feng and Li [30]. They studied volumetric swelling degree of polyester urethane (PULG) during degradation and before degradation in chloroform. They reported that the volumetric swelling degree of the copolymer network increased along with degradation time. This indicated that the polymer chain between crosslink points broke down due to the scission effects of molecules, ions, and radicals in the oxidative degradation solution. Pekel et al. [31] reported that volumetric expansion of the network reduced as a function of increasing crosslink density. Moreover, the volumetric swelling degree of

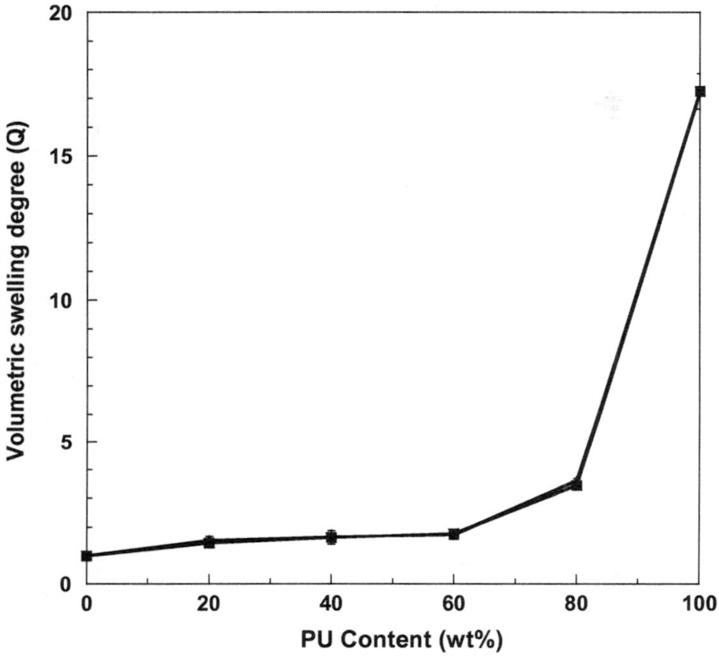

Figure 3.10 Volumetric swelling degree of BA-a/PU alloys at various PU compositions: (●) Traditional cure method, (■) Sequential cure method.

BA-a/PU alloys increased rapidly from 3.5 (BA-a/PU 20/80) to 17 (neat polyurethane) because of their broken network in chloroform. Comparing the swelling values of neat benzoxazine to BA-a/PU 40/60, the volumetric swelling degree of the Ba-a/PU alloys slightly increased with increasing the polyurethane content. Therefore, the benzoxazine structure (more than 40% mass fraction) combined with polyurethane can obstruct their swollen and broken network of polyurethane in chloroform.

3.3 SEQUENTIAL IPN OF POLYURETHANE ACRYLATE AND POLYBENZOXAZINE

Okhawilai et al. [32] prepared the hybrid polymer networks of high-performance UV-curable polyurethane acrylate (PUA) and thermally curable polybenzoxazine (PBA) by sequential cure methods, i.e., UV cure of the PUA and then thermal cure of the PBA component. The network formation between benzoxazine resin and urethane acrylate prepolymers

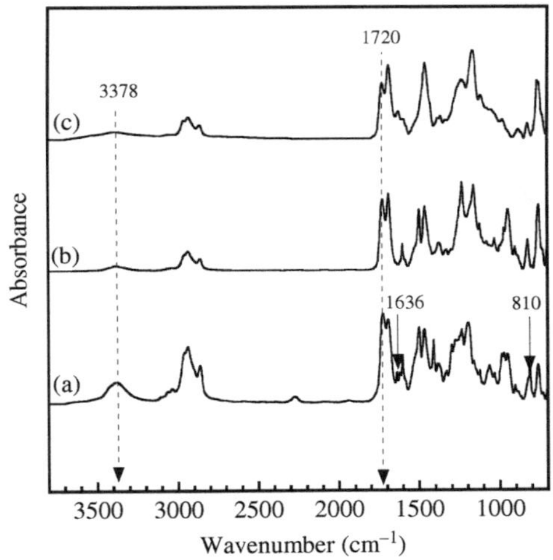

Figure 3.11 FT-IR spectra of PUA/BA-a alloys at 50/50 weight ratio: (a) PUA/BA-a mixture (b) PUA/BA-a after UV cure, (c) PUA/BA-a after UV and thermal cure.

were examined using the FT-IR technique. The main characteristic absorption bands of urethane acrylate prepolymer were presented in Fig. 3.11a, i.e. the bands at 810 cm^{-1} and 1636 cm^{-1} which were assigned to acrylate double bond. In Fig. 3.11b, these absorption bands obviously disappeared. In addition, the disappearance of 1720 and 3378 cm^{-1} assigned to NCO and OH groups also disappeared. This suggested that the urethane acrylate prepolymer converted to PUA after the exposure of UV light at room temperature. It could be implied that the reaction of vinyl group had completely occurred [33,34]. After thermal cure (Fig. 3.11c), the absorption peak of the tri-substituted benzene ring and the backbone of the benzoxazine ring became tetra-substituted. This phenomenon, attributed to the sequential cure method, consisted of two steps: (1) UV cure for polyurethane acrylate network formation first, followed by (2) thermal cure for BA-a/BA-a reactions. Therefore, it was confirmed that the IPN occurred in the system, i.e., the main reactions are the PUA polymerization and the self-polymerization of BA-a.

It has been reported that the system of benzoxazine and urethane resins exhibits interesting and unique synergistic effects on the glass transition temperature [35]. The advantage of sequential cure of the resin mixture is the fast forming of film with highly uniform thickness. An IPN type of both resins provide different curing mechanisms. The storage modulus at rubbery plateau region of this system is presented in

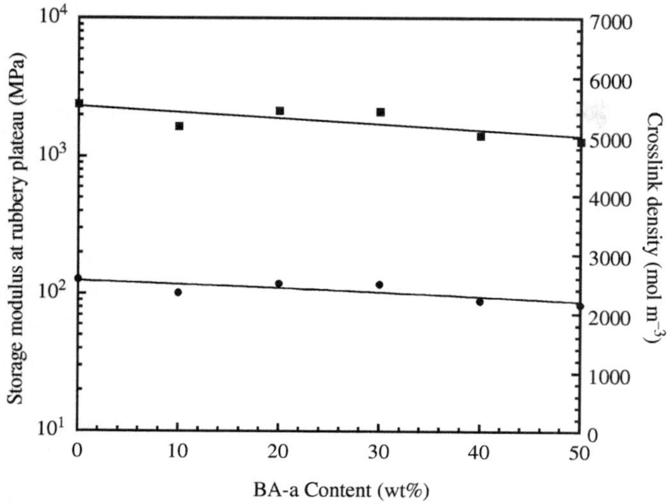

Figure 3.12 Storage modulus at rubbery plateau ($T_g + 50°C$) and crosslink density of PUA/BA-a film alloys at various BA-a compositions: (●) Storage modulus (■) Crosslink density.

Fig. 3.12. The storage modulus in the rubbery plateau region of the polymer alloys was decreased with increase PBA-a content, i.e., the values decreased from 127 MPa to 84 MPa with an addition of the BA-a fraction from 0 to 50 wt%. This result suggests that the increase in the BA-a content in the polymer alloys resulted in a decrease in crosslink density of the fully cured specimens. The crosslink density of a polymer network can be estimated from the theory of rubber elasticity by knowing its rubbery plateau modulus as suggested by Nielson [36,37]:

$$\log\left(\frac{E'}{3}\right) = 293(\rho_x) + 6.0 \quad (3.1)$$

where E' is a storage modulus in a rubbery plateau region, and ρ_x is a crosslink density that is the mole number of network chains per unit volume of the polymers. The crosslink density determined from Nielson's equation of PUA/PBA-a alloys is shown in Fig. 3.12. From the results, the crosslink densities of the PUA/BA-a polymer alloys were slightly decreased from 5554 to 4931 mol m^{-3} with an addition of the BA-a fraction from 0% to 50% by weight. Polybenzoxazine is a well-known thermoset that possesses a relatively low crosslink density compared to epoxy, phenolic, or polyurethane [35]. Some recent works reported the neat PBA-a having a crosslink density value ranging from 4000–4350 mol m^{-3} [38]. Although the rather low crosslink densities observed, polybenzoxazines exhibit high glass transition temperatures as a result of their unique intramolecular [O^{-}...NH^{+}] interactions [18,38].

Loss tangent (tan δ) curves of the DMA thermograms as presented in Fig. 3.13 reveal the glass transition temperature (T_g) and network characteristics of the PBA-a and the PUA/BA-a alloys. From the figure, the T_gs of the alloys were increased with the BA-a content, i.e., from 124°C (neat PUA) to 129°C, 148°C, 162°C, 180°C, and 210°C (at the BA-a contents of 10, 20, 30, 40, and 50 wt% respectively). Interestingly, the T_g of the neat PBA-a was presented at 188°C. Similar synergistic behavior in T_g was also reported in the polymer hybrids between polybenzoxazine and polyurethane [37]. It could be attributed to the complementary property of the two polymers, i.e., the improvement in crosslink density for the polybenzoxazine fraction by the presence of the PUA, while the increase in network rigidity for the polyurethane by the presence of the more rigid polybenzoxazine [18,38].

The loss tangent curves for each composition of polymer alloy reveals only one peak, associated with the α relaxation or glass transition. If the two starting materials present phase separation, two glass transition peaks for each homopolymer would be expected in the tan δ curve. The appearance of only single tan δ peak associated with the glass transition in

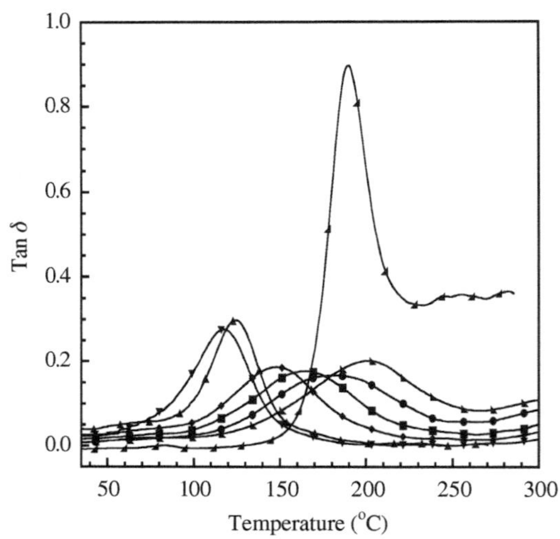

Figure 3.13 Loss tangent (tan δ) of PUA/BA-a film alloys at various compositions: (▼) 100/0, (▲) 90/10, (♦) 80/20, (■) 70/30, (●) 60/40, (◣) 50/50, (◢) 0/100.

the polymer alloys indicates that both resins in fact formed a highly compatible hybrid network without detectable phase separation.

Fig. 3.14 presents the visual appearance of the obtained PUA/PBA-a, which is relatively transparent with the color ranging from light (neat PUA) to dark (in PUA/BA-a = 50/50 and in the neat PBA-a). This optical property also confirms the highly compatible nature of the two polymers in the obtained hybrids.

Thermal stability of PUA/BA-a film alloys under the sequential cure process was investigated by thermogravimetric analysis (TGA) under N_2 atmosphere and the thermograms of the alloys at various BA-a contents are shown in Fig. 3.15. Thermal stability of PUA/BA-a film alloys were evaluated by comparing their degradation temperature (T_d) at 10% weight loss. The thermal decomposition in each alloy was mainly a single stage weight loss. The T_ds of the neat PBA-a and PUA were determined to be 344°C and 368°C, respectively. The decomposition temperatures of PUA/BA-a alloys showed at 90/10, 80/20, 70/30, 60/40, and 50/50

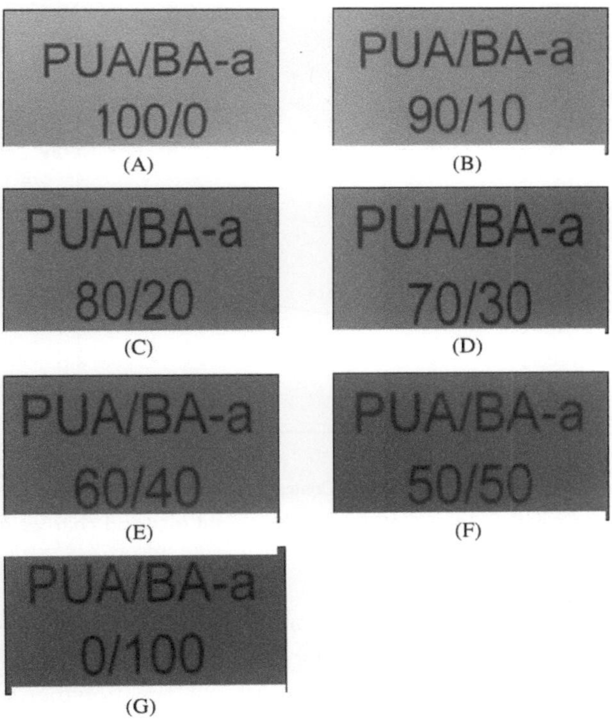

Figure 3.14 Appearance of PUA/BA-a alloy films at various composition: (A) 100/0, (B) 90/10, (C) 80/20, (D) 70/30, (E) 60/40, (F) 50/50, (G) 0/100.

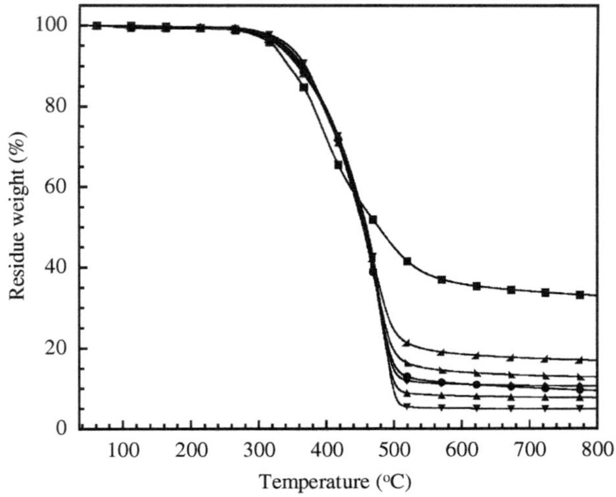

Figure 3.15 TGA thermogram of PUA/BA-a film alloys at various composition: (■) 100/0, (▲) 90/10, (♦) 80/20, (▼) 70/30, (●) 60/40, (◢) 50/50, (◣) 0/100.

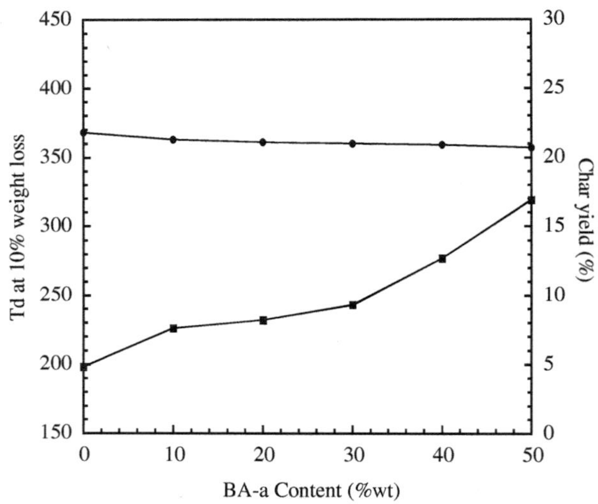

Figure 3.16 Degradation Temperature at 10% weight loss and char yield of PUA/BA-a film alloys at various BA-a contents.

were 363°C, 361°C, 360°C, 359°C, and 357°C, respectively. That means the incorporation of polybenzoxazine could enhance the decomposition temperature of the system. The T_d values of the polymer alloys were found to slightly decrease in a linear manner with increasing BA-a content as presented in Fig. 3.16. Additionally, the T_d values of the obtained mixture were found to be higher than those reported interpenetrating polymer network systems e.g., 301–367°C for blocked PU/epoxy [39], 328–334°C for PBA-a/EPU [40], and 338–341°C for PU/PA [41].

Furthermore, the char yield of PUA/BA-a alloys under the same sequential cure methods, reported at 800°C under N_2 atmosphere, is illustrated in Fig. 3.16. The char yield values of PUA/BA-a alloys at 90/10, 80/20, and 50/50 were reported to be 7.6%, 8.2%, and 16.9%, respectively. The improvement in char formation of the system presents a linear relationship with the amount of the polybenzoxazine as shown in Fig. 3.16. In principle, the higher value of the char yield provides a better fire resistant behavior of the polymer. The enhancement of fire resistant behavior of the hybrid was a result of the presence of greater char forming polybenzoxazine fraction.

3.4 CONCLUSIONS

In this chapter, the examples of hybrid system of polyurethane and thermoset polymers are given including benzoxazine monomer and urethane prepolymer and sequential IPN of polyurethane acrylate and polybenzoxazine. It could be found that the presence of polybenzoxazine (a novel class of phenolic resin) is able to enhance thermal properties and mechanical properties of the original polyurethane, which is beneficial to extend the usage range of the polymer.

ACKNOWLEDGMENTS

Part of the work presented in this book chapter has received financial support from the Ratchadaphiseksomphot Endowment Fund of Chulalongkorn University (CU-58-030-AM) and from Commission of Higher Education, Ministry of Education, Thailand.

REFERENCES

[1] V.R. Sastri, Engineering thermoplastics: acrylics, polycarbonates, polyurethanes, polyacetals, polyesters, and polyamides, in: Plastics in Medical Devices, second ed., Elsevier Inc, Oxford, 2014, pp. 121–173.
[2] D. Randall, S. Lee, The Polyurethanes Books, John Wiley & Sons, British, 2002.
[3] N.B. Shelke, R.K. Nagarale, S.G. Kumbar, Polyurethanes, in: S.G. Kumbar, C.T. Laurencin, M. Deng (Eds.), Natural and Synthetic Biomedical Polymers, Elsevier Science, Burlington, 2014, pp. 123–144.
[4] C. Oppon, P.M. Hackney, I. Shyha, M. Birkett, Effect of Varying Mixing Ratios and Pre-Heat Temperature on the Mechanical Properties of Polyurethane (PU) Foam, Proc. Eng. 132 (2015) 701–708.

[5] K. Piotr, Synthesis methods, chemical structures and phase structures of linear polyurethane, properties and applications of linear polyurethane in polyurethane elastomers, copolymers and ionomers, Prog. Mater. Sci. 52 (2007) 915–1015.
[6] M. Szycher, Szycher's Handbook of Polyurethanes, CRC Press, Boca Raton, FL, 1999.
[7] A. Bezazi, F. Scarpa, Mechanical behaviour of conventional and negative Poisson's ratio thermoplastic polyurethane foams under compressive cyclic loading, Int. J. Fatigue 9 (5) (2007) 922–930.
[8] L.H. Sperling, Interpenetrating Polymer Networks and Related Materials, Plenum Press, New York, 1981.
[9] C. Erbil, E. Kazancioghr, N. Uyanik, Synthesis, characterization and thermoreversible behaviours of poly(dimethyl siloxane)/poly(N-isopropyl acrylamide) semi-interpenetrating networks, Eur. Polym. J 40 (2004) 1145–1154.
[10] L.V. Karabanova, S.V. Mikhalovska, L.M. Sergeeva, S.T. Meikle, M. Helias, W. Lioyd, Semi-interpenetrating polymer networks based on polyurethane and poly (vinyl pyrrolidone) obtained by photopolymerization: Structure-property relationships and bacterial adhesion, Polym. Eng. Sci. 44 (2004) 940–947.
[11] C.H. Chen, M.H. Chen, Synthesis, thermal properties, and morphology of blocked polyurethane/epoxy full-interpenetrating polymer network, J. Appl. Polym. Sci. 100 (2006) 323–328.
[12] J.H. Kim, S.C. Kim, Analysis of reaction injection molding process of polyurethane-unsaturated polyester blends. Part II: Mechanical properties, Polym. Eng. Sci. 27 (1987) 1252–1257.
[13] S.B. Pandit, S.S. Kulkarni, V.M. Nadkarni, Interconnected Interpenetrating Polymer Networks of Polyurethane and Polystyrene. 2. Structure-Property Relationships, Macromolecules 27 (1994) 4595–4604.
[14] J. Yang, M.A. Winnik, D. Ylitalo, J. Devoer, Polyurethane – Polyacrylate Interpenetrating Networks. 1. Preparation and Morphology, Macromolecules 29 (1996) 7047–7054.
[15] F.J. Hua, C.P. Hu, Interpenetrating polymer networks of epoxy resin and urethane acrylate resin: 2. Morphology and mechanical property, Eur. Polym. J 36 (1999) 27–33.
[16] K.P.O. Mahesh, M. Alagar, S.A. Kumar, Mechanical, thermal and morphological behavior of bismaleimide modified polyurethane-epoxy IPN matrices, Polym. Adv. Technol. 14 (2003) 137–146.
[17] S. Rimdusit, C. Jubsilp, S. Tiptipakorn, Alloys and Composites of Polybenzoxazines, Springer, Singapore, 2013.
[18] H. Ishida, D.J. Allen, Physical and mechanical characterization of near-zero shrinkage polybenzoxazines, J. Polym. Sci. B Polym. Phys. 34 (1996) 1019–1030.
[19] T. Takeichi, Y. Guo, T. Agag, Synthesis and characterization of poly(urethane-benzoxazine) films as novel type of polyurethane/phenolic resin composites, J. Polym. Sci. A Polym. Chem. 38 (2000) 4165–4176.
[20] S. Rimdusit, S. Tiptipakorn, C. Jubsilp, T. Takeichi, Polybenzoxazine alloys and blends: Some unique properties and applications, React. Funct. Polym. 73 (2013) 369–380.
[21] S. Rimdusit, M. Sudjidjune, C. Jubsilp, S. Tiptipakorn, Enhanced film forming ability of benzoxazine-urethane hybrid polymer network by sequential cure method, J. Appl. Polym. Sci. (2014) 131 article number 40502.
[22] D.K. Chattopadhyay, K.V.S.N. Raju, Structural engineering of polyurethane coatings for high performance applications, Prog. Polym. Sci. 32 (2007) 352–418.
[23] D.K. Chattopadhyay, D.C. Webster, Thermal stability and flame retardancy of polyurethanes, Prog. Polym. Sci. 34 (2009) 1068–1133.

[24] R. Narayan, D.K. Chattopadhyay, B. Sreedhar, K.V.S.N. Raju, N.N. Mallikarjuna, T.M. Minabhavi, Synthesis and characterization of crosslinked polyurethane dispersions based on hydroxylated polyesters, J. Appl. Polym. Sci. 99 (2006) 368–380.
[25] S. Rimdusit, S. Pirstpindvong, W. Tanthapanichakoon, S. Damrongsakkul, Toughening of polybenzoxazine by alloying with urethane prepolymer and flexible epoxy: A comparative study, Polym. Eng. Sci. 45 (2005) 288–297.
[26] M.T. Huang, H. Ishida, Dynamic mechanical analysis of reactive diluent modified benzoxazine-based phenolic resin, Polym. Polym. Compos 7 (1999) 233–247.
[27] S. Rimdusit, N. Kampangsaeree, W. Tanthapanichakoon, T. Takeichi, N. Suppakarn, Development of wood-substituted composites from highly filled polybenzoxazine-phenolic novolac alloys, Polym. Eng. Sci. 45 (2007) 140–149.
[28] S. Rimdusit, P. Kunopast, I. Dueramae, Thermomechanical properties of arylamine-based benzoxazine resins alloyed with epoxy resin, Polym. Eng. Sci. 51 (2011) 1797–1807.
[29] L.R.G. Treloar, The Physics of Rubber Elasticity, Oxford University Press, London, 1949.
[30] Y. Feng, C. Li, Study on oxidative degradation behaviors of polyesterurethane network, Polym. Degrad. Stab. 91 (2006) 1711–1716.
[31] N. Pekel, F. Yoshii, T. Kume, O. Güven, Radiation crosslinking of biodegradable hydroxypropylmethylcellulose, Carbohydr. Polym. 55 (2004) 139–147.
[32] M. Okhawilai, K. Pudhom, S. Rimdusit, Synthesis and characterization of sequential interpenetrating polymer networks of polyurethane acrylate and polybenzoxazine, Polym. Eng. Sci. 54 (2014) 1151–1161.
[33] F. Wang, J.Q. Hu, W.P. Tu, Study on microstructure of UV-curable polyurethane acrylate films, Prog. Org. Coat 62 (2008) 245–250.
[34] J. Seo, E.S. Jang, F.H. Song, S. Choi, S.B. Khan, H. Han, Preparation and properties of poly(urethane acrylate) films for ultraviolet-curable coatings, J. Appl. Polym. Sci. 118 (2010) 2454–2460.
[35] S. Rimdusit, W. Bangsen, P. Kasemsiri, Chemorheology and thermomechanical characteristics of benzoxazine-urethane copolymers, J. Appl. Polym. Sci. 120 (2011) 3669–3678.
[36] L.E. Nielsen, Mechanical Properties of Polymers and Composites, 1st ed, Marcel Dekker, New York, 1974.
[37] L.E. Nielsen, R.F. Landel, Mechanical Properties of Polymers and Composites, 2nd ed, Marcel Dekker, New York, 1994.
[38] C.H. Lu, Y.C. Su, C.F. Wang, C.F. Huang, Y.C. Sheen, F.C. Chang, Thermal properties and surface energy characteristics of interpenetrating polyacrylate and polybenzoxazine networks, Polymer 49 (2008) 4852–4860.
[39] C. Chen, M. Chen, Synthesis, thermal properties, and morphology of blocked polyurethane/epoxy full-interpenetrating polymer network, J. Appl. Polym. Sci. 100 (2006) 323–328.
[40] H. Yeganeh, M. Razavi-Nouri, M. Ghaffari, Synthesis and properties of polybenzoxazine modified polyurethanes as a new type of electrical insulators with improved thermal stability, Polym. Eng. Sci. 48 (2008) 1329–1338.
[41] S. Simic, B. Dunjic, S. Tasic, B. Bozic, D. Jovanivic, I. Popovic, Synthesis and characterization of interpenetrating polymer networks with hyperbranched polymers through thermal-UV dual curing, Prog. Org. Coat 63 (2008) 43–48.

CHAPTER 4

Role of Nanofillers in Polyurethane Based Blends and Interpenetration Networks

Yao Tong and Mo Song
Loughborough University, Loughborough, United Kingdom

Contents

4.1 INTRODUCTION

Polyurethane is the reaction product between isocyanate and polyol. It is favorable because of its versatility. Although the chemistry of organic isocyanates were discovered by Wurts in 1849 [1], the chemistry and technology of PUs started in 1937 when Otto Bayer and his coworkers started the pioneering work of synthesizing PU polymers [2]. Their initial work investigated the reactions between isocyanates and diamine to produce polyuria; polyurethanes were synthesized in their further research. With the discovery of new materials, many interesting properties of polyurethanes were discovered and the potential of polyurethanes became

Polyurethane Polymers: Blends and Interpenetrating Polymer Networks
DOI: http://dx.doi.org/10.1016/B978-0-12-804039-3.00004-X

greater because the formulas of synthesizing polyurethanes became more versatile. As the technologies and new materials continually advance, the formulations and processing techniques are improved to develop new generation PU based polymers that have more superior properties. The key reaction to produce polyurethane is the reaction between isocyanate group and active hydroxyl group. Apart from this reaction, isocyanate groups can also react with other functional groups because they are very reactive. Reactions can be divided into two main categories: (1) the reactions of isocyanates with compounds containing active hydrogen; and (2) the reactions between isocyanate groups (self-condensation) [3,4]. The pronounced positive character of the C atom in $-N=C=O$ is the main reason for the high reactivity of isocyanate groups towards nucleophilic agents. The large electrophilic character of the carbon atom is imparted from the electronegativity of the oxygen and the nitrogen in $-N=C=O$. In the reaction systems to produce PUs, diisocyanates, polyols, catalysts and even chain extenders can be used. Therefore, the final structure and properties of a PU polymer are highly dependent on the structure and properties of the components used to synthesize the PUs.

Industrial and scientific areas have paid attention to polymer blends because it is widely regarded as the best method to obtain an excellent engineering material in a cost effective way [5,6]. Polymer blends are obtained from mixing two structurally different polymers physically where no chemical reaction takes place between these two polymers. The bonds that bind the two polymers together are only secondary bonds, i.e., hydrogen bonding, London force and dipole interactions [7]. Polymer blends are developed to meet the technical and engineering requirements by combining the advantages of two polymers. The degree of compatibility of polymer blends is the critical factor that determines the final properties. However, other influential factors are blend morphology, chemical composition, crystalline structure, molecular weight, and processing conditions. Incompatible polymer blends usually have poor properties [5–7].

Due to the versatility of polyurethane, it is widely used to blend with other polymers to produce a superior product that can meet the requirements of different applications. Thermoplastic polyurethane (TPU) is the most common PU to be blended because it is a linear copolymer that consists of soft and hard segments, making TPU versatile. Phase separation takes place due to the incompatibility or thermodynamic immiscibility between hard segments and soft segments. The hard segments usually consist of polar materials and they tend to form clusters or aggregate into ordered hard domains due to carbonyl or amino hydrogen bonds. Meanwhile, soft segments form amorphous domains [8]. Composition variables and processing conditions such as properties and the structure of hard segments and soft segments, synthesis methods and thermal history of the PUs are very important for the properties of the final product [4]. Polymer blends contain PU have been widely used in different applications such as membrane applications [9,10], electrical applications [11–13], and damping materials [14].

Interpenetrating polymer network is a term that describes the polymer blends that adopt simultaneous or sequential polymerization of polymer networks to compatabilize different polymers [15]. To an extent, PU based composites can be regarded as polymer blends. The applications of PU based composites are enormous, including engineering materials [16–20], electrical applications [21,22], and coating applications [23]. In some polymer nanocomposites, the polymeric matrixes are polymer blends.

The role of nanofillers in polymer blends is multifunctional. In this chapter, the properties that nanofillers improve for PU based blend nanocomposites are reviewed. The preparation and the characterization techniques for PU based blends/IPNs nanocomposite will be summarized. The polyurethane blends/IPNs used in different applications will be reviewed briefly. Electrical applications, applications for separation purpose, biological applications, shape memory applications and other applications will be reviewed. The preparation methods, structure and the properties of the blends are highlighted.

4.2 PREPARATION AND CHARACTERIZATION OF NANOFILLERS REINFORCED PU BLENDS/IPNs

The properties and characteristics of a polymer blend/IPN nanocomposite highly depend on the preparation methods of the samples. The general methods for preparation are reviewed briefly in this section. In addition, the characterization techniques to characterize nanofiller dispersion, miscibility, and morphology of the blend/IPN nanocomposite are also reviewed.

4.2.1 Preparation of Blends and IPNs

Solvent casting is the simplest method to prepare polymer blends with the aid of a suitable solvent. The general procedure is that the polymers are dissolved in a suitable blend to form a mixture solution. The same solvent is generally used to disperse the nanofillers as they are added into the system. Afterward, the polymers mixture solution and the nanofillers dispersion are mixed together to form a blend composite mixture [15]. Finally, the mixture solution is cast on a smooth plate (petri dish, Teflon mold, etc.) for solvent evaporation. Drop cast is the simplest casting method while doctor blade casting [24] and spin coating [25] can be used if designated film thicknesses are required. The obtained film can be used for any subsequent characterizations. For PU based blends fabricated by this method, the PU used is usually thermoplastic PU [13,26–28], with the PU synthesized directly in the solvent and blended with the other components in some cases [10,29,30]. Proper mixing in the solvent and the subsequent solvent removal process is important because the solvent can significantly change the phase structure even at a very small amount of solvent in the system. In addition, care should be taken when the blend contain high T_g polymers because if the drying temperature is not above the T_g, a misleading 'frozen-in' miscibility will be induced. A immiscible blend may become miscible if a wrong conditioning temperature is selected or freeze drying is adopted [15]. Moreover, the solvent used for blending will affect the phase behaviors of the blends [31]. Solvent casting method is more suitable in the laboratory because the blending process and the drying procedure is time consuming and complicated. It is not cost effective and the production scale is small.

Melt mixing is a common technique for blend preparation that can produce polymer blends on a larger scale. Melt mixing can be performed in two-roll mills [14,32–34], Banbury mixer [15], internal mixers [35–38], and extruders with different screw configurations and sizes [39–43]. In some cases, the PU blends will be mixed first and then they are shaped in the two-roll mills, extruder, or injection machine for subsequent characterizations. Melt mixing method has several advantages: (1) it is solvent free; (2) feasible for large scale sample preparation for standard mechanical properties characterization; (3) compatible to industrial processing. However, the compatibility and the phase structure of the samples produced by melt mixing are not ideal and the performance of the blends is not optimized. Reactive blend was developed to improve the compatibility of PU with other polymers in melt mixing [44–47]. In melt mixing, thermoplastic PUs are directly blended with other polymers while the second component is fed into the mixer first and then polyols, isocyanate, and catalysts are added sequent in reactive processing. An example reactive processing is described by Imer et al. [44]. PLA was added into the mixer first and then diol was added. At last, isocyanate was added into the mixer. The conditions used were 180°C, 50 rpm and 20 min mixing time. Compatibility between PU and PLA was improved by reactive processing and better morphology and properties resulted. For PU blend nanocomposites, the fillers are added into the mixer to mix with the blend directly [48] or mixed with one polymer first and then the other polymer [47]. In addition, reactive processing to produce PU blend nanocomposites was also reported [26].

The blends fabricated by in situ polymerization either simultaneous or sequential are termed interpenetrating polymer networks (IPNs). There is only one polymer that has network structure. Via in situ polymerization, the compatibility of the blends will be improved. IPNs do not necessarily have covalent bonding. Phase separation takes place if the two polymers are immiscible inherently. Improved adhesion between two phases results as the polymerization proceeds due to the interlocking of the polymer chains in the interphase [15]. For IPNs produced from simultaneous polymerization (semi-IPNs), two or more polymerizations take place simultaneous without interfering each other. An example is the preparation of PU/PMMA semi-IPNs [49]. Polyisocyanate, polyol, methyl methacrylate, trimethylpropane trimethacrylate, and the solvent were stirred in a glass reactor initially and then the initiator benzoyl peroxide and tin dibutyl dilaurate were added in sequence. Afterward, the mixture was transferred to a glass mold and allowed the polymerization to start. In terms of IPNs produced from sequential polymerization (sequential IPNs), one polymer was synthesized first and then the second polymer. In the study of Cui et al., the authors polymerized PU in *N,N*-dimethylacetamide (DMA) first and then the monomer of polybenzoxazine was added. The mixture was casted into Teflon mold for curing. The preparation of nanocomposite IPNs is more complicated than pure IPNs. The nanofillers can be added into the system directly or reacted with one of the polymer first before IPNs polymerization [50]. The activities in these area is enormous but the reports concerning PU based IPNs nanocomposites is limited. Tan and his coworkers reported the preparation of acrylonitrile butadiene rubber (NBR) and polyurethane-silica hybrid networks [34]. PU prepolymer was synthesized and mixed with SiO_2 nanoparticles afterwards. The NBR and PU prepolymer coated SiO_2 were mixed in a two-roll mill. The resultant mixture was cured in a standard to generate NBR/(PU-SiO_2) nanocomposites.

4.2.2 Characterization Techniques

The characterization techniques mentioned in this section are the methods generally used to characterize the morphology of the phases, nanofillers dispersion, and the miscibility of the blends and IPNs. The properties of the nanofillers reinforced blends/IPNs are determined by two critical factors: the miscibility of the two polymers in blends/IPNs, and the dispersion of nanofillers in the systems. These two factors can be characterized by several techniques that provide a route to access the properties of the nanofillers reinforced blends and the mechanisms of properties enhancement.

4.2.2.1 Transmission Electron Microscopy

Transmission electron microscopy (TEM) is one of the major techniques used to characterize polymer blends and nanocomposites. An electron beam is employed and high resolution down to 0.2 nm can be achieved. Smaller morphology detail can be obtained compared to SEM and optical microscopy. For polymer blends and nanocomposites, the sample for TEM must be very thin (1–100 nm) to allow electron penetration. The sample is usually prepared by ultramicroscopy and diamond or glass blade cutting in cryogenic temperature. In terms of nanofillers characterization, the particle suspension must be diluted. Water and acetone are the most widely used suspension mediums because of their very low level of hazardousness. Copper grid is used to dip in the suspension and capture the nanofillers for imaging [51]. A high degree of vacuum is required for TEM because the quality of the image will distort in the presence of air [15]. When the electron beam passes through the sample, the electrons are either scattered or unscattered by the nuclei or the electrons of the atom. The intensity distribution of the electrons is shown on the fluorescent screen [15,52]. Low atomic number atoms have limited scattering ability which lead to a poor phase contrast TEM image. In order to get a good image of polymer based materials, a necessary phase contrast of the components is required and the common method to improve phase contrast is staining. The stained medium deposits one of the component in a blend [52]. The most common staining medium is osmium tetraoxide OsO_4 and it is very useful for the polymer contain unsaturation because it can react with the double bond to form osmate ester which has excellent contrast [15]. For PU based blends, IPNs and their nanocomposite, staining was mostly stained by OsO_4 as well, either in liquid or vapor form and the duration varied [17,36,47,52–58]. There are other cases that adopted different staining medium for good phase contrast such as ruthenium teteaoxide [35,59], phosphotungstic acid [60,61] and uranylacetate solutions [62]. Sequential staining by different staining mediums was also adopted by other researchers. For example, Cui et al. stained polyurethane/polybenzoxazine IPNs with osmium tetroxide solution and then with uranium acetate and lead citrate [63]. The standard of selecting a staining medium is not established. Different staining medium may be very effective toward some specific structures. An extensive investigation is needed for the optimized quality of TEM image.

Some example TEM images of different PU based blends, IPNs and their nanocomposite are shown in Fig. 4.1 [17,47,53,62]. From the images, the morphology of the blends and IPNS can be clearly identified characterized by TEM and the dispersion of the nanofillers in the

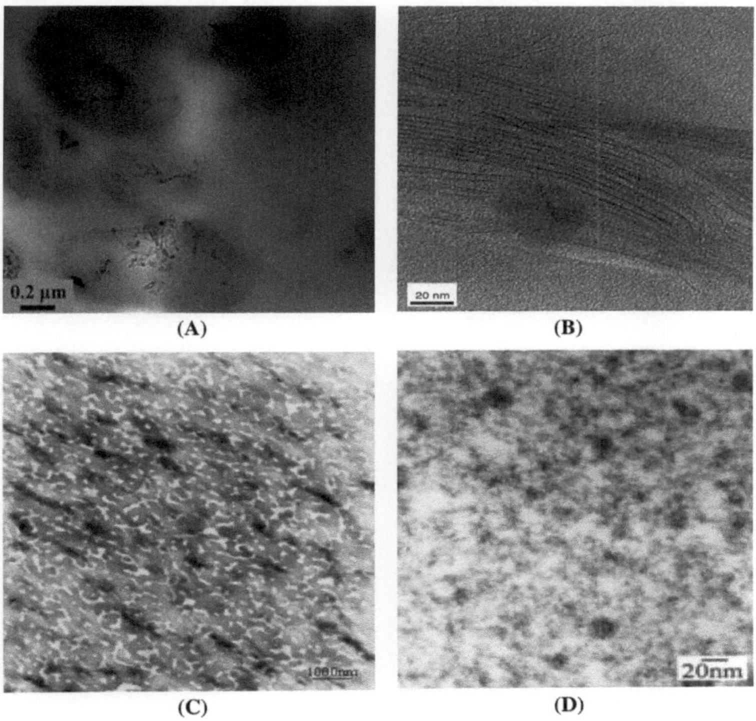

Figure 4.1 TEM images of (A) PLLA/TPU/0.5 wt% CNTs [47]; (B) PU/clay nanocomposite [17]; (C) PU/polyacrylate blend 30/100 by weight [62;] (D) 60:40 PU/VER(EA)IPN [53].

nanocomposite are clear. Different staining methods were adopted to achieve reasonable phase contrast. The example TEM images of nanofillers size measurements and dispersion evaluation are shown in Fig. 4.2 [64–66]. The size and the dimensions of the nanoparticles can be seen clearly under TEM which is beneficial for explaining the mechanism of properties enhancement of nanocomposites.

4.2.2.2 Scanning Electron Microscopy

Scanning electron microscopy (SEM) is a frequently used technique to characterize the surface of materials. It is particularly useful to investigate the morphology of blends/IPNs and the nanofillers dispersion of nanocomposites. When the electron beam hits the specimen, secondary electrons are produced from the release of valence electrons from the atom of the specimen [67]. The secondary electrons are detected and amplified to form an image. The energy of the secondary electrons is very weak and, therefore, only the electrons generated from the surface can be detected

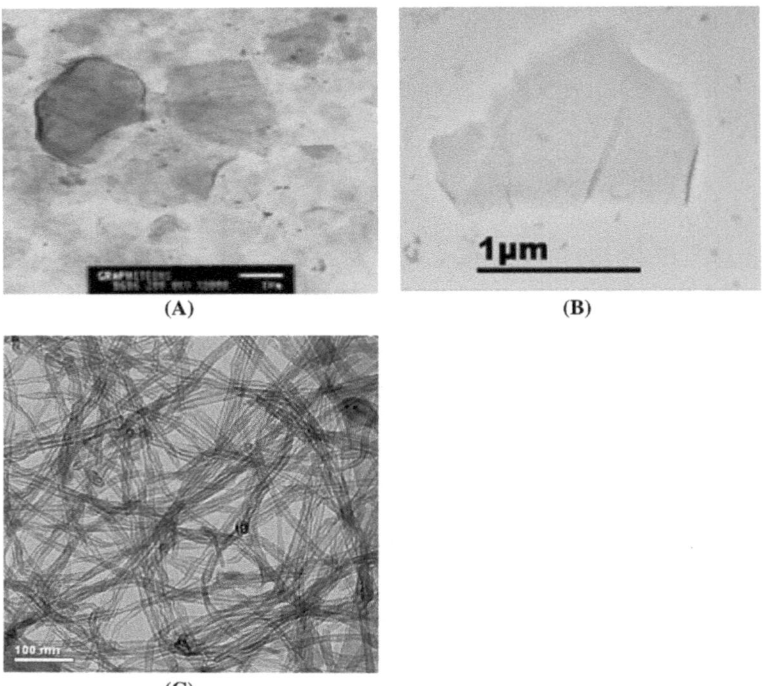

Figure 4.2 TEM images of (A) graphene oxide dispersed in NMP [65]; (B) graphene oxide sheet [64]; (C) carbon nanotubes carpets [66].

clearly. Hence, SEM is a surface-specific technique. The accelerating voltage ranges from 1 kV to 40 kV. A low accelerating voltage is recommended to observe the surface structure clearly. In addition, a high accelerating voltage will melt the polymer which results in a damaged surface. The sample for SEM should be conductive so that the electrons can pass through the sample stage without causing charging which leads to a distorted image. A thin conductive layer (gold or platinum) is usually deposited onto the surface of the polymeric sample. In order to increase the contrast, a staining and etching process may be adopted. For the observation of a cross-section of the sample, cutting by ultramicrotome or razor blade can be used for soft materials. In terms of hard materials, the cross-section is prepared via fracturing and freeze fracture if the materials are soft at room temperature but hard at low temperature. The SEM images of PU based blends, IPNs and nanocomposites are shown in Fig. 4.3 [44,49,68]. Compared to TEM, the resolution is lower but the detail of the surface is clearer. The effect of processing method on the morphology of the PU/PLA blends can be revealed clearly by SEM as

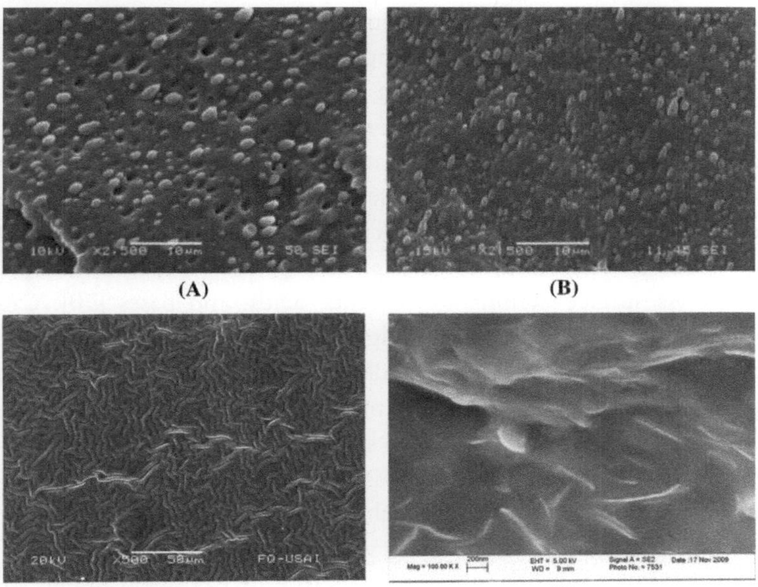

Figure 4.3 SEM images of (A) PU/PLA (70/30 in vol.%) blend prepared by physical blending in an internal mixer; (B) The same blend and equipment as (a) but via reactive processing [44]; (C) PU/PMMA 50/50-3% starch IPN [49]; (D) The fracture surface of PU/graphene oxide nanocomposite [68].

shown in Fig. 4.3A and B. In addition, the depth of the surface is displayed cleared in SEM which facilitates the investigation of the surface topology. Fig. 4.3D was produced from fracturing by liquid nitrogen. The internal structure was shown clearly and the dispersion of nanofillers in the nanocomposite can be evaluated.

4.2.2.3 Optical Microscopy

Compared to TEM and SEM, the resolution of optical microscopy is the lowest and it is only suitable for larger dimensions samples containing polymer blends. Optical microscopy is low cost and it is more convenient for use [69]. As it is convenient, it can be the first technique to evaluate the phase structure of polymer blends before performing any further characterizations. Due to its low resolution and magnification, the applications of optical microscopy for detail phase morphology and nanofillers dispersion are limited. There are various optical microscopy choices, among which transmitted optical microscopy, reflected optical microscopy, and polarized light microscopy are the most widely used. Transmission optical microscopy is suitable for a very thin sample (1–50 μm) as a collimated light beam is passed through the sample for analysis. The PU blends prepared by solvent casting method can prepare a thin sample readily while those prepared by other methods may need microtome for sample preparation. Staining methods, solvent treatment and preferential adsorption of dye or fluorescent species can be used to enhance the contrast between phases [15]. In terms of reflected microscopy, the light is reflected by the sample surface. Therefore, the sample is opaque. The contrast can be improved via surface etching by solvents, acids, and bases [70]. Crossed polarizers are adopted in polarized light optical microscopy and it is usually used to characterize the crystalline structure and the deformation morphologies (crazing, etc.). The sample thickness is similar to those for transmission light microscopy. The samples are usually placed on glass slides and a hot stage microscope is employed for the observation of spherulites growth of the crystallizable components in polymer blends. Example images of optical microscopy are shown in Fig. 4.4 [13,71]. The surface morphologies and crystalline structure of PU based blends can be assessed to an extent. If a detail morphology characterization is required, TEM and SEM are the next techniques to adopt. In terms of nanofillers dispersion, the resolution of optical microscopy is too low to provide a proper assessment in this area.

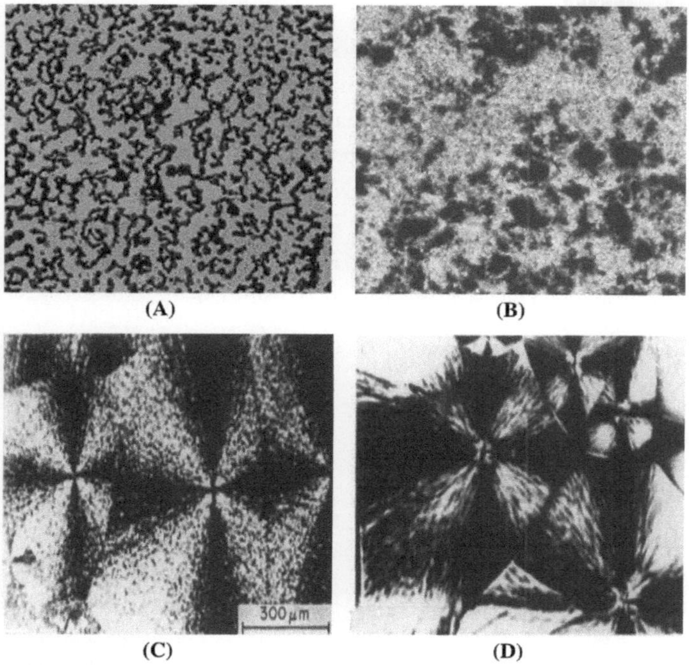

Figure 4.4 Optical microscopy images of (A) PANI/TPU blends 15/85 in wt% prepared by in situ process (bright field); (B) PANI/TPU blends 15/85 in wt% prepared by solution casting (bright field) [13]; (C) PEO/PU blends 40/60 in wt% (crossed polarized light); (D) PEO/PU blends 90/10 in wt% (crossed polarized light) [71].

4.2.2.4 X-ray Techniques

X-rays are generated whenever a beam of high-speed electrons hit a metal target and this usually happens in a vacuum tube [72,73]. In addition, the metal target is always water cooled to prevent melting because most of the dynamic energy generated by the electrons is converted into heat. Nickel-filtered CuK_α radiation (1.54718 Å) is generally used in polymer studies [74]. Small angle x-ray scattering (SAXS) is a technique that measures the electron density variation when an x-ray beam hits the sample at very low angles (0.1–10°) to identify the inhomogeneity on a nanoscale [75]. The shape and the size of polymer chins, characteristic distances of partially ordered materials, pore sizes, and domain sizes can be obtained from SAXS data. SAXS is able to acquire structural information of polymer between 5 nm and 25 nm and the repeat distances in partially ordered systems up to 150 nm. Ultra-small angle X-ray scattering can characterize even larger dimensions. The morphology and phase behaviors of PU in blends and nanocomposites can be characterized by SAXS. The inter-domain repeat distances and domain sizes of the hard segment of PU/clay nanocomposites were characterized by Song et al. [76]. The information of domain sizes and repeat distance can be obtained by resolving the intensity profile of SAXS into one-dimensional correlation function as described by them. The example images of intensity profile and correlation function are shown in Fig. 4.5. The effect of blending

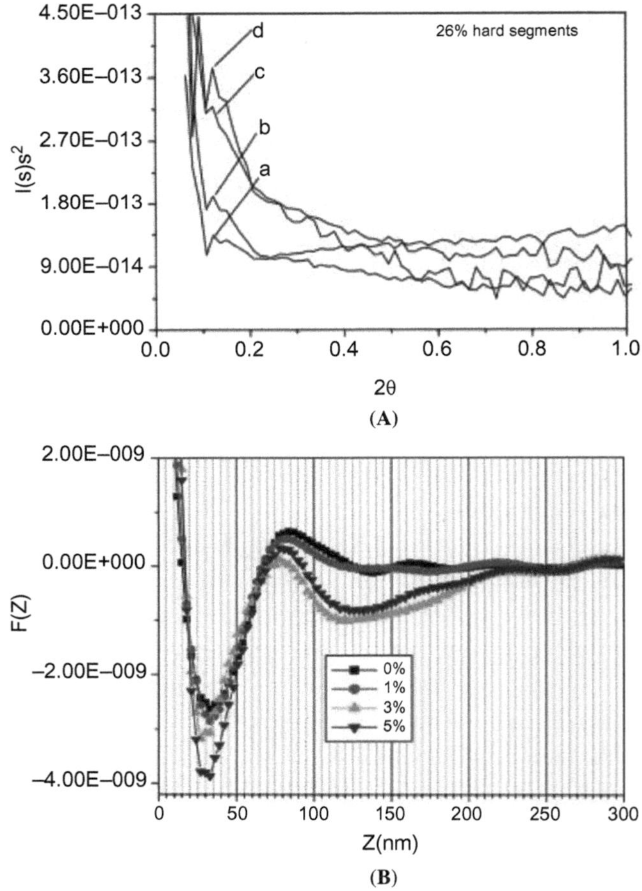

Figure 4.5 (A) Intensity profile of PU (26 wt% hard segment)/clay nanocomposites (clay percentage: a: 0%, b: 1%, c: 3%, d: 5%); (B) Correlation function from intensity profiles of PU (26 wt% hard segment)/clay nanocomposites [76].

polyimide on the phase behaviors of PU was investigated Benson and his coworkers [77]. The domain sizes and hard and soft segments of PU, inter-domain distance, characteristic length, diffuse boundary thickness, and ratios of Porod's invariant were characterized. The morphology and phase behaviors of PU in PU based IPNs can be characterized by SAXS as well [78].

X-ray diffraction is a very effective method to investigate the orderly arrangement of atoms. The information of the arrangement is obtained via the interaction of electromagnetic radiation to interference with structures whilst the sizes of structures compares to the wavelength of the radiation. If the structure is in order, the interferences can sharpen, and, hence, the radiation can be scattered or diffracted under some specific conditions [72]. The reason why diffraction happens is that there are some certain phase relations between two or more waves. The definition of a diffracted beam is a beam that is formed by a large number of scattered rays that can reinforce each another. Therefore, diffraction is a scattering phenomenon which is caused by the cooperation of a large number of atoms without any new interaction between x-rays and atoms [73]. Phases identification, orientation of phases, habit plane of interfaces, crystallography of defects, and order/disorder information are usually obtained via electron diffraction [79]. There are some essential conditions must be met to let diffraction happens, and Bragg's law describes these conditions. The equation of Bragg's law is:$n\lambda = 2d\sin\theta$, where n is the order of diffraction, λ is the wavelength of the radiation, d is the perpendicular distance between the planes that diffraction occur, and θ is the angle between the incident beam and the plane [73,79]. X-ray diffraction is an effective tool to investigate the dispersion state of nanofillers in a composite system especially for layered nanofillers. WXRD can be used to characterize the intercalation capability of polymer [18]. The diffraction peaks shift to lower diffraction angles in intercalated nanocomposites while there are no diffraction peaks in exfoliated nanocomposites. Fig. 4.6 show the WXRD results of PU-organoclay nanocomposites. The results show that polyether polyol molecules have been intercalated into the organoclay layer which led to direction peaks shift to lower angles.

Apart from SAXS and WXRD, there are still other x-ray techniques that can be used to study the properties of PU based blends/IPNs and nanocomposites and they are reviewed in the literature [80].

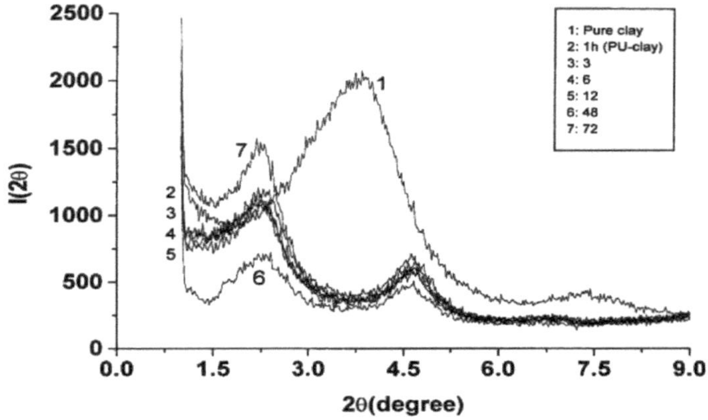

Figure 4.6 WXRD results of PUE-clay composites with different mixing time [18].

4.2.2.5 Fourier-Transform Infrared Spectroscopy

Infrared spectroscopy is one of the most important analytical techniques that can identify the composition of matter. The most significant advantage of infrared spectroscopy is that it can study various states of the substance. Solid surface, solutions, liquids, pastes, powders, films, gases, and other states can be examined through appropriate sampling techniques and many samples that were previously difficult to examine can now be studied because of the development of technology. The instrument for this experiment is Fourier-transform infrared (FTIR) spectroscopy. It is an advanced instrument where the quality of infrared spectra is improved and the time of obtaining the data is minimized. Infrared spectroscopy identifyies the vibrations of the atoms of a molecule. In term of a polymer, infrared spectroscopy examines different groups according to the six different vibrations of the atomic bonding. The peak, which is the energy absorbed in an infrared absorption spectrum, corresponds to the frequency of a vibration of specific part of a sample molecule [81]. Some specific interactions (hydrogen bonding and chemical reactions) in PU based blends/IPNs nanocomposites can be detected by FTIR. The hydrogen bonding can be detected by the shifting in frequency of the absorption peak for the hydrogen bonded unit. An example is the study of polyether blends with 1,3-dimethylurea and a model urethane compound [82]. The N-H absorption peak shifted from 3342 cm^{-1} to 3315 cm^{-1} when the blend contained 90 wt% poly(ethylene oxide) (Fig. 4.7). The composition ratios of PEG based polyurethane and PCL based polyurethane can be estimated via FTIR spectrum by comparing

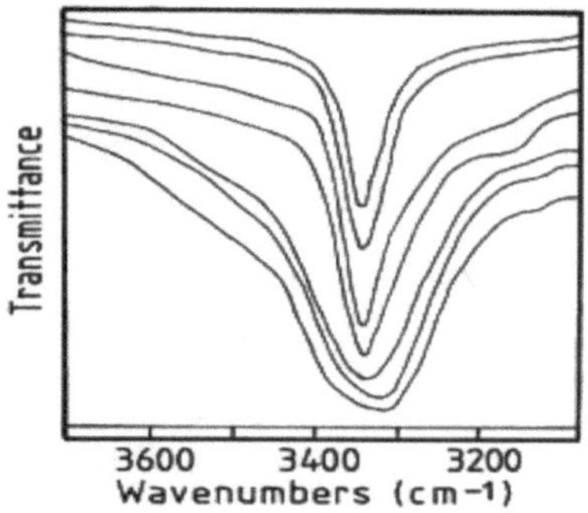

Figure 4.7 N-H region in the FTIR spectrum of DMU and PEO/DMU blends contained different wt% PEO (top to bottom: pure DMU, 10/90, 25/75, 33/67, 50/50, 75/25, 90/10 PEO/DMU) [82].

their signature peaks [83]. There are still many researchers utilizing FTIR to identify the insight of PU based blends/IPNs and their nanocomposites [84–92].

4.2.2.6 Differentially Scanning Calorimetry

Differentially scanning calorimetry (DSC) is a thermal technique that measures the differences in heat flow between a sample and a reference as a function of sample temperature when the two are underwent a controlled temperature changing condition. It is a calorimetric method that can measure the differences in energy between the references and the samples. In addition, DSC was developed to avoid the difficulties of using DTA or quantitatively compensate such as the temperature dependencies of thermal transport and sensor sensitivity [93,94]. There are two types of DSC instruments: power compensation and heat flux mode. The DSC used in this experiment is heat flux DSC. In this type of DSC, heat flows into both the sample and the reference material through electrically heated constant thermoelectric disk. Then heat is transferred into the sample and the reference via the two pans [93]. Heat flux DSC has a similar operative mechanism to a conventional DTA, except the quantitative compensation for the problem areas mentioned above. DSC can determine specific heat, glass transition temperature, melting and crystallization points, heats of fusion and crystallization and heat of reaction. Only small amount of specimen (5–40 mg) is required for each run. The compatibility of the two polymers in a blend can be characterized by the glass transition temperature. The example DSC results of miscible and partially

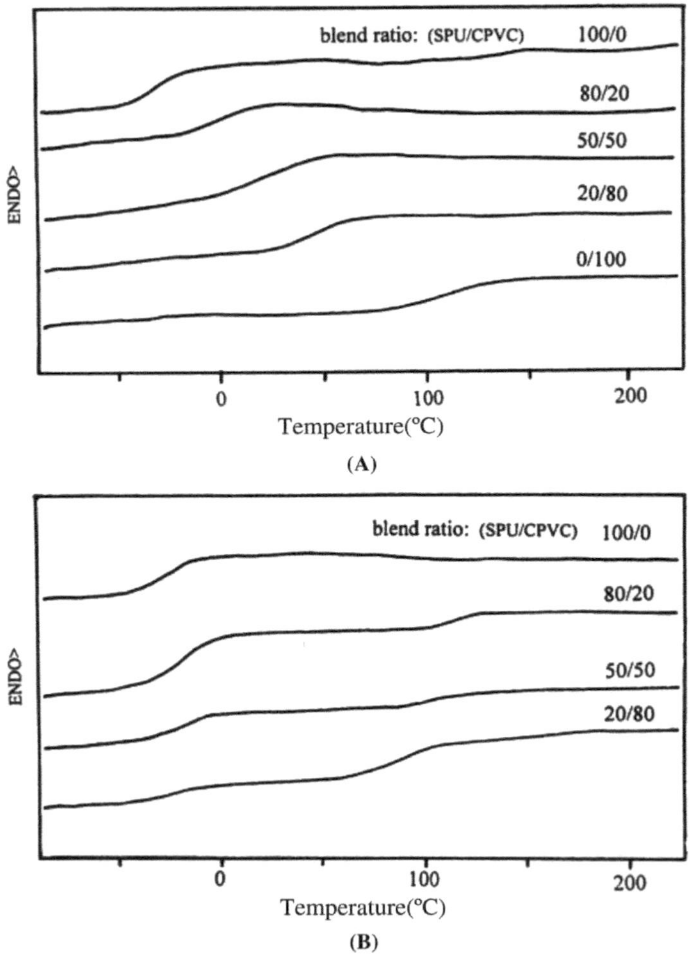

Figure 4.8 (A) Polycaprolactone based PU/chlorinated PVC miscible blend; (B) Polypropylene diol based PU/chlorinated PVC immiscible blend [95].

miscible blends are shown in Fig. 4.8 [95]. A miscible blend only exhibit one glass transition temperature in a DSC curve (Fig. 4.8A) while a partially miscible blend has two glass transition temperatures whose position will change with the varying of the blend's composition. In terms of immiscible blend, there are two distinctive glass transition temperatures which did not change with blend composition (Fig. 4.8 B).

DSC can provide fast analysis to indicate the compatibility of blends but the sensitivity for glass transition temperature or secondary relaxation transitions is lower than a dynamic method such as dynamic mechanical analysis. Modulated temperature DSC (MTDSC) was developed to increase the sensitivity for polymer blend transitions. The simultaneous measurement of the amplitude (modulus) of the complex heat capacity, the heat flow, and the phase angle between heat flow and heating rat (heat flow phase) enables a more detailed study of complicated materials systems, both in quasi-isothermal and nonisothermal conditions. In other thermal analysis techniques that use an oscillating excitation of the materials, dynamic mechanical analysis, dielectric thermal analysis, and dynamic rheometry, e.g., a phase angle is defined between the modulated input and the resulting output signal. The characterization of phase structure of IPNs can be undertaken by MTDSC. The glass transition temperatures of different phases can be identified [96]. By combining MTDSC with SAXS and TEM, a clear understanding of the morphology of PU based blends/IPNs and their nanocomposites may be obtained.

4.3 ENHANCEMENT OF MECHANICAL PROPERTIES

The mechanical properties of PU based blends, IPNs and nanocomposites are widely reported in the literature, comparing the effect of different compositions and nanofillers on properties' enhancement. The properties that the literature most often report are strength (tensile, compressive or flexural), modulus (tensile, bulk, or flexural), toughness, durability, creep resistance, ductility, and wear resistance. Via comparing with the mechanical properties of original polymer, the compatibility of the blends/IPNs and the efficiency of nanofillers can be evaluated.

The mechanisms of improving fracture durability of PU based matrix by organoclay was identified by Jin et al. [16]. Generally, there are two major mechanisms attribute to fracture resistance: energy dissipation and energy absorption as cracks propagate through the materials. During a fatigue test, the deformation of the nanocomposites includes the orientation and stretching of hard segments and nanoplatelets. The total energy required for crack propagation consists of the energy for extending and orienting the hard segment domains and silicate platelets and overcoming the frictions between polymer chains and between polymer chains and nanoplatelets. It was discovered that nanoplatelets showed a reversible deformation behavior where the d-spacing went back to original position when the stretching force was removed (Fig. 4.9). The nanoplatelets and the polymer chains between them acted as nano-springs during

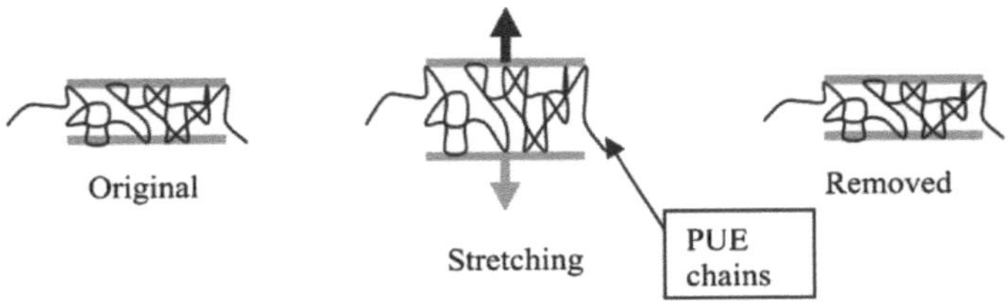

Figure 4.9 Schemes of d-spacing variation during stretching [16].

stretching. These nano-springs could reduce the stress concentration in the materials and prevent the development of fatigue cracks. Therefore, the fatigue durability of the system was improved. Cai and his coworkers investigated the role of graphene oxides in improving hardness and a scratch resistant PU based system [68]. The hardness, modulus, and the scratch resistance of the polymer matrix were improved significantly. Geometrical consequence may be the main reason for the highly improved scratch resistance. The nanofillers acted as rigid and hard particles to resist the scratch of the indenters on the surface of the polymer matrix. Therefore, the scratch depth was reduced. Scratch damage was created difficulty by the indenter when the interspaces among the nanofillers were smaller than the tip size of the indenter. Increasing nanofillers loading led to the formation of thicker nanoflakes instead of reducing the size of interspaces which accounted for the independence of scratch depth toward the graphene oxide loading. Organoclay was added to PU/epoxy based IPNS for improved friction and wear performance [97]. The authors suggested that the addition of nanofillers to improve wear resistance have several advantages: (1) strengthened film can be produced as the nanofillers can blend well with wear particles; (2) the wear loss of composites modified by nanofillers would be milder than the conventional composite because the size of the nanofillers are the same as the segments of the polymer chain surrounding. The observation of the worn surfaces indicated that the friction and wear mechanisms of epoxy resin/PU IPNs were dominated by adhesion and surface fatigue. With the addition of nanofillers, the formation of microploughs was inhibited and the number of microcracks increased which resulted in the reduction of average coefficient of friction. Therefore, the wear loss of epoxy resin/PU IPNs nanocomposites was reduced.

The mechanical properties of PU/NBR blends were improved by the addition of layered double hydroxide (LDH) [98]. Dodecyl sulfate (DS) intercalated LDH reinforced blends exhibited improved tensile strength, elongation at break and modulus than the pure blend. The improvements were attributed to: (1) the strong interfacial interaction between the −OH of DS-LDH and the −CONH and −C≡N polar groups; (2) stress was transferred efficiently by the well-dispersed DS-LDH layers; (3) the synergistic effect of chain slippage and platelet orientation of DS-LDH layers. However, the mechanical properties decreased with increasing nanofillers loading because the presence of aggregated DS-LDH layers in the matrix. Similar functions of nanofillers in mechanical properties enhancement was observed in the graphite oxide nanoplatelets (GONPs) reinforced PU based matrix [65]. The addition of GONPs transferred the stress effectively between the GONPs and the matrix as a result from the chemical bonding. The modulus of PU was increased significantly but the tensile strength and the elongation at break were reduced. Although GONPs were an effective load transfer agent, the crosslinking between PU and GONPS restricted the elastic deformation which led to reduced elongation at a break whilst the presence of nanofillers destroyed the hard segments PUs which resulted in the reduction in tensile strength. The mechanical properties of nanoclay reinforced epoxy modified PU were investigated by Dutta et al. [30]. The tensile strength and scratch hardness were improved. The reinforcing effect induced by intercalated and exfoliated clay increased the length of the crack propagating path during tensile test and reduced the number and the size of voids. In addition, the reinforcing effect was attributed to the increase in crosslinking density resulted from the reaction between −OH and −NCO groups of the polymer matrix with the −OH groups of the bentonite clay during the PU cure. Moreover, a high surface to volume ratio of the nanofillers also facilitated the chemical reactions which increased the crosslinking density of the composites. However, the elongation at break values decreased with increasing clay content due to the increased clay layer restricted the molecular motions of polymer chains.

The utilization of nanofillers for mechanical properties enhancement is promising. Nanofillers act as load transfer medium, crosslinking sites, crack inhibitors, and so on. In some cases, the nanofillers can act as compatibilizer to improve the compatibility between the two components in a blend/IPN. The key factors of optimizing properties are uniform dispersion of nanofillers, good filler-polymer matrix compatibility, and a suitable filler loading. Poor nanofiller dispersion and filler-matrix capability have adverse effect on the mechanical properties of PU based blends/IPNs. If the filler loading exceeds the optimized loading, poor nanofillers dispersion will result. The number of studies of polymer blends/IPNs nanocomposites is enormous but a mature product is still absent from the market.

4.4 IMPROVEMENT OF THERMAL/ELECTRICAL CONDUCTIVITY

Electrical conductivity of polymers is initially very low and the typical value is about 10^{-11} S•m. Therefore, polymers are inherently suitable for insulation materials for wires, cables, and insulating housing. The demand and utilization of electrical conductive polymeric materials are growing in many applications such as corrosion protection [99], antistatic coating [100], and electrodes [101]. PU can blend with intrinsic conductive polymers or nanofillers to form conductive blends, IPNs, or composites. In addition, PU based blends/IPNs can be reinforced by conductive nanofillers to improve electrical conductivity. Ahmed and coworkers studied the effect of MWCNTs in the electrical conductivity and shape memory properties of PU/PS IPNs [29]. Their main results are shown in Table 4.1. The electrical conductivity of the IPNs increased with increasing MWCNTs loading. Due to higher electrical conductivity, the shape recovery was also improved because high electrical conductivity was one of the critical conditions for obtaining good shape memory property. The formation of MWCNTs conductive network in the IPNs system is the reason why the electrical conductivity is improved. The MWCNTs

Table 4.1 Electrical conductivity and shape recovery of PU/PS IPNs nanocomposites [29]

Sample name	Electrical conductivity ($S\ cm^{-1}$)	Shape recovery (%)
PU/PS	0.99×10^{-6}	84
PU/PS/MWCNT 0.05	1.25×10^{-3}	85
PU/PS/MWCNT 0.1	2.87×10^{-2}	86
PU/PS/MWCNT 0.3	3.13×10^{-2}	86
PU/PS/MWCNT 0.5	1.94×10^{-1}	88
PU/PS-NO_2/MWCNT 0.05	3.45×10^{-3}	85
PU/PS-NO_2/MWCNT 0.1	4.88×10^{-2}	88
PU/PS-NO_2/MWCNT 0.3	8.99×10^{-2}	89
PU/PS-NO_2/MWCNT 0.5	9.92×10^{-1}	90
PU/PS-NH_2/MWCNT 0.05	4.38×10^{-3}	92
PU/PS-NH_2/MWCNT 0.1	5.87×10^{-2}	94
PU/PS-NH_2/MWCNT 0.3	8.68×10^{-2}	95
PU/PS-NH_2/MWCNT 0.5	1.08	95

act as a current transport medium. In addition, conducting bridges with IPN conducting domains were formed that contributed to the improved electrical conductivity.

Jin and coworkers studied the mechanisms of electrical conductivity enhancement in polymer composites extensively [102]. They suggested that the particle size of nanofillers was the key factor to improve the electrical conductivity of nanocomposites because the particles with larger particle size were beneficial to reduce the tunneling resistance between discontinuous conductive nanofillers which facilitated the formation of conductive networks and bridges in the composites. Additionally, the adoption of a hybrid filler system and the addition of ionic salt into nanocomposite could enhance electrical conductivity further. They also proposed a theoretical approach to estimate the maximum distances between the nanofillers so that the fillers could form a conductively network. The key equations to estimate the conductance of graphite sheet-PE-graphite sheet (EG-PE-EG) junction are shown below:

The transmission coefficient T' can be expressed in the following equation:

$$T' = \left[1 + \frac{1}{\frac{4\varepsilon}{V_0}\left(1 - \frac{\varepsilon}{V_0}\right)} sh^2 \kappa a\right]^{-1} \tag{4.1}$$

$$\kappa = \frac{\sqrt{2m(V_0 - \varepsilon)}}{h} \tag{4.2}$$

where ε is the energy of electrons, V_0 is the effective potential energy of PE, m is the mass of electrons, s is spin quantum number, h is Planck's constant and a is the distance between graphite sheet.

Considering the hole transport in the PE layer, the transmission coefficient, T, can be obtained as follows:

$$T = \left[1 + \frac{1}{\frac{4\varepsilon}{V_0}\left(1 - \frac{\varepsilon}{V_0}\right)} sh^2 \beta a\right]^{-1} \tag{4.3}$$

$$\beta = \sqrt{2m(-V_0 + \varepsilon)}/h \tag{4.4}$$

Finally, the conductance of graphite sheet-PE-graphite sheet junction can be express in terms of T and T' as the equation below:

$$G = G_0(1 - T' + T)\frac{bc}{(2a + b)(2a + c)} \quad (4.5)$$

where G_0 is the conductance of a graphite sheet and bc is the cross-section area of a graphite sheet.

From Eq. (4.5), the conductance of GS-PE-GS can be estimated when the thickness of PE layer (the distance between the graphite sheets) change. The estimated results are shown in Fig. 4.10. From the estimation plot, the distance between the nanofillers should be close enough to form a conductive network. It implies that the particle size of the nanofillers used for conductive composite should be big enough to form a conductive pathway. Electrical conductance exhibits both Ohm's law and tunneling behavior. A bigger size conductive filler is beneficial for reducing tunneling resistance so that the electrical conductivity of composites can be improved significantly. Although the estimation is based on PE and graphite sheets, it is a pioneering work and it can be the practical guideline for manufacturing electrical conductive composites.

Thermal conductivity is the property of a material to conduct heat. It is an important property for thermal insulation and heat sink applications. The thermal conductivity of PU based blends/IPNs is generally poor. With the addition of nanofillers, the thermal conductivity can be improved significantly. Cai et al. incorporated MWCNT into PU based

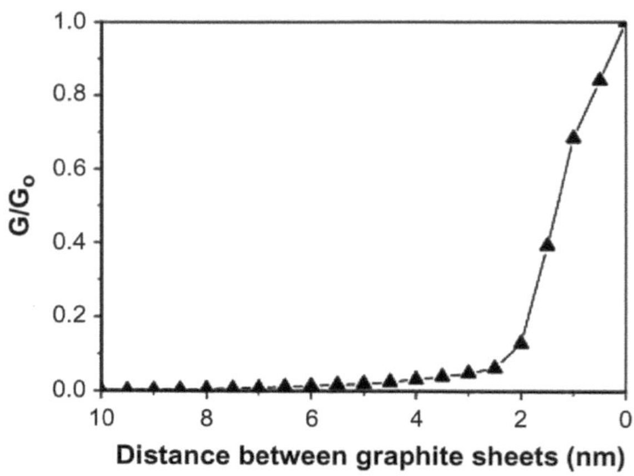

Figure 4.10 Estimated conductance of GS-PE-GS as the function of the thickness of the polymer layer in GS—PE—GS ($e = 2$ eV, $V_0 = 10$ eV).

matrix for thermal conductivity enhancement [103]. Their results are shown in Fig. 4.11. The continuous MWCNTs rich phase could act as a thermal transport tunnel in the composite system. The thermal conductivity of the system was improved significantly. Increasing MWCNTs loading would increase interfacial phonon scattering and reduce thermal conductivity. However, the interfacial phonon scattering could be reduced with the aid of latex technology and PCL crystallites formed around the MWCNTs. The authors concluded that polymer crystallites nucleated at interface and continuous MWCNTs rich phase were the two main factors for the effective reduction of the interfacial phonon scattering. The literature about improving thermal conductivity of PU based polymers with nanofillers is limited but the role of nanofillers in the composite system would be similar to other polymer matrices.

Nanofillers in the PU based blends/IPNs act as electrons or a thermal transportation tunnel. The key to forming conductive composites is the formation of a conductive network and the reduction of tunneling resistance. Utilizing conductive nanofillers with big particle size and blending with intrinsic conductive polymer can be the main route to manufacture conductive PU blends/IPNs nanocomposite. The key to improve thermal conductivity is to reduce the phonon scattering in the system nanofillers.

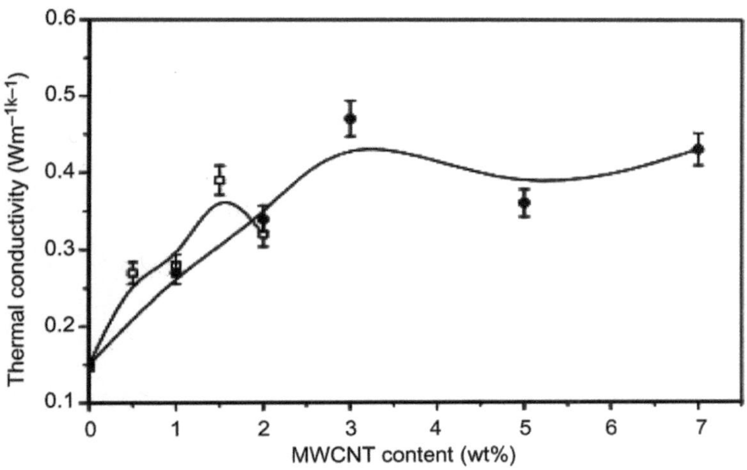

Figure 4.11 Plot of thermal conductivity versus MWCNT content. (■) The samples with consistent 1 wt% sodium dodecyl sulphate (SDS); (●) The samples with consistent weight ratio (1:1) of SDS to MWCNTs [103].

4.5 MODIFICATION OF THERMAL STABILITY

Thermal stability is the stability of a material at high temperatures and it is generally characterized by the thermogravimeric analysis (TGA) technique. The modification of thermal stability can provide PU with a higher working temperature and then can be used in some high temperature applications. The applications range of PU based blends/IPNs can be broadened. Thermal stability of PU was modified by functionalized graphene by Cai et al. [68]. The key results are shown in Fig. 4.12.

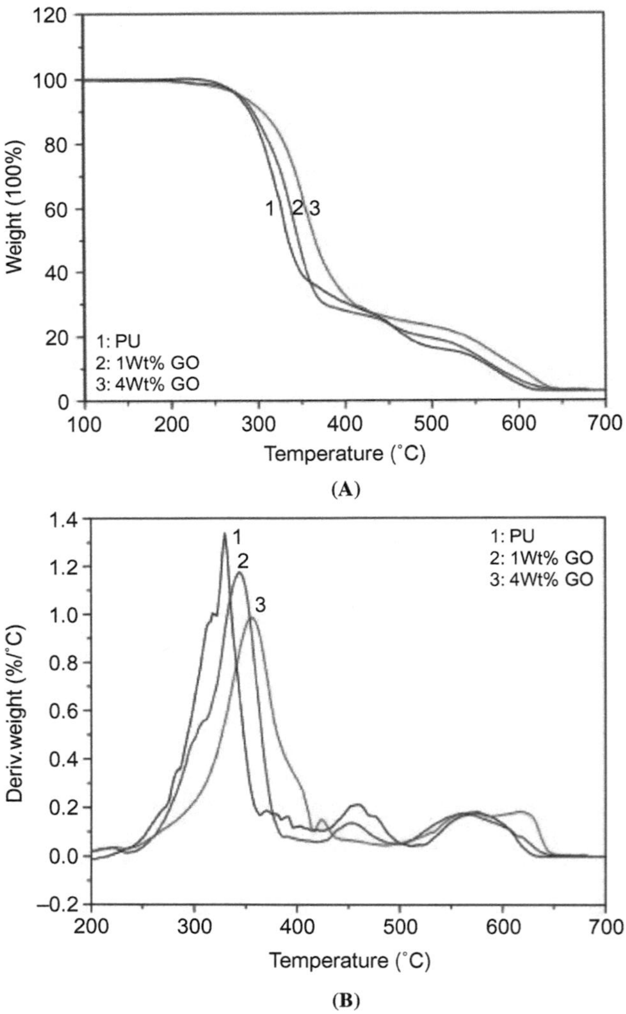

Figure 4.12 Plots of (A) weight loss against temperature; (B) weight loss velocity against temperature of PU/GO nanocomposites [68].

The thermal degradation of PU is two stages of decomposition. Chain scission takes place around 170–200°C due to the breakdown of the urethane bonds. Afterward, the polyol and polyisocyanates start to decompose. The polyisocyanates generally decompose at higher temperature than polyols because the polyisocyanates tend to form carbodiimide and urea. From Figure 4.12, the results show that the main decomposition temperature is increased by 50°C with 4 wt% GO but the decomposition rate is nearly the same. In addition, the onset decomposition temperature of the polyisocyanate increases to 500°C. The authors indicated that GO acted as a barrier to reduce the diffusion of the heat in the polymer. El-Shekeil et al. investigated the effect of fiber content on the properties of kenaf fibers reinforced poly(vinyl chloride) (PVC)/thermoplastic PU (TPU) blend [104]. The degradation mechanism of PVC/TPU/kenaf fiber was three stages of decomposition. Only the onset decomposition temperature of the third stage was improved slightly. The incorporation of kenaf fiber did not show significant enhancement on the thermal stability of PVC/TPU blend. However, the poor adhesion might be the reason of poor thermal stability enhancement.

Dutta and his coworkers incorporated organoclay into an epoxy modified PU for enhanced mechanical properties, cytotoxicity, and biodegradation [30]. The improvement of thermal stability by organoclay was also studied using TGA. The organoclay reinforced composites exhibited one stage decomposition instead of the two stage decomposition of the matrix. The onset decomposition temperature was increased 250°C with 5 wt% organoclay incorporated. The organoclay nanofillers acted as a thermal insulator and mass transport barrier which blocked the path for the volatile products to escape. The thermal stability increased with increased organoclay loading and this was attributed to the clay layers that trapped the polymer chains between them and hence could retard thermal degradation. Therefore, better dispersion of organoclay in the system was beneficial for thermal stability enhancement of the system. An exfoliated state would have the best thermal stability. The thermal stability of LDH reinforced PU/NBR blend was reported by Kotal et al. [98]. The role of LDH in the modification of thermal stability was explained by the authors. The degradation mechanism of nanocomposites is the same as the pure blend. The initial weight loss was accelerated due to the degradation of alkyl chains of the LDH, evaporation of absorbed water in the intercalated layer, and the loss of hydroxide on LDH. The decomposition of the LDH was endothermic which affected the combustion process of the polymer chains trapped within the LDH layers. The charred layer generated in the first step could act as barrier against the emission of volatile degradation product. The thermal stability of the nanocomposite increased with increased LDH loading. In addition, the partially exfoliated LDH tended to block the escape path of the degraded small molecules and, therefore, delayed the catalytic degradation of LDH in the matrix.

The nanofillers which can improve the thermal stability should have one of the following features: (1) Good thermal insulation; (2) Barrier to degraded products which retard the thermal degradation; (3) The decomposition process of the filler can retard the decomposition of the composite. However, excellent compatibility of the nanofillers and the matrix and good dispersion of nanofillers in the system is necessary.

4.6 APPLICATIONS OF PU BASED BLENDS/IPNs

In this section, the application of PU based blends is reviewed extensively while a few examples of the application of IPNs will be mentioned. Several good books or reviews have covered the application of IPNs [105,106].

4.6.1 Separation and Selection Applications

PU based blends are widely used as membranes for separation and selection techniques. The membranes prepared from for liquid and solid separation and selection by Lue et al. tailored the permeability properties of membranes by blending PU and poly(dimethylsiloxane) (PDMS) [9]. The PU used was commercially available PU. The membrane was prepared from solvent casting into a Teflon plate. PDMS prepolymer was dissolved in tetrahydrofuran (THF) first and then PU pallet (a commercial product) and PDMS crosslinkers were added subsequently to form a PDMS/PU solution. The authors indicated that toluene and methanol were diffused more slowly in higher PU content blends and permeability properties could be optimized by blending a certain amount of appropriate polymers. Al Minnath and his colleagues studied the transport properties of substituted benzene through TPU in natural rubber (NR) blends at different temperatures [32]. The TPU was polyether based TPU and the blend membranes were prepared by masterbatch technique using a two-roll mixing mill. TPU and NR were melt blended and then allowed to cure. The maximum solvent uptake and diffusion efficiencies of benzene, toluene and p-xylene decreased with an increase in TPU content and the

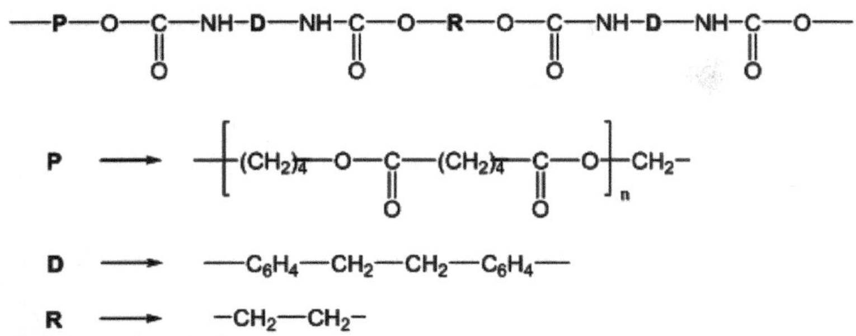

Figure 4.13 The chemical structure of the PU adopted by Zavastin et al. for membrane preparation [107].

diffusivity value decrease if the penetrant molecular size was bigger. Cellulose acetate (CA)/PU membrane used for separation techniques in waste water treatment was developed by Zavastin et al. [107]. The chemical structure of the PU used is shown in Fig. 4.13. The CA/PU blend membranes were prepared by a three step procedure. 20% PU solution in dimethyl formamide and 5% CA solution in acetone were mixed in a mass ratio of 8:7. Then acetone was added into the mixture to obtain a homogeneous suspension. Afterward, the suspension was centrifuged immediately after filtration. The obtained polymeric mixture was processed by a roller machine with adjustable distance to obtain the blend membranes with different thickness. The applicability of the membranes in separation technique was tested by aqueous solutions of new textile dye present in both 5 and 7 forms ((sodium-2-((2-hydroxy-6-sulfonato-naphthalene-1-yl) diazenyl)-6-methoxybenzothiazole-5 and 7-sulfonate) and its complex with iron (III) ions. The membranes had the best results when the aqueous solutions of azo dye complexed with iron (III) ions in molar ratio 3:1.

In terms of gas separation and selection, de Sales and his coworkers utilized a PU-poly(methyl methacrylate) (PMMA) blend as the membrane for gas selection and permittivity [5]. PU was synthesized from MDI (4, 4-diphenylmethane diisocyanate) and poly (tetramethylene glycol). For the blend membrane preparation, PU and PMMA were dissolved in THF and the resultant polymer solution was cast on petri dishes. Their results showed that increasing the PMMA content of the blend led to decreased permeability and diffusion coefficients of CO_2, H_2, O_2, CH_4, and N_2 due to the reduction of free-volume in the blend by adding more glassy components. In addition, an increase in selectivity at low temperature was observed. Semsarzadeh and Ghalei studied the effect of polyethylene oxide—polypropylene oxide—polyethylene oxide triblock copolymer (Pluronic) on the gas permittivity of PU/polyvinyl acetate (PVAc) blend [108]. PU was synthesized from toluene diisocyanate (TDI), polytetramethylene ether glycol (PTMEG), and 1,4-butanediol. The blend membranes were prepared by dissolving PU, PVAc, and Pluronic in DMF and then they were cast on a teflonized glass Petri dish. The addition of Pluronic resulted in improved compatibility between PU and PVAc. The CO_2 solubility increase with PVAc content in the blend and increased permselectivity of pair gas in the blends with different PVAc content was observed. PU blend membranes prepared from blending PU with polyetherimide (PEI) and poly (amide-imide) (PAI) for CO_2 separation was reported by Kim and coworkers [109]. The PU was commercially available thermoplastic PU. PU/PEI blend membranes were prepared by solvent casting method using DMF whilst the PU/PAI blend membranes used NMP as solvent. PEI was immiscible with PU in all composition ranges whilst PAI exhibited partial misciblity with PU. The CO_2 selectivity of PU was improved by blending with PEI and PAI. PAI/PU blend exhibited the highest N_2/CO_2 separation factor which indicated the best CO_2 selectivity was achieved.

Pervaporation is a separation technique has received wide attention due to its energy saving, scalability, small footprint and the feasibility to assemble into other unit operations [110]. Lin et al. studied the possibility of using polyethylene glycol (PEG)/PU blend membranes for FCC gasoline desulphurization by pervaporation [111]. PEG was mixed with PU polymer, maleic anhydride and trimethylamine in NMP initially and then the solution was cast onto polyethersulfone membrane. The blend membranes exhibited higher sulphur enrichment factor (4.03) and permeation flux (2.5 kg/(m^2 h)) which indicated the blends had outstanding thiophene species and sulphur removal effect. Separation of toluene/methanol binary mixture by pervaporation using a polyurethane—poly(dimethylsiloxane) (PDMS) blend was tried initially by Lue and coworkers [112]. Commercially available PU was mixed with PDMS prepolymer and PDMS crosslinker in THF with various PU compositions and then the polymer solution was cast on a Teflon plate for future use. From the results, the PU-PDMS (20/80) blend showed the best toluene selectivity with fair separation efficiency in toluene-lean and azeotropic composites. PU/4-vinyl pyridine (4-VP) IPNS membranes were synthesized for pervaporation separation of water-ethanol mixtures [85]. The PU was synthsized from 4, 4′-di-cyclohexylmethane diisocyanate, hydroxyl terminated polybutadiene (HTPB), and 1,4-butanediol. The IPNs were fabricated via sequential polymerization in a solvent and them the membranes were cast onto a glass plate for subsequent drying. The IPNs had better pervaporation permeability and separation factors than pure PU membrane. The IPN with 8.44 wt% of 4-VP content had the best performance of 4.11 separation factor and 1184 g/m^2h.

Ultrafiltration technique is also a membrane based filtration technique that is used for purifying and concentrating macromolecules (10^3-10^5 Da) or solid solutions, especially protein solutions. The solids or macromolecules in the solution are retained while the water or solvent pass through the semipermeable membrane [113]. The suitability of PU/sulfonated polysulfone (SPS) on ultrafiltration membrane was investigated by Malaisamy et al. [114]. Commercial grade PU was used in this study. The casting solutions consist of PU, sulfonated polysulfone and with or without additive, pore former, and PEG600. The solvent used was DMF. The polymer solutions were cast onto glass plates by a casting blade under specific coating methods and environment. The blend composites that had the best performance were 80/20 and 75/25 (PU/SPS) and the maximum compatible concentration of additive was 7.5 wt%. Pure water flux, pore sizes, and water content were increased while the membrane resistance of the membranes was decreased with the addition of sulfonated polysulfone. The molecular weight cutoff estimated values of the membranes ranged from 19 to 77 kDa. Sivakuma and coworkers developed cellulose acetate/PU blend membrane for ultrafiltration and reserve osmosis applications [115]. Cellulose acetate was blended with PU using DMF as the solvent and polyvinylpyrrolidone (PVP) as the compatibilizer. The blend membranes were cast on glass plates. The results showed that optimized hydrophilic/hydrophobic balance could be achieved by adding PVP and the blend membranes showed better protein flux which indicated they were suitable for aqueous protein separation. Zhou et al. blended castor oil-based PU with cellulose acetate for improved mechanical properties and cold resistivity for reserve osmosis applications [116]. The PU used was synthesized from 2,4-TDI, castor oil and 1,4-butanediol. PU prepolymer was dissolved in THF and then mixed with cellulose acetate/THF solution. Various compositions of cellulose/PU blends were fabricated and cast onto glass plates for subsequent PU cure. Due to strong intermolecular hydrogen bonding, PU was dispersed in cellulose acetate continuous phase to form fine microphase separation domain in the cellulose acetate rich phase. As a result, optimized tensile strength, elongation at break, and cold resistivity were obtained in the blend PU/cellulose acetate (25/75). Moisture vapor permeable films were developed by utilizing the blends of TPU and aliphatic polyester [41]. The TPU used was commercial grade. TPU and aliphatic polyester were dried before blending and they were blended by co-rotating and intermeshing a twin-screw extruder. The blend films were prepared via blow molding. From the results, aliphatic polyester was dispersed uniformly when the content of aliphatic polyester was below 30 wt% but the aliphatic polyester phase domain became irregular above 40 wt% aliphatic polyester because aggregation took place. The fabricated blend exhibited improved adhesive strength and tensile properties but deteriorated moisture vapor permeability and the waterproofness compared to TPU films.

In the separation and selection techniques, the sorption and diffusion properties of the membranes are the critical factors while the permeate flux is determined by the solubility and diffusivity of the permeants [117]. As a result, the solubility and diffusivity of the gas, liquid, and solid are important as well. The PU based blends described were useful potential materials for designed applications but the efficiency of separation and selection may vary with using different gas, solid, or liquid.

4.6.2 Electrical Applications

Blending of PU with intrinsic conductive polymer (ICPs) is usually targeted for improving the mechanical properties of ICPs without compromising too much of the electrical conductivity. The electrical conductivity of PU can be improved by blending with doped poly (o-methoxyaniline) (POMA) as reported by Gonçalves [118]. PU was synthesized from TDI, polypropylene glycol and 1,4-butanediol via a two-step polymerization method. PU and POMA blends were obtained by solvent casting technique using DMF as solvent. The dopants used did not have significant effect on the electrical conductivity. Interconnectivity of conductive domain in the blend was observed. The highest electrical conductivity measured was 10^{-3} S cm^{-1}. Liang and coworkers studied the electrical conductivity and low-frequency noise of PANI/PU blends on different substrates [119]. This kind of blends can be a candidate for metal electrical interference (EMI) shield. Commercially available polyether-polyurethane was blended with sulfonic acid treated PANI in the solvent dichloroacetic acid (DCAA). The resultant solutions were cast on different substrates: Teflon, PVC, and ceramics. The transport mechanism was suited using the noise results. Water-based PU/PANI blends for conductive coating application were fabricated by Chen et al. [120]. Commercially available WPU was dispersed in deionized water and then the DBSA doped PANI and a tackiness agent were put into WPU dispersion to form a uniform polymer solution. The resulting solution was casted on a poly (butylene terephthalate) (PET) substrate. DBSA doped PANI was well-dispersed in WPU and the surface resistance of WPU was decreased from 6.0×10^{10} Ω/sq to 1.5×10^5 Ω sq^{-1} in the blend of WPU/PANI (65/35). The fabricated coatings were suitable for antistatic, electromagnetic interference shielding and electrostatic discharge coating applications. TPU and polyether modified polysiloxane (PEMPS) blend with lithium slats for solid polymer electrolytes applications as reported by Wang and colleagues [121]. The preparation of PU/PEMPS blends and electrolytes is shown in Fig. 4.14. The addition of lithium salt could improve the compatibility between TPU and PEMPS. As a result of induced morphology change from the addition of lithium salt, the tensile, thermal, and ionic conductivity properties were improved significantly. The TPU/PEMPS electrolytes contained LiTFSI had the highest room temperature ionic conductivity up to 2.49×10^{-5} S cm^{-1}. The reason is that the interaction of lithium salt with PEMPS and TPU led to the phase compatibility which allowed the Li ion to easily transfer from one phase to the other. Jeevananda and coworkers studied the effect of adding PANI in PU/PMMA IPNs [122,123]. The PUs were castor oil-based PU where HMDI and TDI were used as the hard segments respectively. The IPNs were synthesized by sequential polymerization and PANI was added as filler. From both studies, the PANI filled IPNs exhibited improved mechanical properties, thermal stability, and especially electrical properties compared to pure PU/PMMA IPNs. PANI acted as reinforcing fillers and conducting bridges at the same time.

Apart from targeting electrical conductivity, PU based blends possessed unique properties for different electrical applications were investigated as well. Chwang et al. developed a polyaniline-dodecyl benzene sulfonic acid/polyurethane (PANI-DBSA/PU) blend with improved dielectric constant [124]. PU was synthesized by toluene diisocyanate (TDI) and polypropylene glycol 700 (PPG-7000) in *N,N*-dimethylacetamide (DMAc). PANI-DBSA/PU blends were prepared by in situ polymerization of PANI-DBSA in PU solution (ISP blend) and mixing PANI-DBAS in NMP with PU solution. The resultant blends were cast on a

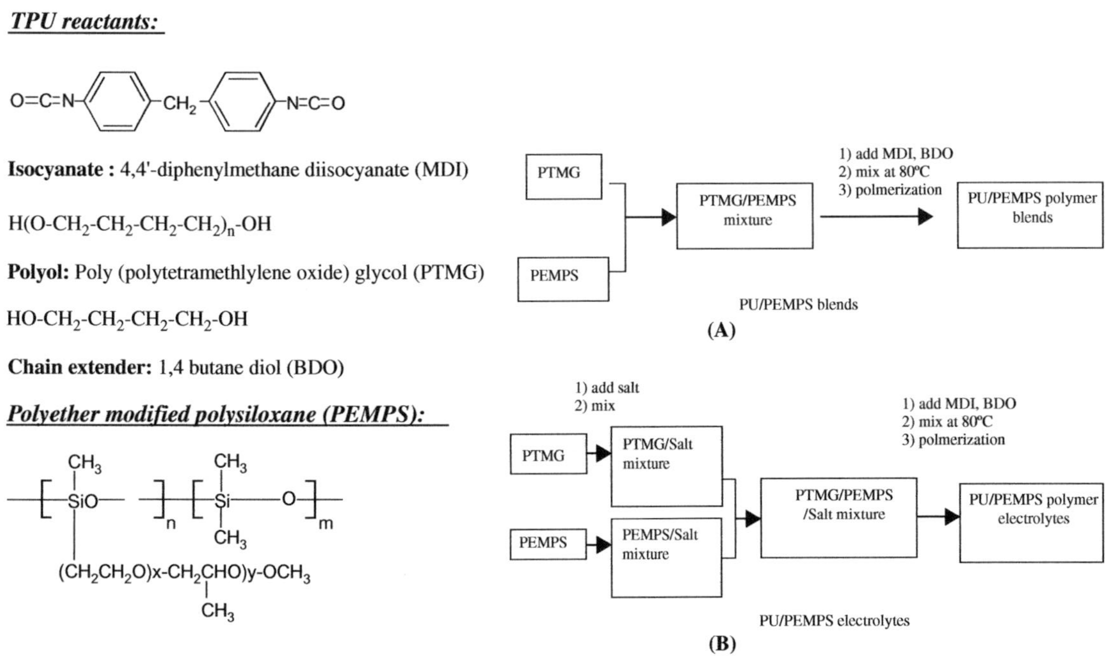

Figure 4.14 The scheme of PU/PEMPS blends and electrolytes preparation used by Want et al. [121].

preheated glass plate and then dried. The ISP blend exhibited a homogeneous structure whilst the SM blend showed inhomogeneous phase segregated with PANI domains. A very high dielectric constant ($\varepsilon_r \sim 1120$) was achieved in the ISP blend containing 17 wt% PANI. A PU/PANI blend with improved electromechanical properties for actuators, sensors and energy conversion were developed by Jaaoh [125]. Commercially available TPU and PANI were blended in DMF or NMF. The obtained solutions were then cast on glass plate to obtain dried film. The results showed that solution cast blend with NMP had superior electric permittivity and Young's modulus which indicated it could be a potential candidate for electromechanical actuator applications. Sulfonated polysulphone (s-PSF)/PU blend membranes for proton exchange membrane fuel cell applications were prepared by Pedicini et al. [24]. 18 wt% Aliphatic polyurethane was mixed with s-PSF in DMAc and then the membranes were cast by the Doctor Blade technique. A slight increase in ionic exchange capacity and lower swelling ratios were observed and the PU blended with PSF sulfonation degree (48%) had the best performance according to the I−V curves.

4.6.3 Biological Applications

Biological materials receive increasing attention because they are vital for medical and medicine applications. The utilization of PU based blends provide materials with some biocompatibility and good mechanical properties. PU based blends used in medical devices, tissue engineering and blood compatible materials are reviewed in this section. Han et al. blended hydrophilic polyurethane (HPU) with poly vinyl alcohol/vinyl butyral copolymer (PVAB) as an outer membrane for a high-performance glucose sensor [126]. Commercially available HPU and PVAB were blended in cyclohexanone with various compositions. A pneumatic dispenser was used to assemble the as-prepared polymer solution on enzyme layers. From the results, the optimized blend was HPU/PVAB (60/40). The fabricated outer membrane presented high oxygen permeability and the ability to control glucose diffusion properly. Li et al. blended citric acid grafted polyurethane with polyethersulfone (PES) to produce a membrane used in blood purification fields [127]. PES and modified PU were first dissolved in the solvent DMAc and then the membranes were cast by spin coater and subsequent processes. From the results, the produced membrane exhibited superior cytocompatibility, biocompatibility, and anticoagulant property than pure PES. The suitability of PU/PCL blends with different ratios used in polymeric endoaortic paving (PEAP) for cardiovascular applications [128]. The PU and PCL used were commercially available and the blends were fabricated by solvent casting method using chloroform as solvent. The blends' stiffness was similar to that of aortic tissue as revealed by biaxial tension testing. Tubes of blends were able to maintain high diameter increases in thermoforming and 50/50 blend tube could maintain 90% of the diameter increased applied at 55°C. In addition, 50/50 blends exhibited a 94% lower degradation rate than pure PCL and the 10/90 blend showed no degradation. The authors indicated that the mechanical properties were adjustable via varying the blend ratios and PEAP consisting of a PCL/PU blend was suitable developing the next generation of endovascular therapy for AAA tissue.

Montini Ballarin utilized electrospinning technique to a bilayer scaffolding structure consisting of poly (L-lactic acid) (PLLA) and bioresorbable segmented polyurethane (SPEU) blends [129]. Aliphatic diisocyanate, aliphatic polyester, and an aromatic chain extender were used to synthesize SPEU. SPEU and PLLA blend were produced by electrospinning and the solvent system was optimized to produce uniform bead-free fibers. The blends showed better thermal, physical, and surface properties than the raw polymer which indicated the produced scaffolding tubular structure is a potential candidate for vascular tissue engineering as bilayered vascular grafts. However, a faster hydrolytic degradation of blends might result from the lower degree of crystallinity of the blends. Eelctrospun nanofibrous scaffolds for tissue engineered tubular grafts were developed by Huang et al. utilizing collagen-chitosan-TPU blends [130]. TPU was chosen to improve the mechanical properties whilst chitosan and collagen were used as bio-mimicking the native extracellular matrices. Commercially available TPU and collagen were dissolved in hexafluoroisopropanol (HFP) while chitosan was dissolved in HFP and trifluoroacetic acid (TFA) mixture. The three solutions were blended in a certain weight ratio. The electrospun nanofibers were collected on a flat collector plate to form a membrane and then the membrane was cured by glutaraldehyde (GTA) vapor. The fabricated scaffolds were flexible while a high tensile strength was maintained. Good biocompatibility was achieved because of the nanofibers alignment that maintained cell morphology.

Lv and coworkers tried to blend PU with heparin for improved blood compatibility [131]. PU was synthesized from IPDI, poly (ethylene oxide-tetramethlene oxide) and 1,4-butanediol via a two-step polymerization method. The obtained PU was dissolved in THF and the heparin was dissolved in water. The two solutions were mixed and then cast on glass plates to form PU/heparin blends. From the results, heparin in the film could keep excellent bioactivity and the film exhibited good blood compatibility in a long time which exceeded the detection range of the clot detection instrument. The blood compatibility of PU could be improved by blending with silk fibroin (SF) as reported by Liu and coworkers [132]. The PU used was a commercially available waterborne PU. SF was dissolved in melted 1-butyl-3-methylimidazolium chloride (BMIMCl) and PU was dissolved in DMF. The two solutions were mixed at various ratios and then the resulting mixtures were cast onto glass plates. In this blending process, the step of dialysis for SF was not necessary because BMIMCI was used. Chemical bonding between PU and TPU and a β-sheet structure of SF in the blend films were observed. With increasing SF content, the hydrophilicity and blood compatibility of the blends were improved. The PU/SF blend was suitable for bioapplications which required good blood compatibility. Highly blood compatible hemodialysis membranes were prepared by blending heparin-mimicking polyurethane (HMPU) with PES [133]. PU was synthesized by MDI and dimethylolpropionic acid (DMPA) and then HMPU was obtained through sulfonation by concentrated sulphuric acid (Fig. 4.15). The HMPU-PES blend membranes were prepared by mixing HMPU and

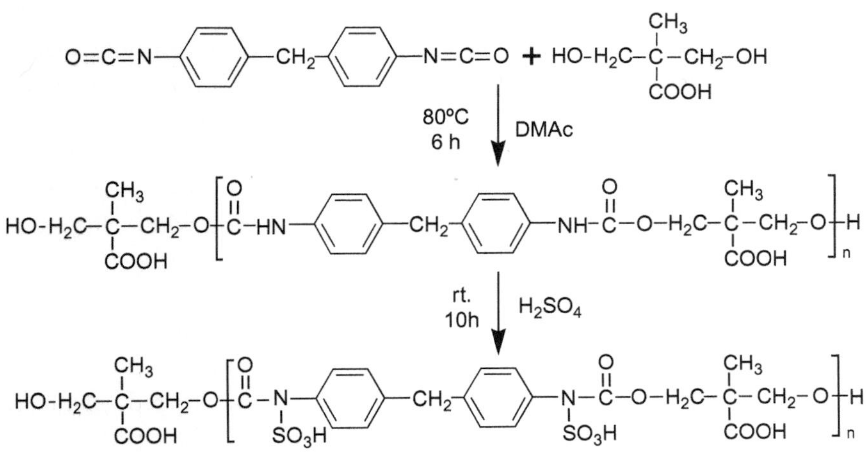

Figure 4.15 The scheme of synthesizing heparin-mimicking polyurethane proposed by Ma et al. [133].

PES in DMAc and then the polymer solutions were cast by spin coating. The blend membranes exhibited lowered thrombin generation activation, suppressed platelet adhesion, good protein antifouling property, good therapeutic effect, prolonged blood coagulation, and decreased protein adsorption than pure PES membranes. This suggested that the HMPU-PES blends had excellent blood compatibility which could be used in blood purification fields.

Liu et al. developed a controlled heparin release system for long lasting antithrombogenicity application based on polyurethane and silk fibroin film [134]. The PU used was commercially available. For the preparation of the blend films, PU was dissolved in DMF first and then heparin sodium and SK were added. The mixture was cast on a glass plate. By controlling the film thickness, SF to PU ratios and the loading amount of heparin, the heparin release rate and the percentage of the cumulative amount of the released heparin was adjusted. The blend films also exhibited excellent blood compatibility according to the results of the coagulation time tests. A similar heparin release system was developed by Yang and coworkers but using biomedical polyurethane (BPU) [135]. Native silk fibroin powder (NSFP) was produced from silk cocoons by extraction and grinding. The obtained NSFP was blended with heparin to obtain heparinized NSFP (H-NSFP). Then the BPU was blended with H-NSFP by a phase inversion technique at room temperature. The results showed that the release of heparin can be controlled effectively by varying the membrane thickness, the amount of heparin contained in the membrane, and the ratios of NSFP and BPU.

Park et al. investigated the possibility of PU blended with poly (ethylene oxide) (PEO)-poly z(polytetramethylene oxide) (PTMO) multi-block copolymer as coating materials for urinary catheters [136]. The PU used in this study contained PTMO as soft segment and 4,4′-methylenediphenylene diisocyanate as the hard segment. The blend coatings with different PU loading were prepared by solvent casting method where THF was used as the solvent. The results showed that the adhesion of *Escherichia coli*, *Proteus mirabilis*, and *Staphylococcus epidermidis* on the surfaces was remarkably reduced by the blend. The blends also exhibited excellent encrustation resistance and the duration of patency was extended up to 20 ± 3.1 h which was much higher than silicone coated catheter. Fenouillot and coworkers investigated the possibility of reactive blending α,ω-aminopropyl polydimethylsiloxane (PDMS) and thermoplastic polyurethane (TPU) for medical devices and the coatings for the settlement of marine organisms [137]. The TPU used was polyester based polyurethane pallet. TPU was melted blended with different siloxane oligomer in a twin-screw extruder and polysiloxane was synthesized during extrusion. The splitting of urethane group of TPU at 200°C allowed the fast reaction between amino-siloxane with the isocyanate, forming a very fine morphology TPU-PDMS copolymer which is beneficial to produce a suitable material for medical devices.

4.6.4 Shape Memory Applications

Shape memory material is a material that can return to its original shape after deformation with the aid of heating. There are five objectives of developing shape memory blends [138]: (1) Shape recovery stress and mechanical properties improvement; (2) Minimizing shape recovery induction time; (3) New blends with shape memory properties; (4) Modifying trigger temperature, mechanical properties and biomedical properties of available shape memory materials; (5) Developing shape memory materials that can be trigger by other external effects such as electric, light, moisture, and magnetic field. In this section, the PU based blends used in shape memory area are reviewed.

Ajili et al. developed a polyurethane/polycaprolacton (PCL) blend as a potential shape memory for cardiovascular stents [35]. The shape memory property was crystallinity-induced. A polyester PU contained MDI and 1,4-butanediol was used and it was blended with PCV via melt blending in an internal mixer. PU/PCL (70/30) blend showed shape recovery in the range of body temperature and excellent biocompatibility which indicated this blend could be used as a stent material. The shape memory effect and miscibility of TPU-phenoxy resin were studied by Jeong and coworkers [28]. The TPU used contained poly (caprolactone) based soft segment and hexamethylene diisocyanate (HDI)/BDO/4,4′dihydroxy biphenyl (DHBP) based hard segment. TPU was mixed with phenoxy resin in DMAc and then the resulting polymer solutions were cast on glass plates. Shape memory effect was obtained when the soft segment/phenoxy resin domain was a reversible phase and hard segment domain had a fixed structure. The T_g of the blends can be designed by varying the ratios of TPU and phenoxy resin. With the increased of the content of hard segment in blend or the decreased of the block length of TPU, the hysteresis in shape recovery was reduced. Similar investigation using TPU and PVC blends was conducted by the same group [139]. The

TPU used was the same as the previous study. The PVC/TPU blend films were produced by solvent casting method using DMAc as solvent. PVC was compatible with the PCL soft segment. From the results, the shape memory property was improved by the increase of hard segment domain. The permanent deformation in the fixed shape which happened along with the recoverable deformation of reversible phase of miscible PVC/PCL segment was more obvious at a rubbery plateau when the modulus ratio of glassy/rubbery state was much lower than 100. The publication of utilization PU based blends in shape memory applications is limited but the reported articles suggested that PU based blends are promising in this area.

4.6.5 PU Based Blends/IPNs for Other Applications

The other potential applications of PU based blends are reviewed in this section. Parnell and coworkers adopted a reactive blending process to produce poly (vinyl chloride) (PVC) and TPU blends [140]. MDI, 1,4-butandiol and poly (butylene adipate) (PBA) were used to synthesize TPU. The reactive processing was performed in an internal mixer by mixing PVC and the TPU reactants and synthesizing TPU at the same time. The produced blends exhibited heterogeneous, multiphase morphologies, and tensile strengths between that of pure PVC and TPU. The fracture toughness of plasticized PVC (PPVC) could be improved by blending with TPU [141]. Two commercial TPUs with different hardness were used for comparison. The PPVC/PU blends with different compositions were fabricated by melt blending in a laboratory-made bench kneader. The blends exhibited improved tensile strength, fire retardancy, impact strength, abrasion resistance and thermal stability but the reduction of tensile modulus and hardness were also observed. PPVC/TPU70 blends showed better fracture toughness improvement than the PPVC/TPU90 blend.

PU based blends can be developed for high impact applications. Huang and coworkers blended thermoplastic polyurethane and PET for improving the impact strengths of PET for high impact applications [6]. Commercially available PET and TPU were blended in a counter-rotating twin-screw extruder. The results showed that good compatibility between TPU and PBT was achieved and an increase about 369.18% in impact strength was achieved by blending with 30 wt% TPU. The impact strength of poly (vinyl butyral) (PVB) was improved by blending with thermoplastic polyurethane as stated by Sincock et al. [61]. PU was synthesized from bases on methylene bis-cyclohexyl diurethane and polytetramethylene ether glycol. Blending of PU and PVB pallets was finished by incorporating certain amount of mixtures into a twin-screw palletizing extruder. Miscibility over a range of compositions was observed and the impact resistance of glass laminates prepared from these blends was improved. Imre et al. blended PU elastomers into PLA for impact modification and the resultant blends could potentially be applied in biomedicine [44]. Poly (tetrahydrofuran) (PTHF), 1,4-butanediol and MDI were used to synthesize PU. A PU/PLA blend was prepared by physical blending and reactive processing which were both performed in an internal mixer. The results showed that the blends produced by reactive processing have superior microstructure and mechanical properties because the formation of a submicron, phase in phase structure. They also suggest that reactive processing is a cost effective, environmental friendly and convenient method.

Packaging applications is also an important aspect of PU based blends applications. Soy protein isolate (SPI) was blended for the first time in waterborne polyurethane for improved properties of SPI films to be used in packaging applications [142]. Poly(butylene adipate) (PBA) based WPU was synthesized from PBA and HDI. Dimethylol propionic acid (DMPA) was added to introduce carboxyl group onto the polymer chains and trimethylamine (TEA) was used to neutralize the carboxyl groups of as-synthesized PU. SPI and WPU were dissolved in deionized water separately and the solutions were mixed and cast onto glass plates to form a blend film. A homogeneous structure of the blend film was formed due to the strong intermolecular interactions between SPI and WPU. The water resistance and mechanical properties of SPI/PU blends were superior to pure SPI films. Tong and coworkers blended crude glycerol-based waterborne polyurethane (CGWPU) with soymeal-derived protein (SMP) for packaging applications [143]. CGWPU was synthesized from glycerol-based polyols, dimethylol propionic acid (DMPA) and isophorone diisocyanate (IPDI). The blends of CGWPU/SMP were compared with the blends prepared from SMP with commercially available polyurethane (MU). The blend preparation scheme is shown in Fig. 4.16. Good compatibility between CGWPU and SMP was achieved because of strong interfacial adhesion and intermolecular interactions. Great improvements in the water resistance, thermal stability, and tensile strength were observed with increasing CWGPU content whilst the elongation at break

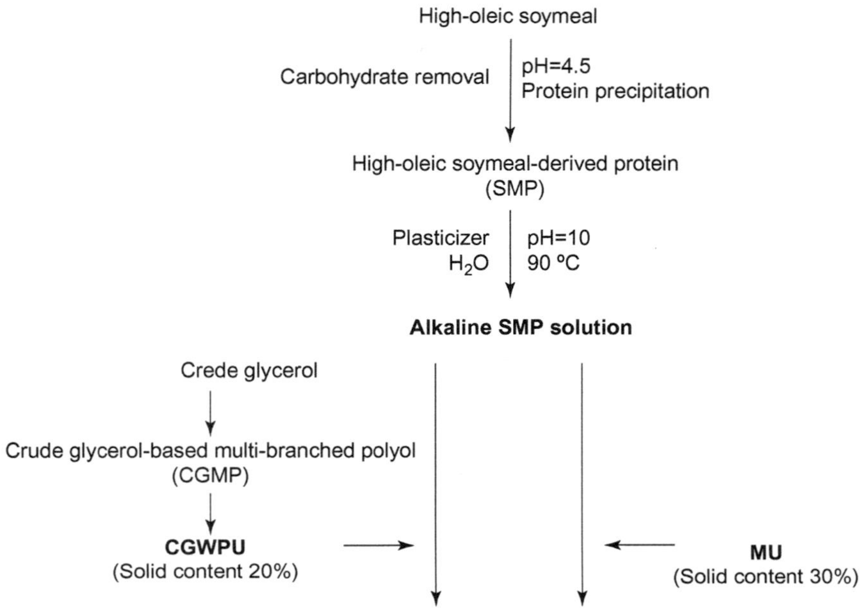

Figure 4.16 The scheme of CGWPU/SMP and MU/SMP blends preparation used by Tong et al. [143].

was decreased. The produced CWGPU/SMP blends had superior properties than MU/SMP blends.

Thermoplastic PU is widely used as coating and adhesive materials. The PU based blends can be readily used in coating and adhesives for improved performance. Water-soluble PU (WPU) was blended with PVA by Wang et al. for potential coating and textile applications [144]. Commercially available WPU and PVA were blended in distilled water and then the mixtures were cast into molds. The toughness of the PVA was improved significantly by blending with WPU. With increasing WPU content, the fracture mechanism transitioned from brittle to ductile according to the results from stress–strain curves and the fractured surface. Severe plastic deformation was also observed in the fracture surface of the blend that contained high WPU content. Song and his colleges blended TPU with olefin block copolymer (OBC) to improve the PU paint adhesion and wettability on OBC substrate [59]. The TPU and OBC were commercially available and the TPU was a polytetramethylene glycol eth-based polyurethane elastomer consisted of MDI, 1,4-butandiol and polyol. TPU and OBC blends were prepared by compounding in a twin-screw extruder or by dry blending. The TPU/OBC blend substrates were fabricated by injection molding. From the results, significant enhancement on paint adhesion and high connectivity of the TPU with the top paint layer were observed in the blends with 25 wt% TPU or greater. Incomplete wetting and gaps in the coating were observed for lower TPU content blend substrates. The proposed mechanism of adhesion was interfacial polyurethane inter-diffusion.

There are some interesting applications where PU based blends can be applied. Santayanon and coworkers blended polyester PU with modified cassava starch to improve its moisture resistance [33]. Cassava starch is a biodegradable material which can be used in coating applications. Propionic anhydride modified starch/PU blends were prepared by a two-roll mill. From the results, miscibility between the starch and the PU was improved by the modification of propionic anhydride and the moisture resistance of the starch was improved by blending PU. However, the blended starch exhibited a slower biodegradation than normal starch. A film forming binder based on caprolactam-modified casein (CA-CPL) and waterborne polyurethane was developed by Ma and coworkers [145]. Film forming binder can be used in textile coatings, leather finishing agents, and packaging applications. CA-CPL/deionized water solution was mixed with WPU aqueous dispersion directly with different weight ratios of CA-CPL to WPU. In terms of leather finishing samples, the mixture was applied on leather through three leather finishing processes according to the finishing formula made by a skilled leather engineer. From the results, poor compatibility between CPL-CA and WPU was observed when the weight percentage of WPU was above 50%. Hydrogen bonding interaction between CPA-CA and WPU was verified. Optimized CPL-CA/WPU blend could provide leather with excellent flexibility and wet rub resistance. PU/polyimide (PI) blends were developed for protection applications by Benson et al. [77]. PU was synthesized from polypropylene glycol, MDI, and ethylene diamine via two-step polymerization. PU/PI blends with various compositions were prepared via solvent casting method using DMF as solvent. The blend films were cast onto glass plates. From the results, the formation of PU hard domains was blocked by the addition of PI and an increase in phase mixing was observed. Improvements in modulus, yield stress, and fracture energy were obtained and those were attributed to the synergistic effect between the nanoscale reinforcement and hydrogen bonding between the hard segment of PU and the PI. Want et al. utilized poly (bispropoxyphosphazene) (LFR) blended with PU for improving thermal stability and fire retardancy [146]. The PU was synthesized by a two-step polymerization from MDI, poly (tetramethylene oxide) oligomer and 1,4-butanediol. LFR was blended with PU during the second step of PU polymerization. Two stages degradation behavior was observed for pure PU and PU/LFR blend. PU/LFR blend exhibited higher degradation rate, lower apparent activation energies of thermal degradation, and improved fire retardancy.

4.7 CONCLUSIONS AND FUTURE PERSPECTIVES

The PU structure, PU based blends/IPNs preparation methods and the properties targeted for different applications have been reviewed in this chapter. The key characterization techniques for blends, IPNs and nanocomposites were detailed. The utilization of PU in polymer blends is versatile and the range of applications is broad. The utilization of nanofillers for mechanical properties enhancement is promising. Nanofillers can act as a load transfer medium, crosslinking sites, crack inhibitors, as compatibilizer and so on. The key factors of optimizing properties are uniform dispersion of nanofillers, good filler-polymer matrix compatibility, and a suitable filler loading. Nanofillers in the PU based blends/IPNs act as electrons or thermal transportation tunnel. The key to form conductive composites is the formation of conductive network and the reduction of tunneling resistance. Utilizing conductive nanofillers with big particle size and blending with intrinsic conductive polymer can be the main route to manufacture conductive PU blends/IPNs nanocomposite. The key to improve thermal conductivity is to reduce the phonon scattering in the system nanofillers. The nanofillers that can improve thermal stability should either have good thermal insulation property or can retard the degradation process of the composite. In addition, excellent compatibility of the nanofillers and the matrix and good dispersion of nanofillers in the system is necessary.

Miscibility is the key factor that affects the final properties of the PU based blend. The selection of the PU used in different studies aim for optimized miscibility of the polymer that the PU blended with. In addition, the preparation methods used are important as well. The preparation methods determine whether a blend with good phase structure, phase domain and interphase can be obtained. Comparing different preparation methods, in situ reaction solvent blending may be the most effective way prepared a blend with molecular scale miscibility. However, the complicated preparation process and time consuming drying processes limit its applications. In terms of scalability and cost effective, in situ reaction solvent blending is not easily used in industry. For solvent blending, the blending procedure is less complicated but the subsequent drying procedure takes a long time as well. Therefore, it is not easy to produce polymer blends on a large scale for commercial applications. Melt blending to produce polymer blends may be the technique that can readily fit into large scale production. However, the miscibility of the resulted blends is usually not ideal and the capital cost for producing blends may be the major limitations of its utilization. The miscibility problems may be resolved by utilizing reactive melt blending where the polymerization of one of the blend's component occurs in the blending mixing (twin-screw extruder for example). Polymer blends have superior properties than pure polymers but the processing of the blends for large scale production is still difficult to achieve. A universal, scalable, cost effective processing technique that can achieve ultimate performance of polymer blend needs to be developed for wider applications of polymer blends.

REFERENCES

[1] A. Wurtz, Justus Liehigs Ann. (1849).
[2] O. Bayer, Polyurethanes Mod. Plast. 24 (1947) 149–152.
[3] R. Narayan, D.K. Chattopadhyay, B. Sreedhar, K.V.S.N. Raju, N.N. Mallikarjuna, T.M. Aminabhavi, Synthesis and characterization of crosslinked polyurethane dispersions based on hydroxylated polyesters, J. Appl. Polym. Sci. 99 (2006) 368–380.
[4] D.K. Chattopadhyay, K.V.S.N. Raju, Structural engineering of polyurethane coatings for high performance applications, Prog. Polym. Sci. 32 (2007) 352–418.
[5] J.A. de Sales, P.S.O. Patrício, J.C. Machado, G.G. Silva, D. Windmöller, Systematic investigation of the effects of temperature and pressure on gas transport through polyurethane/poly(methylmethacrylate) phase-separated blends, J. Memb. Sci. 310 (2008) 129–140.
[6] J. Huang, X. Lu, G. Zhang, J. Qu, Study on the rheological, thermal and mechanical properties of thermoplastic polyurethane/poly (butylene terephthalate) blends, Polym. Test 36 (2014) 69–74.
[7] D. Paul, S. Newmen, Polymer Blends, Academic Press, New York; London, 1978.
[8] R.W. Seymour, S.L. Cooper, Thermal analysis of polyurethane block polymers, Macromolecules 6 (1972) 48–53.
[9] S.J. Lue, J.S. Ou, S.L. Chen, W.S. Hung, C.C. Hu, Y.C. Jean, et al., Tailoring permeant sorption and diffusion properties with blended polyurethane/poly(dimethylsiloxane) (PU/PDMS) membranes, J. Membr. Sci. 356 (2010) 78–87.
[10] M.S. Yen, P.Y. Tsai, P.D. Hong, The solution properties and membrane properties of polydimethylsiloxane waterborne polyurethane blended with the waterborne polyurethanes of various kinds of soft segments, Colloids Surf. A Physicochem. Eng. Asp. 279 (2006) 1–9.

[11] S.M. Yang, H.L. Lee, Conductivity and morphology of the blend of polyaniline and water based polyurethane, Synth. Met. 102 (1999) 1226–1227.
[12] H. Yoshikawa, T. Hino, N. Kuramoto, Effect of temperature and moisture on electrical conductivity in polyaniline/polyurethane (PANI/PU) blends, Synth. Met. 156 (2006) 1187–1193.
[13] D.S. Vicentini, G.M.O. Barra, J.R. Bertolino, A.T.N. Pires, Polyaniline/thermoplastic polyurethane blends: Preparation and evaluation of electrical conductivity, Eur. Polym. J 43 (2007) 4565–4572.
[14] H. Yeganeh, M.R. Mehdizadeh, Synthesis and properties of isocyanate curable millable polyurethane elastomers based on castor oil as a renewable resource polyol, Eur. Polym. J 40 (2004) 1233–1238.
[15] L.M. Robeson, Polymer blends: a comprehensive review, Hanser, Munich; Cincinnati, 2007.
[16] J. Jin, L. Chen, M. Song, K. Yao, An analysis on enhancement of fatigue durability of polyurethane by incorporating organoclay nanofillers, Macromol. Mater. Eng. 291 (2006) 1414–1421.
[17] H. Xia, S.J. Shaw, M. Song, Relationship between mechanical properties and exfoliation degree of clay in polyurethane nanocomposites, Polym. Int. 54 (2005) 1392–1400.
[18] M. Song, D.J. Hourston, K.J. Yao, J.K.H. Tay, M.A. Ansarifar, High performance nanocomposites of polyurethane elastomer and organically modified layered silicate, J. Appl. Polym. Sci. 90 (2003) 3239–3243.
[19] X. Wang, J. Jin, M. Song, Cyanate ester resin/graphene nanocomposite: curing dynamics and network formation, Eur. Polym. J 48 (2012) 1034–1041.
[20] Y. Lin, J. Jin, M. Song, S.J. Shaw, C.A. Stone, Curing dynamics and network formation of cyanate ester resin/polyhedral oligomeric silsesquioxane nanocomposites, Polymer (Guildf) 52 (2011) 1716–1724.
[21] N.-W. Pu, Y.-Y. Peng, P.-C. Wang, C.-Y. Chen, J.-N. Shi, Y.-M. Liu, et al., Application of nitrogen-doped graphene nanosheets in electrically conductive adhesives, Carbon N. Y. 67 (2014) 449–456.
[22] N. Yousefi, M.M. Gudarzi, Q. Zheng, S.H. Aboutalebi, F. Sharif, J.-K. Kim, Self-alignment and high electrical conductivity of ultralarge graphene oxide–polyurethane nanocomposites, J. Mater. Chem. 22 (2012) 12709–12717.
[23] M.-W. Ulrich, Polyurethanes: Coatings, Adhesives and Sealants, Vincentz Network GmbH & Co KG, 2007.
[24] R. Pedicini, A. Saccà, A. Carbone, I. Gatto, A. Patti, E. Passalacqua, Study on sulphonated polysulphone/polyurethane blend membranes for fuel cell applications, Chem. Phys. Lett. 579 (2013) 100–104.
[25] H.L. Wang, C.M. Fu, A. Gopalan, T.C. Wen, Frequency dependent conductivity of the thin film blend of electroluminescent poly(p-phenylene vinylene) with waterborne polyurethane as ionomer, Thin Solid Films 466 (2004) 197–203.
[26] T. Gurunathan, C.R.K. Rao, R. Narayan, K.V.S.N. Raju, Polyurethane conductive blends and composites: synthesis and applications perspective, J. Mater. Sci. 48 (2012) 67–80.
[27] Y. Ti, D. Chen, Mechanical and dynamic mechanical properties of polyurethane/Fe-octacarboxyl acid phthalocyanine blends, Prog. Org. Coatings 76 (2012) 119–124.
[28] H.M. Jeong, B.K. Ahn, B.K. Kim, Miscibility and shape memory effect of thermoplastic polyurethane blends with phenoxy resin, Eur. Polym. J 37 (2001) 2245–2252.
[29] N. Ahmed, a Kausar, B. Muhammad, Shape memory properties of electrically conductive multi-walled carbon nanotube-filled polyurethane/modified polystyrene blends, J. Plast. Film Sheeting. (2015).

[30] S. Dutta, N. Karak, J.P. Saikia, B.K. Konwar, Biocompatible epoxy modified bio-based polyurethane nanocomposites: Mechanical property, cytotoxicity and biodegradation, Bioresour. Technol. 100 (2009) 6391–6397.
[31] A.K. Nandi, B.M. Mandal, S.N. Bhattacharyya, Miscibility of poly (methyl acrylate) and poly (vinyl acetate): incompatibility in solution and thermodynamic characterization by inverse gas chromatography, Macromolecules 18 (1985) 1454–1460.
[32] M. Al Minnath, G. Unnikrishnan, E. Purushothaman, Transport studies of thermoplastic polyurethane/natural rubber (TPU/NR) blends, J. Membr. Sci. 379 (2011) 361–369.
[33] R. Santayanon, J. Wootthikanokkhan, Modification of cassava starch by using propionic anhydride and properties of the starch-blended polyester polyurethane, Carbohydr. Polym. 51 (2003) 17–24.
[34] J.H. Tan, X.P. Wang, J.J. Tai, Y.F. Luo, D.M. Jia, Novel blends of acrylonitrile butadiene rubber and polyurethane-silica hybrid networks, Express Polym. Lett. 6 (2012) 588–600.
[35] S.H. Ajili, N.G. Ebrahimi, M. Soleimani, Polyurethane/polycaprolactane blend with shape memory effect as a proposed material for cardiovascular implants, Acta Biomater. 5 (2009) 1519–1530.
[36] F. Yu, H.-X. Huang, Simultaneously toughening and reinforcing poly(lactic acid)/thermoplastic polyurethane blend via enhancing interfacial adhesion by hydrophobic silica nanoparticles, Polym. Test 45 (2015) 107–113.
[37] M. Maity, B.B. Khatua, C.K. Das, Effect of processing on the thermal stability of the blends based on polyurethane: Part IV, Polym. Degrad. Stab. 72 (2001) 499–503.
[38] J.K. Mishra, S. Roychowdhury, C.K. Das, Effect of the coupling agent's reactivity on the shrinkability of the blends consisting of grafted low-density polyethylene and ethylene acrylic elastomer, Polym. Adv. Technol. 13 (2002) 112–119.
[39] K. Palanivelu, S. Balakrishnan, P. Rengasamy, Thermoplastic polyurethane toughened polyacetal blends, Polym. Test. 19 (2000) 75–83.
[40] K. Palanivelu, P. Sivaraman, M. Dasaratha Reddy, Studies on thermoplastic polyurethane toughened poly(butylene terephthalate) blends, Polym. Test. 21 (2002) 345–351.
[41] X. Wang, X. Luo, X. Wang, Study on blends of thermoplastic polyurethane and aliphatic polyester: Morphology, rheology, and properties as moisture vapor permeable films, Polym. Test. 24 (2005) 18–24.
[42] E.G. Bajsic, A. Pustak, I. Smit, M. Leskovac, Blends of thermoplastic polyurethane and polypropylene. II. Thermal and morphological behavior, J. Appl. Polym. Sci. 117 (2010) 1378–1384.
[43] P. Potschke, J. Pionteck, H. Stutz, Blends of thermoplastic polyurethanes and polyolefins.Pdf, Polymer (Guildf) 43 (2002) 6965–6972.
[44] B. Imre, D. Bedo, A. Domján, P. Schön, G.J. Vancso, B. Pukánszky, Structure, properties and interfacial interactions in poly(lactic acid)/polyurethane blends prepared by reactive processing, Eur. Polym. J. 49 (2013) 3104–3113.
[45] A.S. Vatalis, A. Kanapitsas, C.G. Delides, P. Pissis, Relaxation phenomena and morphology in polymer blends based on polyurethanes investigated by various thermal analysis techniques, Thermochim. Acta 372 (2001) 33–38.
[46] D.J. Duffy, A.M. Heintz, H.D. Stidham, S.L. Hsu, W. Suen, C.W. Paul, The competitive influence of specific interactions and extent of reaction on the miscibility of ternary reactive polymer blends: Model for polyurethane adhesives, Int. J. Adhes. Adhes. 25 (2005) 39–46.
[47] Y. Shi, W. Zhang, J. Yang, T. Huang, N. Zhang, Y. Wang, et al., Super toughening of the poly(l-lactide)/thermoplastic polyurethane blends by carbon nanotubes, RSC Adv. 3 (2013) 26271–26282.

[48] J.K. Mishra, I. Kim, C.S. Ha, Heat shrinkable behavior and mechanical response of a low-density polyethylene/millable polyurethane/organoclay ternary nanocomposite, Macromol. Rapid Commun. 25 (2004) 1851–1855.
[49] M.Á. Garcés-Lara, R. Antonio-Cruz, A.M. Mendoza-Martínez, T.L. Ramírez, A.B. Morales-Cepeda, Effect of initiator and cross-linking agent on a semi-IPN of (PU-Starch)/PMMA, Chem. Eng. Commun. 196 (2009) 1227–1236.
[50] W.S. Chow, Polyamide blend-based nanocomposites: a review, Express Polym. Lett. 9 (2015) 211–232.
[51] D.B. Williams, C.B. Carter, Transmission Electron Microscopy: A Textbook for Materials Science, Springer, United States, 2009.
[52] D. Reifer, R. Windeit, R.J. Kumpf, A. Karbach, H. Fuchs, AFM and TEM investigations of polypropylene/polyurethane blends, Thin Solid Films 264 (1995) 148–152.
[53] C. Qin, Z. Jin, X. Bai, H. Jiang, W. Cai, Compatibility of polyurethane/(vinyl ester resin)(ethyl acrylate) interpenetrating polymer network, Polym. J. 39 (2007) 1365–1372.
[54] T. Trakulsujaritchok, D.J. Hourston, Damping characteristics and mechanical properties of silica filled PUR/PEMA simultaneous interpenetrating polymer networks, Eur. Polym. J. 42 (2006) 2968–2976.
[55] C. Qin, J. Zheng, J. Haijian, T. Qiang, B. Xuduo, The influence of modified VER on PU/VER IPN, Pigment Resin Technol. 37 (2008) 355–361.
[56] L.H. Sperling, J. a Manson, Interpenetrating polymer networks from triglyceride oils containing special functional groups: a brief review1, J. Am. Oil Chem. Soc. 60 (1983) 1887–1892.
[57] C.L. Qin, W.M. Cai, J. Cai, D.Y. Tang, J.S. Zhang, M. Qin, Damping properties and morphology of polyurethane/vinyl ester resin interpenetrating polymer network, Mater. Chem. Phys. 85 (2004) 402–409.
[58] D. Tang, C. Qin, W. Cai, L. Zhao, Preparation, morphology, and mechanical properties of modified-PU/UPR graft-IPN nanocomposites with $BaTiO_3$ fiber, Mater. Chem. Phys. 82 (2003) 73–77.
[59] J. Song, A. Batra, J.M. Rego, C.W. MacOsko, Polyethylene/polyurethane blends for improved paint adhesion, Prog. Org. Coatings 72 (2011) 492–497.
[60] H. Liu, J. Song, S. Shang, Z. Song, D. Wang, Cellulose nanocrystal/silver nanoparticle composites as bifunctional nanofillers within waterborne polyurethane, ACS Appl. Mater. Interfaces 4 (2012) 2413–2419.
[61] T. Sincock, D. David, Miscibility and properties of poly(vinyl butyral) and thermoplastic polyurethane blends, Polymer 33 (1992) 4515–4521.
[62] C. Wang, F. Chu, C. Graillat, A. Guyot, C. Gauthier, J.P. Chapel, Hybrid polymer latexes: Acrylics-polyurethane from miniemulsion polymerization: Properties of hybrid latexes versus blends, Polymer 46 (2005) 1113–1124.
[63] Y. Cui, Y. Chen, X. Wang, G. Tian, X. Tang, Synthesis and characterization of polyurethane/polybenzoxazine-based interpenetrating polymer networks (IPNs), Polym. Int. 52 (2003) 1246–1248.
[64] X. Wang, J. Jin, M. Song, An investigation of the mechanism of graphene toughening epoxy, Carbon 65 (2013) 324–333.
[65] D. Cai, K. Yusoh, M. Song, The mechanical properties and morphology of a graphite oxide nanoplatelet / polyurethane composite, Nanotechnology 20 (2009) 085712–085716.
[66] R. Sanjinés, M.D. Abad, C. Vâju, R. Smajda, M. Mionić, A. Magrez, Electrical properties and applications of carbon based nanocomposite materials: an overview, Surf. Coatings Technol. 206 (2011) 727–733.
[67] JEOL Ltd, Scanning Electron Microscope A To Z - Basic knowledge for using the SEM.

[68] D. Cai, J. Jin, K. Yusoh, R. Rafiq, M. Song, High performance polyurethane/functionalized graphene nanocomposites with improved mechanical and thermal properties, Compos. Sci. Technol. 72 (2012) 702–707.
[69] G. Allen, J.C. Bevington, C. Booth, C. Price, Comprehensive Polymer Science: The Synthesis, Characterization, Reactions & Applications of Polymers / Vol.1, Polymer Characterization, Pergamon, Oxford, 1989.
[70] J.I. Kroschwitz, Polymers: Polymer Characterization and Analysis, Wiley, New York, 1990.
[71] Q. Guo, H. Xu, D. Ma, S. Wang, Miscible blends of poly(ethylene oxide) and an amorphous copolyester-polyurethane, Eur. Polym. J. 26 (1990) 67–71.
[72] F.W. Billmeyer, Textbook of Polymer Science, 3rd ed, Wiley, New York; Chichester, 1984.
[73] B.D. Cullity, Elements of X-ray diffraction, Upper Saddle River, NJ, 3rd ed, Prentice-Hall International, 2001.
[74] L.E. Alexander, X-ray Diffraction Methods in Polymer Science, Wiley-Interscience, New York, 1969.
[75] O. Glatter, O. Kratky, Small Angle X-ray Scattering, Academic Press, London, 1982.
[76] M. Song, H.S. Xia, K.J. Yao, D.J. Hourston, A study on phase morphology and surface properties of polyurethane/organoclay nanocomposite, Eur. Polym. J. 41 (2005) 259–266.
[77] R.S. Benson, M.W. Lee, D.W. Grummitt, Nanostructured polymers: a molecular interpretation of nanoscale reinforcement in polyurethane/polyimide blends, Nanostructured Mater. 6 (1995) 83–91.
[78] M. Song, D.J. Hourston, F.U. Schafer, A study of the morphology of polyurethane-polystyrene interpenetrating polymer networks by means of small angle X-ray scattering, modulated-temperature differential scanning calorimetry, and dynamic mechanical thermal analysis techniques, J. Appl. Polym. Sci. 79 (2001) 1958–1964.
[79] D.J. Dyson, X-ray and Electron Diffraction Studies in Materials Science, Maney for the Institute of Materials, Minerals, and Mining, London, 2004.
[80] P.S. George, Polymer Characterization Techniques and Their Application to Blends, American Chemical Society, Washington, DC, 2003.
[81] S. Barbara, Infrared Spectroscopy: Fundamentals and Applications, Analytical Techniques in The Science, John Wiley & Sons, Chichester, 2004.
[82] E. Yilgör, E. Yurtsever, I. Yilgör, Hydrogen bonding and polyurethane morphology. II. Spectroscopic, thermal and crystallization behavior of polyether blends with 1,3-dimethylurea and a model urethane compound, Polymer (Guildf) 43 (2002) 6561–6568.
[83] D. Sarkar, J.C. Yang, N. Klettlinger, S.T. Lopina, Blends of L-tyrosine based polyurethanes for biomaterial applications, Express Polym. Lett. 1 (2007) 724–733.
[84] H. Xia, M. Song, Z. Zhang, M. Richardson, Institutional Repository Microphase separation, stress relaxation and creep behavior of polyurethane nanocomposites, J. Appl. Polym. Sci. 103 (2007) 2992–3002.
[85] S.L. Huang, J.Y. Lai, HTPB-H12MDI based polyurethane IPN membranes for pervaporation, J. MembR. Sci. 115 (1996) 1–10.
[86] M. Liu, R.V. Gregory, A blend of polyurethane and a urethane substituted polythiophene in the β-position, Synth. Met. 69 (1995) 349–350.
[87] K. Pielichowski, I. Hamerton, Compatible poly(vinyl chloride)/chlorinated polyurethane blends: thermal characteristics, Eur. Polym. J. 36 (2000) 171–181.
[88] P.C. Rodrigues, L. Akcelrud, Networks and blends of polyaniline and polyurethane: Correlations between composition and thermal, dynamic mechanical and electrical properties, Polymer (Guildf) 44 (2003) 6891–6899.

[89] J. Huang, J.W. Zou, P.R. Chang, J.H. Yu, A. Dufresne, New waterborne polyurethane-based nanocomposites reinforced with low loading levels of chitin whisker, Express Polym. Lett. 5 (2011) 362−373.
[90] M. Kotal, S.K. Srivastava, B. Paramanik, Enhancements in conductivity and thermal stabilities of polypyrrole/polyurethane nanoblends, Inorg. Mater. 115 (2011) 1496−1505.
[91] K.K.S. Hwang, S.B. Lin, S.Y. Tsai, S.L. Cooper, Properties of polyurethane oligomeric blends versus high molecular weight block copolymers, Polymer 25 (1984) 947−955.
[92] H. Harani, S. Fellahi, M. Bakar, Toughening of epoxy resin using synthesized polyurethane prepolymer based on hydroxyl-terminated polyesters, J. Appl. Polym. Sci. 70 (1998) 2603−2618.
[93] D.A. Skoog, F.J. Holler, T.A. Nieman, Principles of Instrumental Analysis, 5th ed, Thomson Learningm Inc., Belmont, 1997.
[94] P.K. Gallagher, Thermoanalytical Instrumentation, Techniques, and Methodology, in: E.A. Turi (Ed.), Therm. Charact. Polym. Mater, 2nd ed, Academic Press, New York; London, 1997.
[95] W. Wu, X. Luo, D. Ma, Miscibility of the segmented polyurethane blends with chlorinated poly(vinyl chloride), Eur. Polym. J. 35 (1999) 985−990.
[96] D.J. Hourston, M. Song, Applications of modulated temperature differential scanning calorimetry to polymer blends and related sytems, in: M. Reading, D.J. Hourston (Eds.), Modulated Temperature Differential Scanning Calorimetry, Springer, Dordrecht, 2006.
[97] E. Passaglia, M. Bertoldo, S. Coiai, S. Augier, S. Savi, F. Ciardelli, Nanostructured polyolefins / clay composites: role of the molecular interaction at the interface, Polym. Adv. Technol. 19 (2008) 560−568.
[98] M. Kotal, S.K. Srivastava, A.K. Bhowmick, Thermoplastic polyurethane and nitrile butadiene rubber blends with layered double hydroxide nanocomposites by solution blending, Polym. Int. 59 (2010) 2−10.
[99] C.K. Tan, D.J. Blackwood, Corrosion protection by multilayered conducting polymer coatings, Corros. Sci. 45 (2003) 545−557.
[100] J.R. Potts, D.R. Dreyer, C.W. Bielawski, R.S. Ruoff, Graphene-based polymer nanocomposites, Polymer 52 (2011) 5−25.
[101] J. Ge, H. Yao, W. Hu, X. Yu, Y. Yan, L. Mao, et al., Facile dip coating processed graphene/MnO2 nanostructured sponges as high performance supercapacitor electrodes, Nano Energy (2012) 1−8.
[102] J. Jin, Y. Lin, M. Song, C. Gui, S. Leesirisan, Enhancing the electrical conductivity of polymer composites, Eur. Polym. J. 49 (2013) 1066−1072.
[103] D. Cai, M. Song, Latex technology as a simple route to improve the thermal conductivity of a carbon nanotube/polymer composite, Carbon 46 (2008) 2107−2112.
[104] Y.A. El-Shekeil, S.M. Sapuan, M. Jawaid, O.M. Al-Shuja'a, Influence of fiber content on mechanical, morphological and thermal properties of kenaf fibers reinforced poly(vinyl chloride)/thermoplastic polyurethane poly-blend composites, Mater. Des. 58 (2014) 130−135.
[105] L.H. Sperling, Interpenetrating Polymer Networks and Related Materials, Springer Science & Business Media, New York, 2012.
[106] R. Satgurunathan, Interpenetrating Polymer Networks, Pergamon Press, Oxford, 1987.
[107] D. Zavastin, I. Cretescu, M. Bezdadea, M. Bourceanu, M. Drăgan, G. Lisa, et al., Preparation, characterization and applicability of cellulose acetate-polyurethane blend membrane in separation techniques, Colloids Surf. A Physicochem. Eng. Asp. 370 (2010) 120−128.

[108] M.A. Semsarzadeh, B. Ghalei, Characterization and gas permeability of polyurethane and polyvinyl acetate blend membranes with polyethylene oxide-polypropylene oxide block copolymer, J. Membr. Sci. 401–402 (2012) 97–108.
[109] M.-J. Kim, B. Sea, K.-H. Youm, K.-H. Lee, Morphology and carbon dioxide transport properties of polyurethane blend membranes, Desalination 193 (2006) 43–50.
[110] H. Matuschewski, U. Schedler, MSE - modified membranes in organophilic pervaporation for aromatics/aliphatics separation, Desalination 224 (2008) 124–131.
[111] L. Lin, Y. Kong, K. Xie, F. Lu, R. Liu, L. Guo, et al., Polyethylene glycol/polyurethane blend membranes for gasoline desulphurization by pervaporation technique, Sep. Purif. Technol. 61 (2008) 293–300.
[112] S.J. Lue, J.S. Ou, C.H. Kuo, H.Y. Chen, T.H. Yang, Pervaporative separation of azeotropic methanol/toluene mixtures in polyurethane-poly(dimethylsiloxane) (PU-PDMS) blend membranes: correlation with sorption and diffusion behaviors in a binary solution system, J. Membr. Sci. 347 (2010) 108–115.
[113] M. Cheryan, Ultrafiltration Handbook, Technomic Publishing Co. Inc., Basel, 1986.
[114] R. Malaisamy, D.R. Mohan, M. Rajendran, Polyurethane and sulfonated polysulfone blend ultrafiltration membranes, J. Colloid Interface Sci. 254 (2002) 129–140.
[115] M. Sivakumar, R. Malaisamy, C.J. Sajitha, D. Mohan, V. Mohan, R. Rangarajan, Ultrafiltration application of cellulose acetate-polyurethane blend membranes, Eur. Polym. J. 35 (1999) 0–4.
[116] Q. Zhou, L. Zhang, M. Zhang, B. Wang, S. Wang, Miscibility, free volume behavior and properties of blends from cellulose acetate and castor oil-based polyurethane, Polymer (Guildf) 44 (2003) 1733–1739.
[117] S.J. Lue, W.W. Chen, S.Y. Wu, L.D. Wang, C.H. Kuo, Vapor permeation modeling of multi-component systems using a poly(dimethylsiloxane) membrane, J. Membr. Sci. 311 (2008) 380–389.
[118] D. Gonçalves, a Waddon, F.E. Karasz, L. Akcelrud, Conducting blends of soluble polyurethane and poly(o-methoxyaniline), Synth. Met. 74 (1995) 197–199.
[119] C. Liang, G. Leroy, J. Gest, L.K.J. Vandamme, J.L. Wojkiewicz, 1/f noise in polyaniline/polyurethane (PANI/PU) blends, Synth. Met. 159 (2009) 1–6.
[120] C.H. Chen, Y.T. Kan, C.F. Mao, W.T. Liao, C.D. Hsieh, Fabrication and characterization of water-based polyurethane/polyaniline conducting blend films, Surf. Coatings Technol. 231 (2013) 71–76.
[121] S. Wang, K. Min, Solid polymer electrolytes of blends of polyurethane and polyether modified polysiloxane and their ionic conductivity, Polymer 51 (2010) 2621–2628.
[122] T. Jeevananda, Siddaramaiah, Synthesis and characterization of polyaniline filled PU/PMMA interpenetrating polymer networks, Eur. Polym. J. 39 (2003) 569–578.
[123] T. Jeevananda, M. Begum, Studies on polyaniline filled PU/PMA interpenetrating polymer networks, Eur. Polym. J. 37 (2001) 0–5.
[124] C.-P. Chwang, C.-D. Liu, S.-W. Huang, D.-Y. Chao, S.-N. Lee, Synthesis and characterization of high dielectric constant polyaniline/polyurethane blends, Synth. Met. 142 (2004) 275–281.
[125] D. Jaaoh, C. Putson, N. Muensit, Deformation on segment-structure of electrostrictive polyurethane/polyaniline blends, Polymer 61 (2015) 123–130.
[126] J.H. Han, J.D. Taylor, D.S. Kim, Y.S. Kim, Y.T. Kim, G.S. Cha, et al., Glucose biosensor with a hydrophilic polyurethane (HPU) blended with polyvinyl

alcohol/vinyl butyral copolymer (PVAB) outer membrane, Sensors Actuat. B Chem. 123 (2007) 384–390.
[127] L. Li, C. Cheng, T. Xiang, M. Tang, W. Zhao, S. Sun, et al., Modification of polyethersulfone hemodialysis membrane by blending citric acid grafted polyurethane and its anticoagulant activity, J. Membr. Sci. 405–406 (2012) 261–274.
[128] J.H. Ashton, J. a M. Mertz, J.L. Harper, M.J. Slepian, J.L. Mills, D.V. McGrath, et al., Polymeric endoaortic paving: mechanical, thermoforming, and degradation properties of polycaprolactone/polyurethane blends for cardiovascular applications, Acta Biomater. 7 (2011) 287–294.
[129] F. Montini Ballarin, P.C. Caracciolo, E. Blotta, V.L. Ballarin, G.A. Abraham, Optimization of poly(l-lactic acid)/segmented polyurethane electrospinning process for the production of bilayered small-diameter nanofibrous tubular structures, Mater. Sci. Eng. C 42 (2014) 489–499.
[130] C. Huang, R. Chen, Q. Ke, Y. Morsi, K. Zhang, X. Mo, Electrospun collagen-chitosan-TPU nanofibrous scaffolds for tissue engineered tubular grafts, Colloids Surf. B Biointerfaces 82 (2011) 307–315.
[131] Q. Lv, C. Cao, H. Zhu, A novel solvent system for blending of polyurethane and heparin, Biomaterials 24 (2003) 3915–3919.
[132] X. Liu, C. Zhang, W. Xu, H. Liu, C. Ouyang, Blend films of silk fibroin and water-insoluble polyurethane prepared from an ionic liquid, Mater. Lett. 65 (2011) 2489–2491.
[133] L. Ma, B. Su, C. Cheng, Z. Yin, H. Qin, J. Zhao, et al., Toward highly blood compatible hemodialysis membranes via blending with heparin-mimicking polyurethane: study in vitro and in vivo, J. Membr. Sci. 470 (2014) 90–101.
[134] X.Y. Liu, C.C. Zhang, W.L. Xu, C.X. Ouyang, Controlled release of heparin from blended polyurethane and silk fibroin film, Mater. Lett. 63 (2009) 263–265.
[135] H. Yang, H. Xu, H. Liu, C. Ouyang, W. Xu, A novel heparin release system based on blends of biomedical polyurethane and native silk fibroin powder, J. Control. Release 152 (2011) e106–e108.
[136] J.H. Park, Y.W. Cho, I.C. Kwon, S.Y. Jeong, Y.H. Bae, Assessment of PEO/PTMO multiblock copolymer/segmented polyurethane blends as coating materials for urinary catheters: in vitro bacterial adhesion and encrustation behavior, Biomaterials 23 (2002) 3991–4000.
[137] F. Fenouillot, F. Méchin, F. Boisson, P. Alcouffe, T. Pokropski, T. Kallel, et al., Coarsening of nanodomains by reorganization of polysiloxane segments at high temperature in polyurethane/α,ω-aminopropyl polydimethylsiloxane blends, Eur. Polym. J 48 (2012) 284–295.
[138] Q. Meng, J. Hu, A review of shape memory polymer composites and blends, Compos. Part A Appl. Sci. Manuf. 40 (2009) 1661–1672.
[139] H.M. Jeong, J.H. Song, S.Y. Lee, B.K. Kim, Miscibility and shape memory property of poly (vinyl chloride)/ thermoplastic polyurethane blends, J. Mater. Sci. 6 (2001) 5457–5463.
[140] S. Parnell, K. Min, Reactive blending of thermoplastic polyurethane in situ with poly(vinyl chloride), Polym. Eng. Sci. 45 (2005) 876–887.
[141] C.-S. Ha, Y. Kim, W.-K. Lee, W.-J. Cho, Y. Kim, Fracture toughness and properties of plasticized PVC and thermoplastic polyurethane blends, Polymer (Guildf) 39 (1998) 4765–4772.
[142] M. Zhang, F. Song, X.L. Wang, Y.Z. Wang, Development of soy protein isolate/waterborne polyurethane blend films with improved properties, Colloids Surf. B Biointerfaces 100 (2012) 16–21.
[143] X. Tong, X. Luo, Y. Li, Development of blend films from soy meal protein and crude glycerol-based waterborne polyurethane, Ind. Crops Prod. 67 (2015) 11–17.

[144] J. Wang, L. Ye, Structure and properties of polyvinyl alcohol/polyurethane blends, Compos. Part B Eng. 69 (2015) 389–396.
[145] J. Ma, Q. Xu, D. Gao, J. Zhou, J. Zhang, Blend composites of caprolactam-modified casein and waterborne polyurethane for film-forming binder: miscibility, morphology and properties, Polym. Degrad. Stab. 97 (2012) 1545–1552.
[146] P.S. Wang, W.Y. Chiu, L.W. Chen, B.L. Denq, T.M. Don, Y.S. Chiu, Thermal degradation behavior and flammability of polyurethanes blended with poly(bispropoxyphosphazene), Polym. Degrad. Stab. 66 (1999) 307–315.

CHAPTER 5

Polyurethane Ionic Blends as the Electrically Conductive Coatings

Piotr Król[1], Bożena Król[1], Marek Zenker[2] and Jan Subocz[2]
[1]Rzeszów University of Technology, Rzeszów, Poland
[2]West Pomeranian University of Technology, Szczecin, Poland

Contents

5.1 INTRODUCTION

Very wide applicability of polyurethene (PU) results from the fact that their performance properties can be widely modified by selecting appropriate raw materials, catalysts, and auxiliary compounds, by employing various production methods and/or by employing various methods for further processing and/or for shaping the final products. Resulting from their specific microphase structure, which is formed by rigid chain segments and flexible chain segments, polyurethanes offer very good elasticity with reasonably high mechanical strength and abrasion resistance at the same time, and also controllable hardness. Polyurethanes can be available both as relatively rigid plastomers and as flexible elastomers with compact or foamed structures. There are two inhibiting factors for applicability of polyurethanes: their limited stability at temperatures above 90°C, and their flammability, in particular for foamed polyurethanes [1,2].

PUs are produced in the form of foamed plastics, structural elastomers and coating elastomers, adhesives, leather-like materials, and auxiliary agents.

Polyurethane Polymers: Blends and Interpenetrating Polymer Networks
DOI: http://dx.doi.org/10.1016/B978-0-12-804039-3.00005-1

Flexible cellular PU elastomers became inherent in comfort-providing elements, especially in furniture making and automotive industry. On the other hand, rigid foamed PUs can be converted into lightweight elements with structural stability and superior thermal insulation performance (closed-cell foams) and/or acoustic insulation performance (open-cell foams). One should also note that thermoplastic elastomers (PUT) are also available. These can be processed as typical thermoplastics due to thermal restoration of their allophanate groups [3–7].

A newer trend in application of PUs is the production of ionomers and of the practical significance are both anionomers and cationomers. These make the feedstock for waterborne polyurethenes that need no external emulsifiers and become environmentally-friendly lacquers, coating, adhesives, and sealants [8,9]. The combination of emulsions of PU ionomers and emulsions of other polymers, like poly(vinyl alcohol) or poly(vinyl acetate), yields interpenetrating polymer networks (the so-called semi-IPNs) with very good adhesion to metallic and ceramic substrates, and with improved resistance to water [10].

5.2 POLYURETHANE ANIONOMERS

Polyurethane anionomers include the group of linear polyurethanes or those with lower density of cross-linking (PU) which have very small numbers of acidic functional groups, e.g., COO^- or HSO_3^-, in their chains [11]. In general, the structure of a PU anionomer chain, like that of any elastomer, is composed of blocks and it comprises both rigid urethane segments (derived from diisocyanates and diamines) and flexible polyether or polyester segments (Formula 5.1):

$$-R_1-NH-CO-O-\underset{\displaystyle COO^-}{\underset{|}{R}}-O-OC-NH-R_1-NH-CO-NH-R_3-NH-OC-NH-R_1-NH-CO- \qquad (5.1)$$

where

R_1—aromatic, alicyclic or aliphatic radical derived from diisocyanate employed in the synthesis,

R_2—aliphatic chain derived from polyol, e.g., oligo-oxypropylene glycol (Formula 5.2):

$$-[O-CH_2-CH(CH_3)]_n-O- \qquad (5.2)$$

R_3—aliphatic radical, e.g., $-(CH_2)_6-$ from diamine employed to extend the isocyanate prepolymer chain, built-in into the chain through applicable urea structures.

The COO^- groups, which make principal elements of PU anionomers, are located in the element R and they are derived in most cases from some additional substrates, e.g., dihydroxyacids (Formula 5.3):

$$R-\underset{CH_2OH}{\overset{CH_2OH}{\underset{|}{\overset{|}{C}}}}-COOH \quad (5.3)$$

but they can also be introduced in other locations within the PU chain, resulting from chemical modifications of that chain [12].

Yet, the actual structure of a PU chain can be much more complex and it is dependent not only on the type of employed raw materials but also on the conditions adopted for the polyaddition process, e.g., which feedstocks were charged earlier and which were charged later, stoichiometry of reaction, process temperature and time, as well as possible presence and amount of polyaddition catalyst used. The process conditions are responsible for the fact that PU macromolecules can comprise also biuret groups, allophanate ones, carbodiimide or azaheterocyclic isocyanurate structures, or oxazolidone ones. The majority of additionally created structures are polar in their nature. This adds to the nature of carboxyl, urethane and ether or ester groups which are typical for anionomers. Hence, the polarity of resultant PUs becomes considerable and those materials offer better solubility in water (as compared to insoluble nonionic polyurethanes) and in polar organic solvents, e.g., *N,N*-dimethylacetamide, dimethylformamide, acetonitrile, dimethylsulfoxide. When a PU anionomer is not completely water soluble, it is usually capable of forming aqueous emulsions or dispersions, and it can form specific polymer gels in the presence of organic solvents [13]. This should be accounted for by enhanced polar interactions of the dipole–dipole or dipole–induced dipole type and by ionic interactions in the *polymer–solvent* and *polymer–polymer* system. Moreover, an anionomer has a higher tendency to form hydrogen bonds as compared to nonionic PUs. It is for the above reasons that the molecular weight specifications of ionomers have to be analyzed by the GPC method with the use of polar solvents, while a medium-polarity eluent like tetrahydrofuran is sufficiently good for linear polyurethanes [14–16].

One can expect that such strong interactions between polar functional groups in PU anionomers will notably affect the supermolecular structure, arrangement of phases and separation of flexible and rigid segments present in PUs. These issues, in turn, are decisive for their electrical properties: improved conductivity and higher dielectric constant of those materials. The surface energy of the anionomers considered can be modified by introducing functional groups and structural segments with hydrophobic performance (nonpolar hydrogen groups and low-polarity ether groups) and hydrophilic performance (polar acid groups, amine groups, and ammonium groups) [17]. PU elastomers that have some features of ionomers were found to offer improved thermal stability and improved mechanical properties, and even higher potential for thermodynamic miscibility with nonpolar polymers, e.g., poly(p-phenylene vinylene), or with polar ones, e.g., poly(styrene-acrylic acid) [18,19].

For the above reasons, polyurethane ionomers draw attention of numerous researchers from the 1980s [12,20,21]. The systems were synthesized which contained carboxyl and sulfo groups [22–26], and with carbon or fluorocarbon trunk chains [27]. The literature reports confirm continual interest in that group of polymers, applicable for example as waterborne and environmentally-friendly lacquers and adhesives wherein no additional emulsifiers are required [28–31]. This is even more interesting since the intermediates used for their manufacture—urethane-isocyanate prepolymers—are not water compatible and they need to be accurately reacted with polyols in the aqueous medium first. Urethane-isocyanate anionomers, on the other hand, demonstrate the ability to react selectively in the aqueous medium with diamines, e.g., with 1,6-hexamethylenediamine. This process extends the PU chain but no hydrolysis of isocyanate and no release of CO_2 takes place, which is the case when foamed plastics are produced in situ [32,33]. In case of classic linear PUs containing strongly hydrophilic urethane and hydroxyl groups, it is also possible to produce emulsions but additional surfactants are required to make those emulsions stable, and the dispergation process becomes effective only after high shear rates are employed [34].

We employed polyurethane anionomers in our previous research in the form of aqueous dispersions, as the so-called temporary binders which made it possible to mold the oxide-type ceramic materials into products. Attempts were also made to evaluate surface energy of so- produced aqueous dispersions [16,17].

5.2.1 Methods of the Anionomer Synthesis

In this chapter a multistage process of the polyurethane anionomer synthesis, which is based on the previously worked out prepolymer method [18], is presented. The process is presented in details in References [15,18,19].

At the first stage, the urethane-isocyanate prepolymer is prepared from a selected pair: diisocyanate and polyol. At further steps, it is reacted with 2,2-bis-(hydroxymethyl)propionic acid (DMPA), and finally its COO^- groups are neutralized with triethylamine, and the prepolymer chain is extended in the aqueous dispersion medium with 1,6-hexamethylenediamine. What is critical for the adopted synthesis method is the need to provide adequate molar ratios for the introduced reacting substances at successive reaction stages [15,19].

Stage I—Synthesis of isocyanate prepolymer (AB_2)

The essence of the stage I is illustrated in Formula 5.4:

$$\begin{array}{ccccccc} \text{HO}-\text{A}-\text{OH} & + & 2\,\text{OCN}-\text{B}-\text{NCO} & \rightarrow & \text{OCN}-\text{B}-\text{NH}-\text{CO}-\text{O}-\text{A}-\text{O}-\text{CO}-\text{NH}-\text{B}-\text{NCO} \\ (\text{A}) & + & (\text{B}) & \rightarrow & (\text{AB}_2) \end{array} \quad (5.4)$$

where: A—segment derived from polyester or polyether as polyoxypropylene glycol (Rokopol 7 P, $M_n = 450$) from Chemical Factory 'Rokita S. A.' in Brzeg Dolny (Poland), the product was dried under vacuum in nitrogen, at 120°C during 2 h) (Formula 5.5).

$$\text{HO}-[\underset{\displaystyle \text{CH}_3}{\underset{|}{\text{CH}}}-\text{CH}_2-\text{O}]\text{n}-\text{H} \quad (5.5)$$

where B—segment derived from diisocyanate structure.

In the quoted work [19] segments B derived from the following diisocyanates:

2,4- and 2,6-Tolylene diisocyanate (TDI) which was a mixture of 2,4-TDI and 2,6-TDI isomers at the ratio of 80 and 20%, respectively (Formula 5.6);

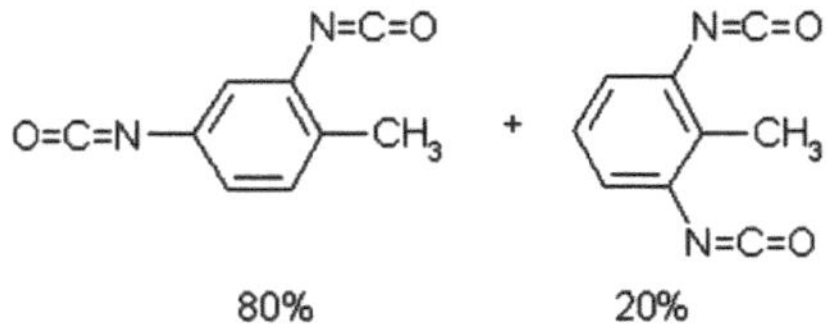

(5.6)

4,4′-Methylenebis(phenyl isocyanate) (MDI) (Formula 5.7);

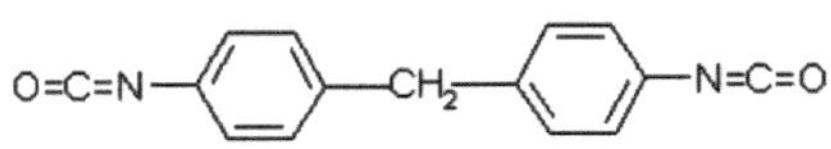

(5.7)

Hexamethylene-1,6-diisocyanate (HDI) (Formula 5.8). All the isocyanate reagents were used as purchased.

$$O{=}C{=}N(CH_2)_6N{=}C{=}O \tag{5.8}$$

Isophorone diisocyanate [5-isocyanato-1-(isocyanatomethyl)-1,3,3-trimethylcyclohexane] (IPDI) (Formula 5.9).

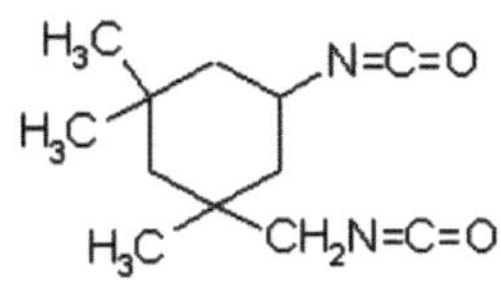

(5.9)

4,4′-Methylenebis(cyclohexyl isocyanate) (HMDI) (Formula 5.10).

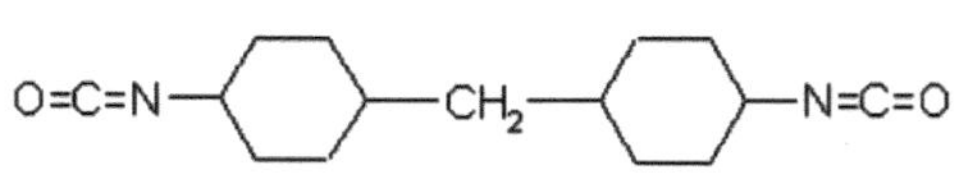

(5.10)

Urethane-isocyanate prepolymer was obtained by adding molten polyoxypropylene glycol drop by drop to the flask where successively selected diisocyanates had been placed previously. The temperature of polyoxypropylene glycol was maintained at 30–100°C by means of an IR radiator. In case of MDI and HMDI diisocyanates, the reactions were conducted in the presence of the diazabicyclo-[2,2,2]-octane (DABCO) catalyst. The specific conditions for individual reactions have been detailed in Table 5.1.

Table 5.1 Conditions for synthesis of polyurethane anionomers [29]

Anionomer no	Stage no	Type of diisocyanate	Type of polyol	Type of ionogenic reactant	Content of –NCO groups[a] (wt%) Theoret.	Content of –NCO groups[a] (wt%) Experim.	Type of extension agent	Molar ratio of groups (– NCO: – OH)	Dry matter content (wt%)
1	1	Synthesis of isocyanate prepolymer ($T = 80°C$, $t = 20$ min)			10.52	9.90	–	2:1	–
		TDI	Rokopol 7P ($M = 450$)	–					
	2	Synthesis of isocyanate ionomer ($T = 60°C$, $t = 20$ min)			2.72	2.30	–	2:1	–
		–		DMPA					
		Neutralization of ionomer by means of TEA ($T = 20°C$, $t = 15$ min, pH = 7.5)							
	3	Extension of isocyanate ionomer and emulsification in water ($T = 20°C$, $t = 20$ min)					HMDA	1:1	20.0
	4	Evaporation of solvents from emulsion and formation of coating ($T = 110°C$, $t = 60$ min)							
2	1	Synthesis of isocyanate prepolymer catalyst DABCO (0.1 wt%) ($T = 30°C$, $t = 20$ min)			5.09	3.50	–	3:2	–
		MDI	Rokopol 7P ($M = 450$)	–			–		
	2	Synthesis of isocyanate ionomer ($T = 50°C$, $t = 60$ min)			2.44	1.19	–	2:1	–
		–		DMPA					
		Neutralization of ionomer by means of TEA ($T = 0°C$, $t = 15$ min, pH = 7.5)							

(*Continued*)

Table 5.1 (Continued)

Anio-nomer no	Stage no	Type of diisocyanate	Type of polyol	Type of ionogenic reactant	Content of–NCO groups[a] (wt%) Theoret.	Content of–NCO groups[a] (wt%) Experim.	Type of extension agent	Molar ratio of groups (– NCO: – OH)	Dry matter content (wt%)
	3	Extension of isocyanate ionomer and emulsification in water ($T=20°C$, $t=60$ min)					HMDA	1:1	27.0
	4	Evaporation of solvents from emulsion and formation of coating ($T=110°C$, $t=60$ min)							
3	1	Synthesis of isocyanate prepolymer catalyst DABCO (0.1 wt%) ($T=100°C$, $t=40$ min)			4.98	6.60	–	3:2	–
		HMDI	Rokopol 7P ($M=450$)	–					
	2	Synthesis of isocyanate ionomer ($T=80°C$, $t=30$ min)			2.39	2.00	–	2:1	–
		–		DMPA					
		Neutralization of ionomer by means of TEA ($T=80°C$, $t=10$ min, pH = 7.5)							
	3	Extension of isocyanate ionomer and emulsification in water $T=80°C$, $t=30$ min					HMDA	1:1	22.0
	4	Evaporation of solvents from emulsion and formation of coating ($T=110°C$, $t=60$ min)							
4	1	Synthesis of isocyanate prepolymer catalyst DABCO (0.5 wt%) ($T=90°C$, $t=120$ min)			5.36	4.30	–	3:2	–
		IPDI	Rokopol 7P ($M=450$)	–					

	2	Synthesis of isocyanate ionomer $T = 80°C$, $t = 30$ min			2.57	1.80	–	2:1	–
		–		DMPA					
		Neutralization of ionomer by means of TEA ($T = 80°C$, $t = 10$ min, pH = 7.5)							
	3	Extension of isocyanate ionomer and emulsification in water $T = 70°C$, $t = 30$ min					HMDA	1:1	24.0
	4	Evaporation of solvents from emulsion and formation of coating ($T = 110°C$, $t = 60$ min)							
5	1	Synthesis of isocyanate prepolymer $T = 100°C$, $t = 90$ min			5.98	6.0	–	3:2	–
		HDI	Rokopol 7P ($M = 450$)	–					
	2	Synthesis of isocyanate ionomer ($T = 100°C$, $t = 30$ min)			2.85	1.8	–	2:1	–
		–		DMPA					
		Neutralization of ionomer by means of TEA ($T = 80°C$, $t = 10$ min, pH = 7.5)							
	3	Extension of isocyanate ionomer and emulsification in water ($T = 80°C$, $t = 30$ min)					HMDA	1:1	26.0
	4	Evaporation of solvents from emulsion and formation of coating ($T = 110°C$, $t = 60$ min)							

[a]That determination involved a well-known method, and dibutylamine was used in the tests. Excess of unreacted amine was titrated with the HCl solution and bromophenol blue was used as an indicator.

Stage II—Synthesis of urethane-isocyanate anionomer and its ammonium salt

Urethane-isocyanate anionomer which contained −COOH groups was synthesized with the use of 2,2-bis(hydroxymethyl)propionic acid (DMPA) (Formula 5.11);

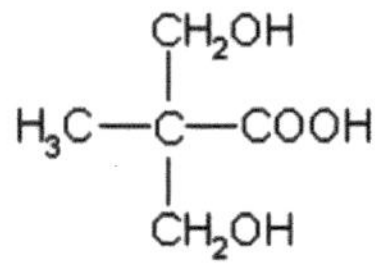

(5.11)

as a chain extender for the AB_2 prepolymer (Formula 5.12)

$$2AB_2 + X \rightarrow BABYBAB \quad (5.12)$$

where: X = DMPA and Y = segment derived from built-in DMPA.

DMPA was added to the reactor in the form of solution in DMF (1:7). The reaction was continued at 50−100°C over 20−60 min, and then the concentration of −NCO groups was checked again and it was compared to the theoretical value (Table 5.1). Afterwards, the urethane-isocyanate macro-anionomer (with the carboxyl groups in its structure) is subjected to the reaction (structural fragment Y) with triethylamine (TEA) which yields the triethylammonium cation. That reaction can be represented by Formula 5.13):

$$\underset{\displaystyle \text{COOH}}{\text{BABYBAB}} + N(C_2H_5)_3 \longrightarrow \underset{\displaystyle \text{COO}^{(-)}\,[NH(C_2H_5)_3]^{(+)}}{\text{BABYBAB}} \quad (5.13)$$

The macro-anionomer produced in that way was capable of forming aqueous dispersions. The obtained isocyanate-urethane prepolymer with the number of −COOH groups known from stoichiometric calculations was reacted with the equimolar amount of TEA. The amine was added at 20−80°C until the level of pH = 7.5 was reached.

Under these conditions the free −NCO groups of this macro-anionomer react much slower with the carboxylate groups than with water and $-NH_2$ groups coming from 1,6-hexamethylenediamine (HMDA) which is put in during the Stage III. Therefore additional creation of the possible amide crosslinks with some part of—COO^- groups which are present in the segments coming from DMPA can be neglected.

Stage III—Extension of urethane macro-anionomer
The macromolecules of isocyanate anionomers which were not converted at stage III must be extended at the next stage by means of HMDA. The reaction is carried out in aqueous dispersion, taking advantage of higher reactivity of isocyanate groups with amino groups that isocyanate groups and water (Formula 5.14):

$$\underset{\substack{|\\ COO^{(-)}[NH(C_2H_5)_3]^{(+)}}}{nBABYBAB} + n\,H_2N-(CH_2)_6-NH_2 \longrightarrow -[NH-(CH_2)_6-NH-\underset{\substack{|\\ COO^{(-)}[NH(C_2H_5)_3]^{(+)}}}{BAHBYBHA}]_n-NH- \qquad (5.14)$$

The major amount of HMDA (90%) was added in the form of solution in THF, at room temperature. The mixture was then subjected to mechanical dispergation during another 30 min, at room temperature: intense agitation was applied with the dropwise addition of redistilled water. THF was subsequently recovered from the obtained emulsion by means of distillation and water is evaporated under mild conditions from the aqueous dispersion which have been applied to the perfectly smooth and chemically inert surface, e.g., Teflon, to produce the polymer film. After evaporation of water, polymer films were produced that were used in further investigations.

Bearing in mind that the assessment of the extent of polar interactions in anionomers is critical for giving explanation to the observed physical-chemical and mechanical properties, the research program intended to identify instrumental methods capable of evaluating said properties in relation to the ionomer structure was launched.

Moreover, as revealed in GPC tests, considerable differences in reactivity of −NCO groups present in diisocyanate feedstocks resulted in diversified average molecular weights of the synthesized ionomers. Hence, one can expect that strong interactions within rigid urethane segments and flexible polyol-ether segments, resulting from structural polarity, ionic interactions and intermolecular hydrogen bondings, would affect the morphological features of the condensed phases of anionomers. These factors may cause the polymer chains organizing in the supermolecular structures with the defined crystallographic order, which occurs in the separate phase regions made from rigid and flexible segments. During creating such an ordering the way of polymer coating formation plays the important role. They can consequently affect the thermal and electrical properties of the new coating materials obtained from polyurethane anionomers [28].

5.2.2 Electrical Properties of Polyurethane Anionomer Coatings

More and more attention has been paid recently to the synthesis of ionomers, in which the structure makes the controlling factor for the electrical properties, so that those products are not considered typical polymeric dielectric materials but rather the materials which are electrically conducting to some extent. That situation is in particular applicable to polyurethane plastics synthesized in the reaction of diisocyanates, polyols as well as numerous amine-type modifiers and chain extenders. So, for example, considerable polarity and increased ionic conductivity in polyurethane anionomers creates the possibility of obtaining new materials with nonlinear properties, applicable in electronics and optics [20–22]. Irrespective of that, the measurements of electrical properties, and in particular of dielectric permittivity and ionic conductivity, can be utilized to study chemical structures and the arrangement of supermolecular structures in urethane polymers [23,24]. The presence of ionic structures, derived from selected types of polyols and tertiary amines which have been additionally incorporated into the polymer chains, modifies further the physical-mechanical properties and surface properties of the items produced of polyurethane plastics [25]. Those structures also affect the interactions resulting from the presence of hydrogen bonds formed with the participation of polar urethane groups, which improves the mechanical properties and thermal stability of the polymer [26]. Moreover, the polyurethane ionomer additives improve thermodynamic miscibility of polymer blends wherein polar polymers, polyolefins, and polystyrene are used, which is essential for easier processing of those materials by means of injection molding [27].

Having in mind the above questions in the present part of the work it was considered advisable to obtain polyurethane anionomers with the use of aromatic, alicyclic, and aliphatic diisocyanates, commonly applicable in polyurethane manufacturing processes, and to study the impact of their chemical and supermolecular structures on the observable electrical properties.

5.2.2.1 *Methods for Determination of Electrical Properties*

The coatings for tests of electrical properties were obtained by application of aqueous dispersions of synthesized anionomers, with the polymer content of about 30 wt%, to 9 × 9 cm pieces of steel sheet. The water was then evaporated under conservative conditions, in the drier at 110°C over about 1 h, and the polymer coating samples were subjected to seasoning at ambient temperature over 10 days minimum [28].

Measurements of electrical properties were taken at ambient temperature $T = 20°C$ with the use of a steel electrode with the diameter of 71 mm. The following properties of the produced coatings were determined (Formula 5.15):

- Volume resistivity (Ω mm):

$$\rho_V = \frac{U \cdot S}{I \cdot d} \tag{5.15}$$

where *U*—value of DC voltage, *I*—steady value of current flowing through a sample, *S*—surface of electrode, *d*—sample thickness,

which was then re-calculated into specific conductivity (S/cm^{-1}) (Formula 5.16)

$$\gamma = \frac{1}{\rho} \tag{5.16}$$

- The real part of complex relative permittivity:

$$\varepsilon = \frac{C_x}{C_0} \tag{5.17}$$

where C_x—capacitance as measured, C_0—electrode capacity arranged in air, calculated from the Formula 5.18:

$$C_0 = \varepsilon_0 \frac{S}{d} \tag{5.18}$$

where ε_o—permittivity of vacuum, which amounts to $8.854 \cdot 10^{-12}$ F/m.
- Dielectric dissipation factor tan.
- Imaginary part of complex relative permittivity (Formula 5.19):

$$\varepsilon'' = \varepsilon'' \tan\delta \tag{5.19}$$

As the sample surface was not perfectly even, the electrode could not adhere to it precisely, which resulted in air gaps at certain locations and the size of that gap reached 1/3 of coating thickness maximum. That factor gave a considerable contribution to increased errors in our measurements. Vacuum application of silver electrodes was not advisable as that could damage the air micro-bubbles present in the coating layer.

Because of that reason, our findings for volume resistivity have the burden of errors of the order of tens percent, which means that in practice only their order of magnitude can be evaluated properly. The shapes of permittivity curves and dielectric dissipation factor curves, on the other hand, can be declared close to their real profiles, although the values for permittivity, if close to 1, are most probably affected by the presence of air bubbles (for air $\varepsilon = 1$) in the coating and by the air gap between the electrode and the polymer surface. Table 5.2 provides the measured maximum and minimum values for the sample thickness, and the sample capacitance C_o:

$$C_o = \frac{\varepsilon_o \cdot S}{d} \tag{5.20}$$

where: S = surface of electrode; d = max. thickness of coatings, which is determined by the sample thickness and size of electrode.

Also, volume current intensity (I) and volume resistivity have been specified in the Table 5.2. Resistivity was measured at the voltage of 10 V. The maximum values were taken for calculations since the electrode was touching the coating at the places with the highest thickness.

The parameters for permittivity (designated as: ε'—real part, ε''—imaginary part) and the dielectric dissipation factor (tg δ) were found from the measurements taken at the sinusoidal voltage with the maximum value of 10 V, within the frequency range of from 20 Hz to 1 MHz.

5.2.2.2 Interpretation of the Electrical Properties of the Synthesized Polyurethane Anionomers

The supplementary information on the nature of physical interactions which are decisive for the polarization phenomena taking place in the studied polyurethane anionomers, and which add to the analyzed structural parameters, can be obtained on the basis of electrical properties. The most typical in this regard is the value of the real part of complex permittivity ε'. As seen from the charts presented in Fig. 5.1. the values of ε' for synthesized PU anionomers, as measured within the frequency range 20 Hz–1 MHz, change very little, although the biggest changes were observed for anionomer no. 5 within 20–25. The chain structure for that polymer was found to be, in accordance with the adopted structural criteria, the most polar one. The much bigger decline in ε' observed for that polymer speaks for the limited capacity of dipoles in that anionomer to follow the orientation of the fast oscillating electric field. This should be

Table 5.2 Electrical properties of synthesized anionomers [29]

Sample no	Type of diisocyanate	Max. thickness of coating (mm)	Min. thickness of coating (mm)	Capacitance C_o (C)	I (A)	Volume resistivity ($\Omega \cdot$ mm)	Specific conductivity (Ω^{-1} cm^{-1})	ε' (1000 Hz)	ε'' (1000 Hz)	tan δ (1000 Hz)
1	TDI	0.34	0.12	$1.03 \cdot 10^{-10}$	$3.00 \cdot 10^{-9}$	$4.0 \cdot 10^{13}$	$2.5 \cdot 10^{-13}$	2.66	0.19	0.07
2	MDI	0.43	0.30	$8.15 \cdot 10^{-11}$	$8.50 \cdot 10^{-10}$	$1.0 \cdot 10^{14}$	$1.0 \cdot 10^{-13}$	1.19	0.07	0.06
3	HMDI	0.30	0.12	$1.17 \cdot 10^{-10}$	$1.00 \cdot 10^{-7}$	$1.0 \cdot 10^{12}$	$1.0 \cdot 10^{-11}$	2.57	0.44	0.17
4	IPDI	0.43	0.25	$8.15 \cdot 10^{-11}$	$1.00 \cdot 10^{-7}$	$9.0 \cdot 10^{12}$	$1.0 \cdot 10^{-12}$	1.60	0.10	0.06
5	HDI	0.34	–	$1.03 \cdot 10^{-10}$	$8.00 \cdot 10^{-6}$	$1.0 \cdot 10^{10}$	$1.0 \cdot 10^{-9}$	10.50	4.84	0.46

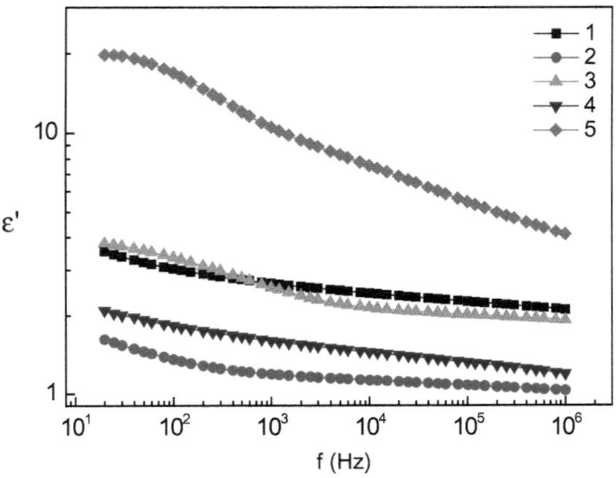

Figure 5.1 Relation between real part of electrical permittivity ε and frequency [28]. 1-anionomer synthesized from TDI; 2-anionomer synthesized from MDI; 3-anionomer synthesized from HMDI; 4-anionomer synthesized from IPDI; 5-anionomer synthesized from HDI.

accounted for by the superior deformability of the aliphatic structure in the electric field. When the frequency of oscillating field increases, the dipoles fail to keep pace of those changes, which consequently is responsible for the drop of ε'. The ε' values for other anionomers are clearly lower, they equal to a few units only, and they vary insignificantly only in the specified scope of frequencies. Except for the much lower value for anionomer no. 2 (synthesized with the use of the aromatic diisocyanate MDI)—probably because of the errors resulting from numerous air bubbles in the coating—the values of ε' for other anionomers correspond to the sequence, which is defined by the polarity of their chemical structures. Except for sample no. 5, the profiles of the obtained dispersion curves for ε' should be considered typical for polar polymers. For example, in poly(urethane-imides) modified with diphenylsilanediol, the values of ε' amounted from 3.9 to 3.1 at the frequency of 1.4 kHz [26]. In the case of aromatic polyurethanes that have poly(urethane-isocyanurate) interpenetrating networks (IPN), the values of ε' fell within 4–8 for the frequency range of 10^{-1}–10^6 Hz, depending on the content of the polyurethane component in the composition (0–100%). Similarly, those values were dependent on the frequency to a slight extent only [29].

In order to compare the effect of the urethane structure type on ε', those values were specified in Table 5.2 for the frequency of 1000 Hz. The electrical properties can be also affected by the nature of phase arrangement in the studied anionomers, adding to the effects of chemical structures. The supermolecular structures of synthesized anionomers show some phase order which is clearly seen in the SAXS dissipation curves [29]. In particular, this refers to ionomers obtained from cycloaliphatic diisocyanates HMDI, IPDI and aliphatic HDI. The structural order in aromatic anionomers obtained from TDI and MDI turned out surprisingly to be inferior, despite their relatively high polarity. It is this factor which, adding to polarity and aromatic nature of the chemical structures, can also affect the specific conductivity observed for the investigated anionomers. In case of polar crystalline polymers, like polyacrylonitrile and polyamide 6,6, the electrical permittivity amounts to 3.5 and 4.0, respectively, while their specific conductivity is of the order of $10^{-15}\ \Omega^{-1}\mathrm{cm}^{-1}$ [30,31]. As can be observed in Table 5.2, the values of volume resistivity measured for the tested samples differ greatly, even within 4 orders of magnitude. Consequently, the values of specific conductivity are of the order of $10^{-13}\ \Omega^{-1}\ \mathrm{cm}^{-1}$ for aromatic anionomer coatings no. 1 and 2, in the order of 10^{-11} and 10^{-12} for cycloaliphatic anionomers no. 3 and 4, and much more—of the order of 10^{-9} for the aliphatic anionomer no. 5. Much higher conductivity of the synthesized anionomers, even in relation to polar polymers like mentioned polyacrylonitrile, results from the mobility of COO^- anions and positive counter-ions created by protoned incorporation of TEA to those structures. This accounts for a considerable share of the ionic conductance in polyurethane anionomers. That is confirmed by the findings presented in [47], according to which introduction to the isocyanate prepolymer, produced from IPDI and polyethylene glycol, ionic groups derived from DMPA, yielded the water soluble polyurethane gels with the conductance at the temperature range 0–50°C increasing from 10^{-6} to $10^{-5}\ \Omega^{-1}\ \mathrm{cm}^{-1}$.

Interesting conclusions result, on the other hand, from the measurements of the dielectric loss factor (Fig. 5.2). The lowest values of $\tan\delta = 0.06-0.07$ at 1000 Hz were obtained for anionomers no. 1, 2, and 4, which suggests that neither aromatic components in the structure nor its polarity do not contribute much to that phenomenon, but it is rather the scope of conductivity resulting from the chemical structure and supermolecular structure of anionomers which is important. The dielectric loss factor for anionomers obtained with the use of HMDI and HDI is much higher and it amounts to 0.17 and 0.46, respectively, which results from high conductivity of those materials. Polarizability of their structures, on the other hand, is decisive for increased real part of the complex electrical permittivity.

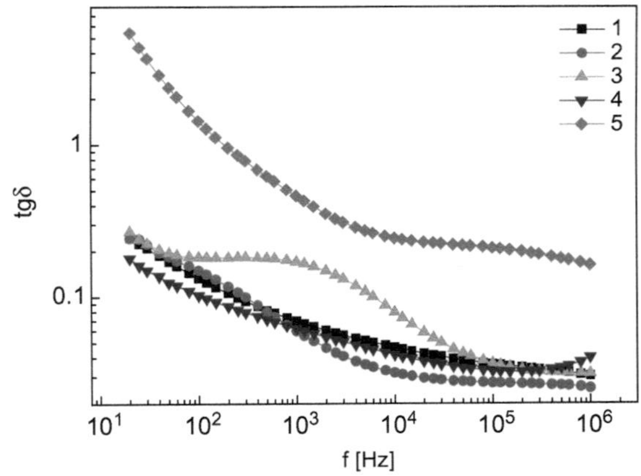

Figure 5.2 Relation between dielectric [28]. Designations for curves like in Fig. 5.1.

The values of tg δ define the ability of the polymer material to convert the electric energy into thermal energy and they also decline at higher frequencies. For the measurements taken at 20°C, a poorly distinguished relaxation peak can be observed for every sample at the frequency range of $f = 1000-5000$ Hz which represents the so-called α-relaxation connected with the glass transition of the rigid urethane-urea segments at the temperature T_{g2}. As discussed above, complex phase transitions can be observed at the temperatures above 0°C which are connected with the glass transition of the rigid urethane segments, and which are responsible for considerable broadening of the range of T_{g2} [25]. The peak which represents the α-relaxation can be pretty clearly seen in the dispersion curve for anionomer no. 3, synthesized from HMDI, for which also the DSC thermogram shows clearly a single phase transition at $T_{g2} = 50$°C.

The effect of β-relaxation type Johari-Goldsteina [32], which is always present in DSC thermograms of all synthesized anionomers within −70°C to −60°C, and which is connected with the movements within soft ether-type segments, can not be found in the recorded curves for dispersion ε''. The probable reason is that it can be possibly observed at 20°C for much higher frequencies only. For example, in case of segmented PU ionomer synthesized from MDI and polyoxyethylene glycol, the peak for β-relaxation in the dispersion curve ε'' could be observed at 20°C not earlier but above 10^8 Hz [29].

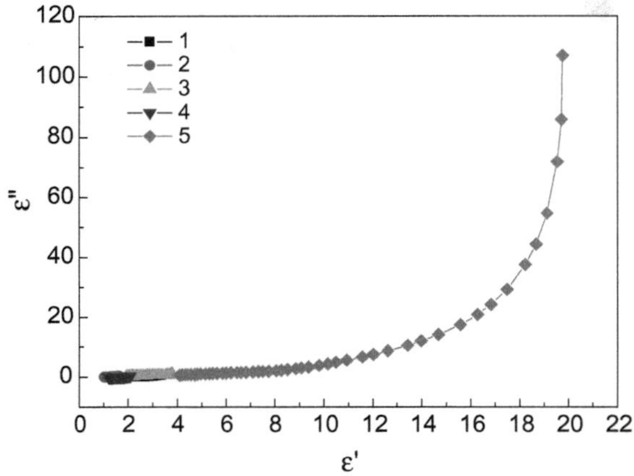

Figure 5.3 Cole–Cole diagram for anionomer no. 5 [28].

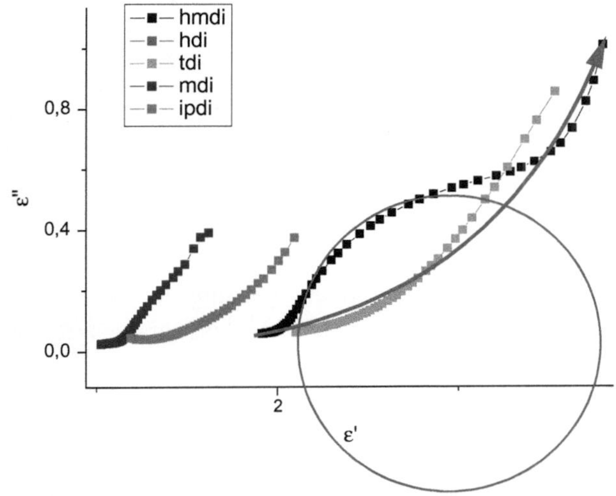

Figure 5.4 Cole–Cole diagrams for anionomers no. 1–4. The designations for curves are the same as in Figure 5.1.

Figs. 5.3 and 5.4 present the charts for the relation $\varepsilon'' = f(\varepsilon')$ which refer to anionomers 5 and 1–4, respectively. These are the totals for the relations resulting from the Debye relaxation model and for the strongly outlined ionic conductivity. The conductivity is clearly dominant in the samples 1, 2, 4, and 5. Hence, the typical Cole–Cole semicircles or those deformed after incorporation of the correction factors to the Debye equation, are not visible in practice, except for the diagram for the sample no. 3, in which the

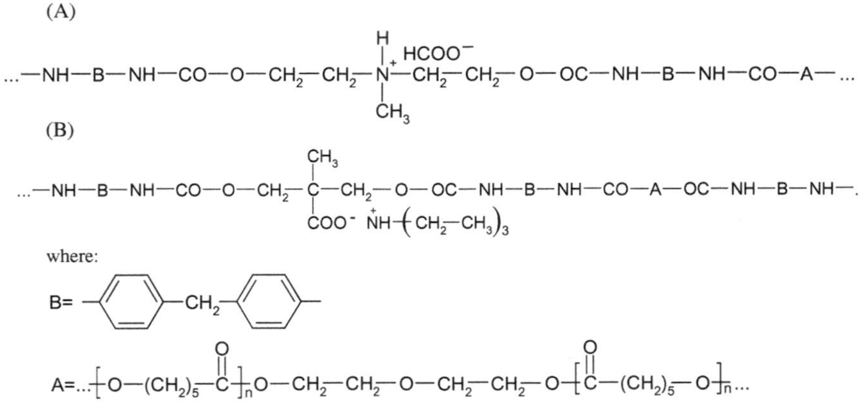

Figure 5.5 Ionic structure in polyurethane cationomer (A) and anionomer (B).

additional semicircle and the conductivity curve were demonstratively plotted (Fig. 5.5). The actual profiles for the Cole–Cole diagrams $\varepsilon'' = f(\varepsilon)$ make the total for both the dispersion curves. The chart for the sample no. 5, Fig. 5.4, was subjected to re-scaling because of the units applicable to the ε'' axis. So, both the polarization effects and ionic conductivity can be observed in every sample. However, conductivity is dominant in most cases and it is hard for that reason to analyze the polarization phenomena in the synthesized anionomers. The information on that subject must hence be derived from the reports on the analysis of Cole–Cole diagrams for classical polyurethanes, i.e., typical dielectric materials [33].

5.3 POLYURETHANE CATIONOMERS

Polyurethane cationomers offer properties which are just as interesting, and result from an even wider variety of chemical structures, specific to linear polyurethanes. The latest research pays much less attention to polyurethane cationomers than to polyurethane anionomers, although the former can offer—as we are going to present here—new applicability as the so-called intelligent materials for which far-reaching changes in physical properties (electrical, optical, and surface properties) are possible, depending on structural modifications. This subject was already discussed in References [34,35] in which the manufacturing method, structure and properties were presented for polyurethane cationomers obtained from 2,4- and 2,6-tolylene diisocyanate

(2,4- and 2,6-TDI) or 4,4′-methylenebis(phenyl isocyanate) (MDI), from selected polyether polyols or polyester polyols, and from difunctional tertiary amines, e.g., *N*-methyl or *N*-butyldiethanoloamine, which were incorporated into chains and reacted with formic acid to yield stable alkylammonium cations. Thus obtained cationomers in the form of aqueous dispersions were employed as polyurethane lacquer coatings. Those applications have long been mentioned in research reports, but in our opinion the results obtained so far are not so satisfactory as those for polyurethane anionomers [36,37]. Aqueous polyurethane cationomers were produced also in a multistage polyaddition process of hexamethylene-1,6-diisocyanate (HDI), poly(oxypropylene) diol ($M = 400$), and *N*-methyldiethanoloamine (*N*-MDEA) as an ionogenic compound and isocyanate-urethane prepolymer chain extender at the same time. The obtained linear polyurethane was converted to cationomer in the quarternization reaction of amino groups with dimethyl sulphate (VI), acetic acid or hydrochloric acid [38]. Some Schiff bases, such as 2-chloromethyl-*N*-(2-hydroxybenzilidene) aniline, 3-(11-bromoundecanoyloxymethyl)-*N*-(2-hydroxy-benzilidene) aniline and 4-chloromethylphenyl-(carbamoyloxymethyl-2-hydroxy-benzil39 idene), especially synthesized for that purpose, were also employed as quarternization reagents [39].

A strong potential can be seen here for structural modifications of cationomers, not only by changes in types of diisocyanates and polyols used—as is the case for anionomers—but also by incorporating various dihydroxyamines in chains of tertiary urethane-isocyanate prepolymers, and—as a subsequent stage—converting them in various methods to alkylammonium compounds. So, this method of production of, for instance, elastomer polyurethane coatings with additionally interesting optical performance becomes feasible. For example, a polymer material in the form of the photosensitive poly(triazene-urethane) cationomers was obtained in the polyaddition reaction of 2,4-TDI and poly(tetramethylene oxide)diol (PTMO) ($M = 2000$), when *N*-MDEA and difunctional triazene compounds were used as chain extension elements. They additionally played the role of UV chromophores: 1-(m-hydroxymethyl)phenyl-3-(2-hydroxyethyl)-3-methyltriaz(1)ene and l-p-nitrophenyl-3,3-di(2-hydroxyethyl) triaz(1), and the quaternization reaction of those phenyltriazenes involved the use of benzyl chloride. It was found under UV exposure conditions that the photolysis rate constant for the so produced ionomer film was lower than that for a nonionic polymer with the similar structure of urethane segments [40]. An original group of poly(urethaneimide) cationomers was obtained in the reaction of MDI, PTMO, *N*-MDEA and

4,4′-hexafluoroisopropylidene-bis-phthalic anhydride, when the counterions X^- employed were: methyl sulfate, tosylate, triflate, chloride, bromide and iodide, with their alkyl groups (C_1-C_6) [41].

5.3.1 Polyurethane Cationomer Coatings Modified by Graphene

Since the first publication on graphene over 10 years ago a large number of publications have arisen regarding the use of the allotropic forms of carbon as a nanofiller in polymer composites [42,43]. Polymer/graphene coatings, compared to the traditional polymer coatings, show superior mechanical, thermal, gas barrier, and electrical properties. However, the improvement in the physicochemical properties of these nanocomposites depended on the distribution of graphene layers and polymer bulks as well interfacial bonding between graphene layers and polymers matrix. Pristine graphene is not compatible with polymer structure and does not form homogeneous composites [44,45]. Therefore, numerous methods of chemical functionalization of the graphene were developed from which the greatest attention was paid to the processes of oxidation and synthesis of graphene oxide [46–48]. This is advantageous in terms of mechanical properties, but causes a considerable deterioration of electrical properties of the coating. Unlike graphene, graphene oxide is electrically isulating, and does not allow its application for the production of conducting nanocoposites. So it is worthwhile still making new attempts to apply nonocovalent methods for graphene modification as a special sonification in organic solvents or in ionic liquids [44].

In recent years, attention has also been paid to the possibility of polyurethane materials modified by graphene or his functionalized derivatives. New reinforced polyurethane elastomers are being sought in the automotive industry, and as varnishes are very useful in the anticorrosive protection of steel and concrete, as well as a conducting nanocoposites in electronic [49–51].

In Reference [52], the authors described the possibility of producing graphene-reinforced waterborne nanocomposite coatings by the sol–gel method based on chemical modified graphene and polyurethane anionomers produced from isophorone diisocyanate (IPDI), polyoxypropylene glycol and dimethylolpropionic acid. In [53,54] the authors presented the possibility of modifying anionomer polyurethane dispersion produced from IPDI, polycaprolactone diol and dimethylol butanoic acid by the graphite oxide nanofillers.

Bearing in mind our former achievements concerning the production of waterborne polyurethane cationic dispersion [55], we made attempts to check how the presence of graphene noncovalent functionalized in THF will influence on the properties of polyurethane coating. Previously, we paid particular attention to the reduction of the hydrophilicity of the coatings prepared from polyurethane cationomers [56]. It was possible, but often has been connected with worsening their mechanical properties. The use of graphene proved to be very useful here, but not without complications in obtaining a physically homogeneous composition. Herre, we are combining the application of graphene with the earlier described possibilities of improvement of the electrical properties of coatings of this type [28]. In work undertaken previously [57] we have limited ourselves to describe a synthesis method and examination of thermal, mechanical, and surface properties of coatings obtained from polyurethane cationomers modified by graphene.

In the first step of the polyaddition process the AB_2 prepolymer was prepared by analogy extending synthesis as in the case anionomer (see Formula 5.4).

In the second stage of the polyaddition process, suspension of graphene in THF was introduced to the resulting mixture of MDI and AB_2 prepolymer. Appropriate suspensions were being obtained in the process of sonication at 50°C during 30 min in ELMASONIC P ultrasonic baths at frequency of 80 kHz.

In step 3, the NMDA (X) amine and DBTL catalyst was introduced to the reaction system, and then the AB_2 prepolymer extension reaction was continued at 50°C for a further 30 min. which shows the following reaction (Formula 5.21):

$$nBAB + nX \rightarrow -(XBABXB)_n- \tag{5.21}$$

The final content of NCO groups in the resulting mixture no longer exceeded 0.25 wt%. Alkylammonium cations were produced by neutralization of tertiary amino groups. For this purpose the pure HCOOH was introduced to the reaction mixture.

At stage 4, as in the synthesis of anionomers, redistilled water with a slight amount HMDA amine was added under intensive agitation conditions. This stage was intended not only to produce the water dispersion (38–42 wt%) but cationomer chains with the residual −NCO groups were subjected to extension at the same time in the reaction between groups −NCO and water (Formulas 5.22 and 5.23):

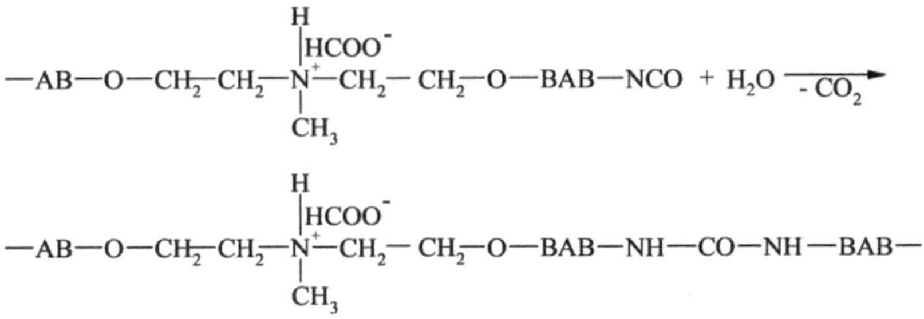

(5.22)

or in the reaction between residual groups −NCO and HMDA;

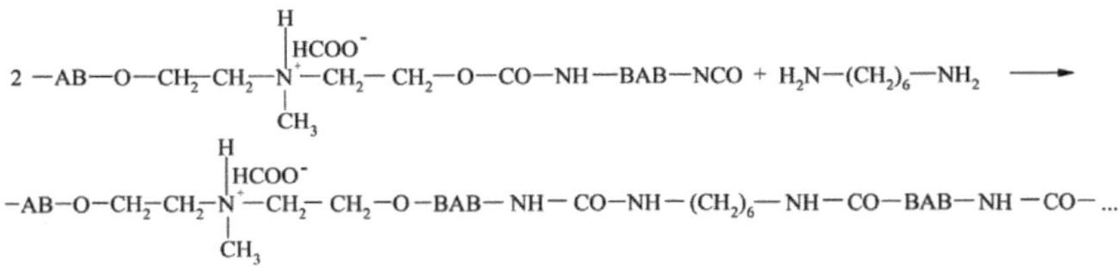

(5.23)

The resulting dispersion of waterborne polyurethane was degassed in ultrasonic field.

The reference films were prepared by covering poly(tetrafluoroethylene) (PTFE) or glass plates by the polyurethane cationomers dispersions. Next, the films were placed in a vacuum drier, at 80°C, over 6 h, and the process of film formation was completed by exposure to ambient air during 10 days. In this way polymer films containing respectively 0 (PU-0 sample); 0.1 (PU-0.1); 0.5 (PU-0.5); 1 (PU-1); 1.5 (PU-1.5) or 2 wt% (PU-2) of the graphene were received [58].

5.3.2 Electrical Properties of the Polyurethane Cationomer Coatings with Graphene

So far there have been many descriptions of modifications of polymer electric properties, for typical dielectrics after admixture introduction of metal nanoparticles like Au, Ag, polyaniline, carbon black or specially prepared multiwalled carbon nanotubes (MWCNT) [59–61]. Limiting the discussion to polyurethanes, which are of interest to this chapter, it is worth noticing that there is the possibility of electric and magnetic properties changes of waterborne polyurethane nanocomposite based on anionomer polyurethane after introduction of Fe_3O_4 particles modified by oleic acid, which were dispersed in PPG glycol under ultrasonic bath. The addition of 0–3.5 wt% of Fe_3O_4 caused the loss of the volume resistivity of obtained polymer foils from the value of 1.04×10^{14} to $5.96 \times 10^4\ \Omega \cdot cm$ and surface resistivity from the value 10^{13} to $10^5\ \Omega\ cm^{-2}$ [62]. On the other hand, the presence of GO initially modified with 4,4′-MDI added in the amount of 1.5 to 3% to thermoplastic polyurethane synthesized from MDI, poly(tetramethylene ether) glycol and 1,4-butanediol had no influence on the conductivity, however it resulted in significant increase of dielectric constant (ε') from the value of approx. 8 to 50 and dielectric loss constant (ε'') from the value of 0.1 to 0.9 measured by the frequency of 0.01 Hz. The measurements conducted in conditions of slow increase of frequency up to 10^6 Hz in the case of these nanocomposites containing above 1.5% GO showed nonlinear loss of constants values ε' and ε'', while these constants in the case of PU without admixtures had values 5 and below 0.1 and changed with frequency insignificantly [63]. In general, electric properties of polyurethanes are important, when they are applied as corrosion protection coatings [64], electrochromic actual devices [65] and nonlinear electric or piezoelectric sensors [66], therefore at present there are many research tasks conducted on improving polyurethane layers' electric properties. The material in this chapter is consistent with the research on developing new ecological water-thinnable PU acrylic topcoats, which would allow one to obtain thin polymeric films in an electrically conductive form; so-called intelligent materials with electrical properties that would change when influenced by frequency and intensity of the applied electric field.

In [58] we have examined following the electrical properties of the resulting coatings involving graphene:

Volume resistivity

The measurements of volume resistivity were taken in accordance with the standards:

- *IEC 93:1980 Methods of test for volume resistivity and surface resistivity of solid electrical insulating materials*
- *ASTM D 257-99 Standard test methods for DC resistance or conductance of insulating materials.*

Dielectric properties

The measurements of dielectric properties were taken in the system containing three electrodes at temperature of 20°C and in the frequency range of 10^{-4}–$5 \cdot 10^3$ Hz. The relaxation processes and the basic dielectric parameters were defined on the basis of frequency domain spectroscopy (FDS). The system DIRANA of OMICRON was used to conduct measurements. The software Origin v.8.0 and WinFit of Novocontrol was used to present and analyze the results of the measurements. The generic complex equation Havriliak–Negami (H–N) equation was applied to set basic relaxation functions' parameters in frequency, which described the changes of dielectric permittivity in frequency function (Formula 5.24) [67,68]:

$$\varepsilon(\omega) = -\mathrm{j}\left(\frac{\sigma_0}{\varepsilon_0 \omega}\right)^N + \sum_{k=1}^{2}\left(\frac{\Delta\varepsilon_k}{(1+(j\omega\tau_k)^{\alpha_k})^{\beta_k}} + \varepsilon_{\infty k}\right) \tag{5.24}$$

where:

k—the total number of partial: ($k = 1$, for low frequency component HN1 and $k = 2$ for high frequency component HN2)

N—coefficient of conductivity character

$\Delta\varepsilon = \varepsilon_o - \varepsilon_\infty$ polarizability, the difference between static permittivity (low frequency) (ε_o) and optical high frequency permeability (ε_∞)

$\Delta\varepsilon_k$—polarizability,

τ_k—relaxation time

$\omega = 2\pi \cdot f$—circular frequency, f—vibration frequency

α, β—constants of the equation H-N,

σ_o—parameter of AC conductivity

The most important properties of the obtained films were described [57] and they are presented in Table 5.3. It is relevant, because the electrical properties are interpreted in the context of a chemical structure and the physical properties of the obtained materials. From the data presented in Table 5.4, it can be inferred that the volume resistivity of the tested polymer matrix is approximately $6 \times 10^8\ \Omega$ m. It is an amount lower than thermoplastic polyurethane called Tradename WHT–1570 made from polyester and MDI as a standard elastomer with $T_g = -40$°C, its volume resistivity is more than $10^{11}\ \Omega \cdot$m [61]. The previously tested polymer film that was obtained from anionomer polyurethane synthesized from MDI, polyoxypropylene diol ($M = 450$) and 2,2, bis (hydroxymethyl)propionic acid (DMPA) had a volume resistivity of $10^{11}\ \Omega \cdot$m [28]. The volume resistivity of the currently tested film is much smaller even without the graphene admixture. The chemical structure of the

Table 5.3 Physicomechanical properties of the synthesized PU films [57]

Sample no.	Graphene content in the PU film (wt%)	Glass transition of soft segments (°C)	Glass transition of hard segments (°C)	Thermal properties (by TG analysis)			Mechanical properties			Surface free energy (0.001 J m^{-2})	The statistical parameters of the surface roughness by AFM height[a] sensor			
		T_{g1} inflection point	T_{g2} inflect	$T_{5\%}$	T_{max}	ash (%)	σ_r (MPa)	ε_r (%)	E (MPa)	γ_S	Surface area (μm^2)	R_a (nm)[b]	R_q (nm)[c]	R_{max} (nm)[d]
		by conventional DSC												
PU-0	0	−49.7	23.9	115	497	4.67	5.51	145	162	45.3	2500	28.4	35.2	361
PU-0.1	0.10	−49.4	30.9	115	495	5.40	68.07	471	334	42.3	2500	65.7	94.4	887
PU-0.5	0.50	–	–	116	499	6.57	5.82	130	85	41.4	–	–	–	–
PU-1.0	1.0	−45.2	32.3	172	495	6.80	38.65	508	267	38.7	–	–	–	–
PU-1.5	1.5	−47.9	37,3	180	491	7.50	13.17	462	194	39.4	–	–	–	–
PU-2	2.0	−43.4	38.6	182	496	11.24	58.35	509	383	38.4	2500	15.4	32.0	657

Mean: the average of all *Z* values within the enclosed area.

[a]Surface area: the total area of examined sample surface (the three-dimensioned area of a given region expressed as the sum of the area of all the triangles formed by three adjacent data points).

[b]R_a (mean roughness): the mean value of the surface relative to the center place.

[c]R_q (R_{ms}): the standard deviation of the *Z* values within the given area.

[d]R_{max} (max height): the difference in height between the highest and lowest points on the surface relative to the mean plane.

Table 5.4 The volume resistivity results of polymer films [67]

Sample no.	**Multiplicity of the measurement**	**The measuring voltage $U = 100$ V**	
		Volume resistivity ρ_s ($\Omega \cdot$ m)	**The value average value, ρ_s ($\Omega \cdot$ m)**
PU-2	1 2 3	$2.6 \cdot 10^5$ $1.0 \cdot 10^5$ $6.6 \cdot 10^5$	$3.4 \cdot 10^5$
PU-1.5	1 2 3	$7.6 \cdot 10^6$ $8.0 \cdot 10^6$ $9.0 \cdot 10^6$	$8.2 \cdot 10^6$
PU-1.0	1 2 3	$8.8 \cdot 10^7$ $8.9 \cdot 10^7$ $9.0 \cdot 10^7$	$8.9 \cdot 10^7$
PU-0.5	1 2	$1.5 \cdot 10^8$ $1.2 \cdot 10^8$	$1.4 \cdot 10^8$
PU-0.1	1 2 3	$2.3 \cdot 10^8$ $2.2 \cdot 10^8$ $2.8 \cdot 10^8$	$2.4 \cdot 10^8$
PU-0	1 2 3	$6.4 \cdot 10^8$ $6.3 \cdot 10^8$ $5.4 \cdot 10^8$	$6.0 \cdot 10^8$

polymer and its phase construction have significant impact on decrease of resistivity. The tested PU is segmented polar polyurethane cationomer. In the hard segments of the PU there are ionic groups: alkylammonium cations in the main chain and formate anions acting as counter anion what is presented on Fig. 5.5A. The phase construction of cationomer is crucial, it allows to form the elastomer film. The elastomer is constructed of flexible segments ($T_{g1} = -49.7$°C) and hard segments ($T_{g2} = -23.9$°C), which gives it good mechanical properties and considerable polarity with high surface energy of 45.3×10^{-3} J m^{-2}. In the chain of anionomers analyzed in [28] there were clusters of ion but, contrary to the cationomers, the anions were less mobile since they originated from DMPA acid built-in the main chain Fig. 5.5B. It can be inferred that conductivity in the analyzed cationomer is definitely iobic, and the formate anions have the pivotal role in transporting of charges. It should be emphasized that 2% concentration graphene results in the decrease in the volume resistivity of three orders of magnitude 6.0×10^8 $\Omega \cdot$m for the pure material to $3.4 \cdot 10^5$ $\Omega \cdot$m. The surface of approximately 1% of the graphene admixture where the sudden decrease of resistivity is probably caused by the exceed of percolation threshold is noticeable. It should be noticed that a little smaller amount of percolation threshold (0.5–1%) was observed in polyurethane composites modified by multiwalled carbon nanotubes (MWCNT), where the significant decrease in resistivity from 2.5 $10^{12}\ \Omega \cdot m$ to 5–6 orders of magnitude was observed [61].

Significantly more cross-linked structure polyurethane/hyperbranched poly(urea-uretane) doped the similar amount of MWCNTs did not show any significant changes in resistivity [70]. It can be claimed that, apart from the content of grapheme, the line structure of the cationomer's chain is the crucial factor in helping decrease resistivity.

Having analyzed the obtained FDS characteristics (Fig. 5.6A,B) it can be seen that the increase of the graphene concentration causes the increase in the real value of the part of electric permittivity ε' (capacitance) and imaginary part of electric permittivity ε'' (dielectric losses) over the whole measurement range of frequency. The obtained parameters ε' and ε'' for the polyurethane cationomers which have not been doped are approximately 1–2 orders of magnitude higher than the ones registered in the case of the polyurethane anionomers [28] as well as doped polyurethanes 0.5–3% oxide graphene [63] that confirms the higher polarity and ionic activity within the chains of cationomers and is compatible with the significant increase of conductivity of nanocomposites by graphene, which was noncovalent functionalized in THF and was observed.

On the basis of changes of dielectric loss factor tg δ it has been stated that the occurrence of those two relaxation processes (Fig. 5.6C). They are characterized by the local maximums on the graph tg δ in the low frequency range from 0.1 Hz and in the range HF over 10 Hz. The high frequency process is probably related to the so-called α-relaxation effects occurring in the area of hard urethane segments, which is especially visible in the DSC and DMTA research in the temperature T_{g2} in the extent of 25–40°C. The low frequency relaxation β (type Johari-Goldstein [32]) is related to the processes occurring in the flexible polyester segments and is registered to be below 0°C in the DSC measurements. In the researched cationomers the range is $T_{g1} = -40°C$ to $-50°C$. The relaxation Maxwell-Wagner [69] occurring in the border of phases composite surface of the samples and the air cannot be excluded, however, it is less likely regarding its roughness of surface of the researched polymer films (Table 5.3).

The parameters of FDS characteristics, estimated with H–N equation for tested polyurethane films are given in Table 5.5. The quantitative

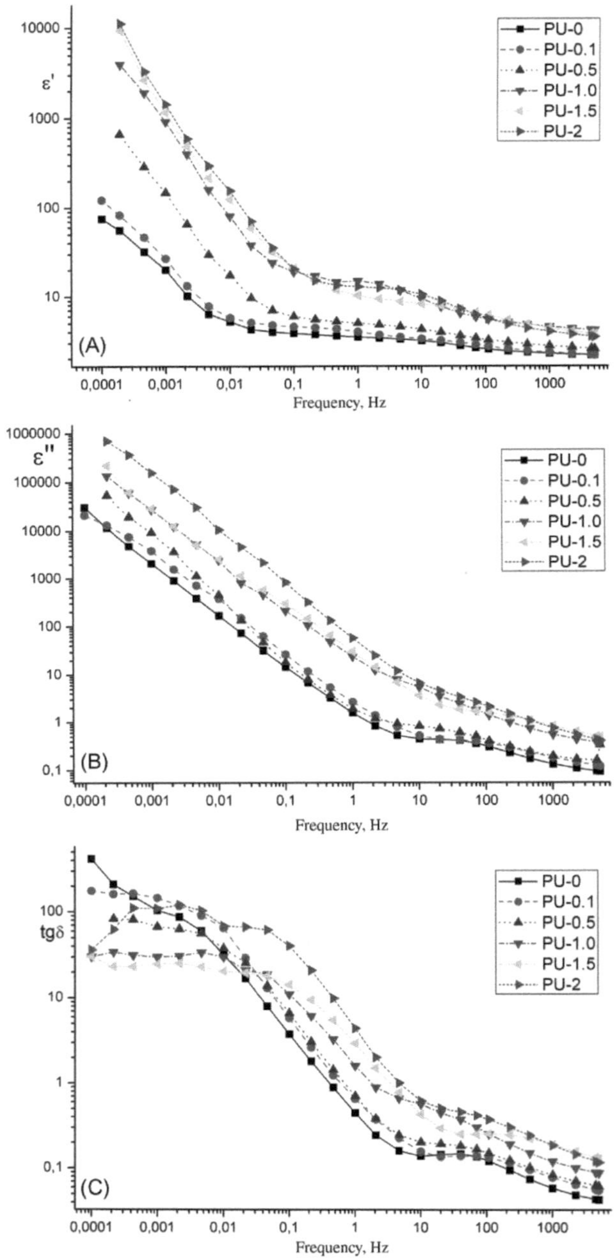

Figure 5.6 Changes of the real part of permittivity ε′ (A), imaginary part of permittivity ε″ (B) and of dielectric losses coefficient tan δ (C) in the frequency domain [58].

Table 5.5 FDS parameters determined by the Havriliak–Negami equation (1) (for $k=1$ and $k=2$) for the examined polyurethane films [67]

Sample no.	σ_o $(\Omega^{-1}\,m^{-1})$		For H–N 1 equation		For H–N 1 equation	
PU-0	σ_o	1.01e-12	$\Delta\varepsilon$	8.8	$\Delta\varepsilon$	1.5
			τ	579	τ	5.4e-3
	N	1	ε_∞	1.07	ε_∞	1.45
			α	0.82	α	0.59
			β	1	β	1
PU-0.1	σ_o	1.59e-12	$\Delta\varepsilon$	141	$\Delta\varepsilon$	2.2
			τ	616	τ	10.6e-3
	N	1	ε_∞	1.07	ε_∞	1.43
			α	0.84	α	0.5
			β	1	β	1
PU-0.5	σ_o	2.49e-12	$\Delta\varepsilon$	1.3	$\Delta\varepsilon$	2.2
			τ	643	τ	4.2e-3
	N	1	ε_∞	1.30	ε_∞	1.44
			α	0.86	α	0.65
			β	1	β	1
PU-1	σ_o	5.03e-12	$\Delta\varepsilon$	5038	$\Delta\varepsilon$	8.5
			τ	716	τ	13e-3
	N	1	ε_∞	1.71	ε_∞	1.45
			α	0.891	α	0.46
			β	1	β	1
PU-1.5	σ_o	1.19e-11	$\Delta\varepsilon$	1e5	$\Delta\varepsilon$	5.5
			τ	3700	τ	1.4e-3
	N	1	ε_∞	2.16	ε_∞	1.45
			α	0.91	α	0.6
			β	1	β	1
PU-2	σ_o	4.46e-11	$\Delta\varepsilon$	1e5	$\Delta\varepsilon$	7.14
			τ	3770	τ	3.3e-3
	N	1	ε_∞	2.26	ε_∞	1.37
			α	0.912	α	0.78
			β	1	β	1

analysis of the dielectric response of samples with addition of graphene was performed with H–N equation, taking into consideration AC conductivity parameter σ_o and an analysis of observed relaxation processes in tested samples was conducted.

The parameter $N=1$ occurring in the equation H–N allows us to conclude that in the tested films, the ionic conductivity prevails. The increase of the graphene concentration in the range from 0.1–2% results in logarithmic increase of AC conductivity parameter σ_o (conductivity) from 1e−12 Ω^{-1} m^{-1} for the for the undoped material, to 4.5e−11 Ω^{-1} m^{-1} for 2% of admixture. It is the process related to the increase of the concentration of ions and the change of structure as well. The admixture of graphene significantly influences the structural relaxation of the material that is observed in prolongation of the relaxation time and the rapid increase of the parameter value τ when amount of admixture exceeds 1% is visible (Fig. 5.7A). This effect is probably related to exceeding percolation threshold, which in the case of other tested carbon

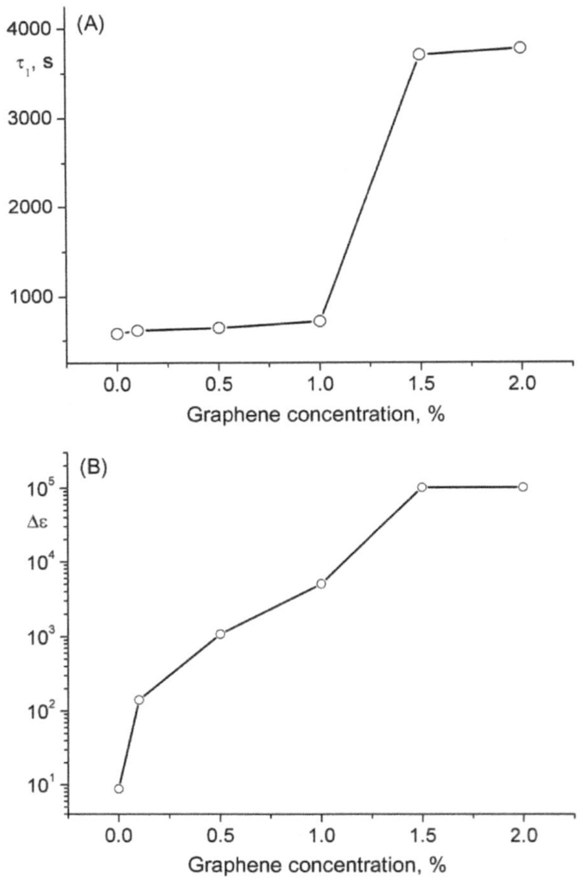

Figure 5.7 Changes of time constant τ_1 (A), polarizability $\Delta\varepsilon$ (B), α_1 parameter (C) and optical permeability parameter ε_∞ (D) of low frequency relaxation process taking place in tested cationomers with various contents of graphene [58].

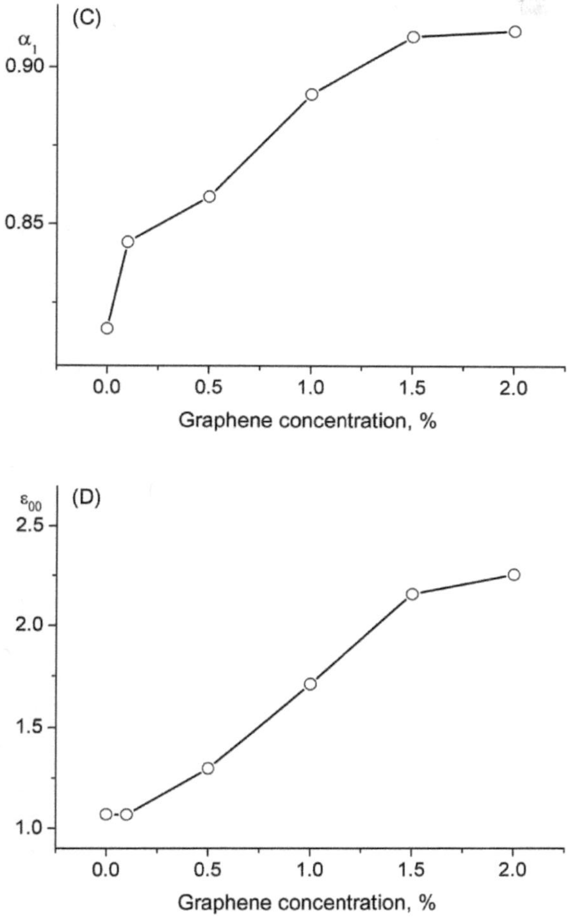

Figure 5.7 (Continued)

nanostructures (MWCNT, fullerene) or expanded graphene may take place at low concentration of value of 1% [61,70]. The abovementioned conclusions are confirmed by recorded changes of polarizability of tested samples, which after exceeding percolation threshold, had parameter $\Delta\varepsilon$ practically set on maximal value (Fig. 5.7B). In Fig. 5.7C changes of α_1 parameter are presented (low frequency relaxation process) for various concentration of graphene in tested polyurethane cationomers. The increase of admixture amount causes the parameter α_1 to reach value of 1, which confirms Debye character of relaxation (Fig. 5.7D) and with very high value of polarizability and high time constant it allows to suppose that ions present in polymer chains are highly mobile when they are exposed to external electric field.

5.4 CONCLUSIONS

The research has shown that polymer films made from waterborne cationomer polyurethane is characterized with a lower volume resistivity than that of layers made from polyurethane and polyurethane anionomer produced from the same diisocyanates and similar in structure polyols. These cationomers have higher values of the real part of permittivity ε' (capacity) and imaginary part of permittivity ε'' (dielectric losses), which can be explained by the high polarity and the phase structure. In the relaxation phenomena in the tested material there are probably dominant interactions of the *dipole–dipol*e type, therefore a strong frequency dispersion of constants ε' i ε'', especially in the range of 10^{-4} to 10^{-1} Hz and significant increase of relaxation time (with the increase of graphene contents) especially at low frequency, are observed. The admixture of graphene in amount 0.5–2 wt% causes the increase of conductivity and percolation threshold is set on the level of approx. 1 wt%. The values of constants ε' and ε'' clearly grow with the increase of graphene amount with the similar shape of frequency dispersion curves as for PU without admixture. There has been the large decrease of PU resistivity in the presence of graphene, which was noncovalent functionalization in THF solvent in the field of ultrasound, which gives a significant advantage over nonconducting polyurethane compounds obtained with oxide graphene.

REFERENCES

[1] O. Bayer, Angew. Chem. 71 (1947) 26.
[2] G. Oertel, Polyurethane Handbook, Hanser Publishers, Munich, 1994.
[3] D.J. Martin, G.F. Meijs, P.A. Gunatillake, S.J. McCarthy, J. Appl. Polym. Sci. 64 (1997) 803.
[4] H.D. Kim, J.H. Huh, E.Y. Kim, C.C. Park, J. Appl. Polym. Sci. 69 (1998) 1349.
[5] H.D. Kim, T.J. Lee, J.H. Huh, D.J. Lee, J. Appl. Polym. Sci. 73 (1999) 345.
[6] M.S. Sánchez-Adsuar, E. Papon, J.J. Villenave, J. Appl. Polym. Sci. 76 (2000) 1596.
[7] M.S. Sánchez-Adsuar, E. Papon, J.J. Villenave, Int. J. Adhes. Adhes. 20 (2000) 291.
[8] E. Žagar, M. Žigon, Polymer 40 (1999) 2727.
[9] M.A. Peréz-Limiñana, F. Aràn-Aís, Int. J Adhes. Adhes. 25 (2005) 507.
[10] M. Potoczek, M. Heneczkowski, M. Oleksy, Ceram. Internat. 29 (2003) 29.
[11] Xin Wei, Yu Xuehai, J. Polym. Sci. Part B Polym. Phys. 35 (1997) 225.
[12] G. Robila, T. Buruiana, E.C. Buruiana, Eur. Polym. J. 35 (1999) 1305.

[13] S. Nomula, S.L. Cooper, J. Colloid Interface Sci. 205 (1998) 331.
[14] P. Król, B. Pilch-Pitera, Polymer 44 (2003) 5075.
[15] P. Król, B. Król, Polimery 49 (2004) 530.
[16] P. Król, B. Król, S. Pikus, K. Skrzypiec, Colloid Surf. A Physicochem. Eng. Aspects 259 (2005) 35.
[17] P. Król, B. Król, J. Eur. Ceram. Soc. 26 (2006) 2241.
[18] Hsing-Lung wang, Ten-Chin Wen, Mater. Chem. Phys. 82 (2003) 341.
[19] L. Shuyan, R. Xiaozhen, L. Yinpei, X. Xiangming, Eur. Polym. J. 34 (1998) 1899.
[20] G.N. Mahesh, S. Ramedh, N.,C. Gautam Ram, G. Radhakrishnan, Polymer 37 (1996) 1657.
[21] W. Hsing-Lung, W. Ten-Chin, Mater. Chem. Phys. 83 (2003) 341.
[22] Ch,A. Jayasuriya, S. Tasaka, N. Inagaki, Polymer 39 (1997) 455.
[23] W. Ten-Chin, D. Yu-Lin, M. Digar, Eur. Polym. J. 38 (2002) 1039.
[24] W. Ten-Chin, W. Yeong-Jyh, Ch Tsung-Tien, Y. Chien-Hsing, Polymer 40 (1999) 3979.
[25] G. Georgoussis, A. Kyritsis, P. Pissis, V. Savelyev Yu, Eur. Polym. J. 1999 (2007) 35.
[26] H. Deligöz, T. Yalcinyuva, S. Özgümüs, Eur. Polym. J. 41 (2005) 771.
[27] Ch,P. Papadopoulou, N.,K. Kalfoglou, Polymer 40 (1999) 905.
[28] P. Król, B. Król, L. Subocz, P. Andruszkiewicz, Colloid Polym. Sci. 285 (2006) 177.
[29] P. Król, B. Król, S. Pikus, M. Kozak, Colloid Polym. Sci. 285 (2006) 177.
[30] "Encyclopedia of Polymer Science and Engineering" Vol. 5, J. Wiley-Interscience Publication, New York, 1990, p. 507.
[31] D.W. Van Krevelen, Properties of Polymers, Third, Completely Revised Edition, Elsevier Amsterdam, Lausanne, 1995, p. 321.
[32] G.,P. Johari, M. Goldstein, J. Chem. Phys. 53 (1970) 2372.
[33] R.,J. Sengva, K. Kulvinder, R. Chaudhary, Polym. Int. 49 (2000) 599.
[34] P. Król, B. Król, Polimery 49 (2004) 615.
[35] P. Król, B. Król, Polimery 52 (2007) 427.
[36] Y. Meng-Shung, T. Hsiang-Chin, H. Po-Da, J. Appl. Polym. Sci. 100 (2006) 2963.
[37] U. Siwoń, J. Drobionka, Pol. Patent No. 335466, 1999.
[38] M. Sudipta, N. Krishnamurti, J. Appl. Polym. Sci. 62 (1997) 1993.
[39] E.C. Buriana, M. Olaru, M. Olaru, M. Strat, G. Strat, B.C. Simionescu, Polym. Int. 54 (2005) 1296.
[40] E.C. Buriana, V. Niculescu, T. Buriana, J. Appl. Polym. Sci. 88 (2003) 1203.
[41] M. Awkal, A. Jonquieres, C.R. Lochon, Polymer 47 (2006) 5724.
[42] K.S. Novoselov, A.K. Geim, S.V. Morozov, D. Jiang, Y. Zhang, S.V. Dubonos, Science 304 (2004) 666–669.
[43] R.J. Young, I.A. Kinloch, L. Gong, K.S. Novoselov, Comp. Sci. Technol. 72 (2012) 1459–1476.
[44] T. Kuilla, S. Bhadra, D. Yao, N.H. Kim, S. Bose, J.H. Lee, Prog. Polym. Sci. 35 (2010) 1350–1374.
[45] S. Kugler, T. Spychaj, Polimery 58 (2013) 58 93-99 (part 1) and 177-179 (part 2).
[46] J. Bian, H.L. Lin, F.X. He, X.W. Wei, I.-T. Chang, E. Sancaktar, Compos. Part A 47 (2013) 72–82.
[47] J. Seppala, N.D. Luong, eXpress Polym. Lett. 8 (2014) 373.
[48] R.K. Layek, A.K. Nandi, Polymer 54 (2013) 5087–5103.
[49] H. Quan, B. Zhang, Q. Zhao, R.K.K. Yuen, R.K. Li, Compos. Part A 40 (2009) 1506–1513.
[50] S.K. Yadav, J.W. Cho, Appl. Surf. Sci. 266 (2013) 360–367.
[51] U. Khan, P. May, A. O'Neill, J.N. Coleman, Carbon 48 (2010) 4035–4041.
[52] X. Wang, W. Xing, L. Song, H. Yan, G.H. Yeoh, Surf. Coat. Technol. 206 (2012) 4778–4784.

[53] vS.H. Yoon, J.H. Park, E.Y. Kim, B.K. Kim, Colloids Polym. Sci. 289 (2011) 1809–1814.
[54] Y.J. Kim, B.K. Kim, Colloids Polym. Sci. 292 (2014) 51–58.
[55] P. Król, B. Król, R. Stagraczyński, K. Skrzypiec, J. Appl. Polym. Sci. 127 (2013) 2508–2519.
[56] P. Król, B. Król, J.B. Lechowicz, Colloids Polym. Sci. 292 (2014) 1051–1059.
[57] P. Król, B. Król, Pielichowska, K.M. Špirková, Colliods Polym. Sci. 293 (2015) 421.
[58] P. Król, B. Król, M. Zenker, J. Subocz, Colliods Polym. Sci. 293 (2015) 2941.
[59] E. Hrehorova, V.N. Bliznyuk, A.A. Pud, V.V. Shevchenko, K.Y. Fatyeyeva, Polymer 48 (2007) 4429.
[60] I. Novák, I. Krupa, I. Chodák, J. Mater. Sci. Lett. 21 (2002) 1039.
[61] J. Chen, Z.-X. Zhang, W.-B. Huang, J.-l Li, J.-H. Yang, Y. Wang, et al., J. Mater. Des. 69 (2015) 105.
[62] S. Zhang, Y. Li, L. Peng, Q. Li, S. Chen, K. Hou, Compos. Part A 55 (2013) 94.
[63] K.M. Sadasivuni, D. Ponnamma, B. Kumar, M. Strankowski, R. Cardinaels, P. Moldenaers, et al., Compos. Sci. Technol. 104 (2014) 18.
[64] H. Wei, D. Ding, S. Wei, Z. Guo, J. Mater. Chem. A 1 (2013) 10805.
[65] J. Song, J. Li, J. Xu, H. Zrng, Nano Lett. 14 (2014) 6298.
[66] M.C. Moreno-Bondi, F. Navarro-Villoslada, E. Benito-Pena, J.L. Urraca, Curr. Anal. Chem. 4 (2008) 316.
[67] S. Havriliak Jr, S.J. Havriliak, Dielectric and Mechanical Relaxation in Materials. Analysis, Interpretation, and Application to Polymers, Hanser Publischers, Munich, Vienna, New York, 1997.
[68] P. Ortiz- Serna, M. Carsí, B. Redondo-Foj, M.J. Sanchis, M. Culebras, C.M. Gõmez, et al., J. Appl. Polym. Sci. 132 (2015) 1.
[69] F. Kremer, A. Schönhals, Broadband Dielectric Spectroscopy, Springer–Verlag, Berlin Heidelberg, 2002.
[70] J.-C. Zhao, F.-P. Du, X.-P. Zhou, W. Cui, X.-M. Wang, H. Zhu, et al., Compos. Part B 42 (2011) 2111.

CHAPTER 6

Blends and IPNs of Polyurethane Polymers With Block Copolymers

Thomas E. Müller
RWTH Aachen University, Aachen, Germany

Contents

6.1 MOLECULAR ARCHITECTURE OF POLYURETHANES

Polyurethanes [1,2], which are composed of organic units linked by urethane groups (Scheme 6.1), are produced commonly by reacting polyisocyanates with polyols. Due to environmental concerns, isocyanate-free routes have more focus from researchers [3–6]. The properties of polyurethanes are determined essentially by the nature of the organic units in between the urethane groups and the density of crosslinking points connecting the polymer chains. The crosslinking density usually is determined by the average functionality of the polyisocyanate and the polyol used in polyurethane synthesis. Short units in combination with high crosslinking density give rigid polyurethane polymers. Long, flexible units, commonly introduced via the polyol, provide elasticity to the polyurethanes. In combination with low crosslinking density polyurethane, materials are obtained that can be quite stretchy. With increasing crosslinking density, the toughness to the polyurethanes materials increases. Long units and intermediate crosslinking density give polyurethane polymers that are useful for making foams. Thus, the properties are essentially determined by the type of polyisocyanate and polyol used to make them. The three-dimensional connectivity generated by the use of a polyisocyanate and polyol with average

Polyurethane Polymers: Blends and Interpenetrating Polymer Networks
DOI: http://dx.doi.org/10.1016/B978-0-12-804039-3.00006-3

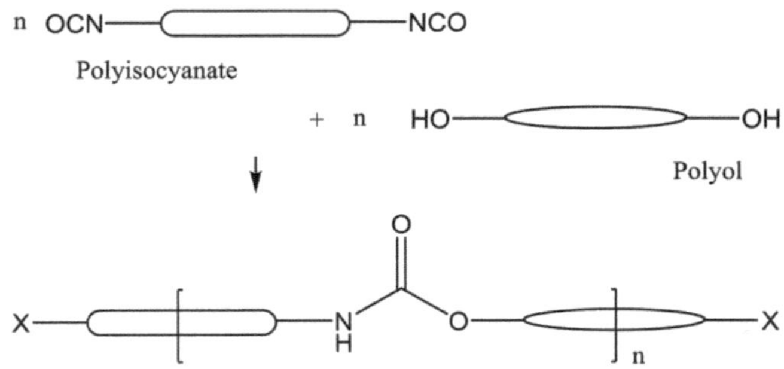

Scheme 6.1 Most common route to the preparation of polyurethanes by reacting polyisocyanates with polyols.

functionality $F > 2$ renders polyurethanes thermosetting polymers that do not melt when heated. Nevertheless, thermoplastic polyurethanes with linear connectivity are also available through the use of a diisocyanate in combination with a diol.

6.2 POLYURETHANES MADE WITH SUSTAINABLE POLYOLS

Similar to most polymeric materials, polyurethane production relies on petroleum oil-derived feedstock for its major components of isocyanates and polyols. Striving for developing more sustainable materials, the development of renewable feedstocks for polyurethane manufacturing becomes highly desirable. Recent developments are polyols derived to a certain extent from carbon dioxide [7,8]. The polyols are obtained by copolymerization of epoxides and carbon dioxide [9–12]. Depending on the catalyst and the reaction conditions, the products can be either polyethercarbonates or alternating polycarbonates. Also bio-based raw materials have been introduced as renewable feedstock [13,14].

Polyethercarbonates are characterized by the presence of carbonate as well as ether linkages. Bimetallic [15] or nanoparticular zinc catalysts [16,17], such as double metal cyanide catalysts [18,19], have been shown to be suitable for the nonalternating copolymerization of epoxides and CO_2. By the use of an appropriate alcohol starter, the functionality of these polyethercarbonate polyols is controlled. Alcohol starters with low molecular weight tend to inhibit double metal cyanide catalysts due to chelating interaction with neighboring zinc centers. This effect is less pronounced in the case of oligomeric alcohol starters. Building up of the polyethercarbonate chain gives rise to polyols with multiblock-structure (Scheme 6.2) [18].

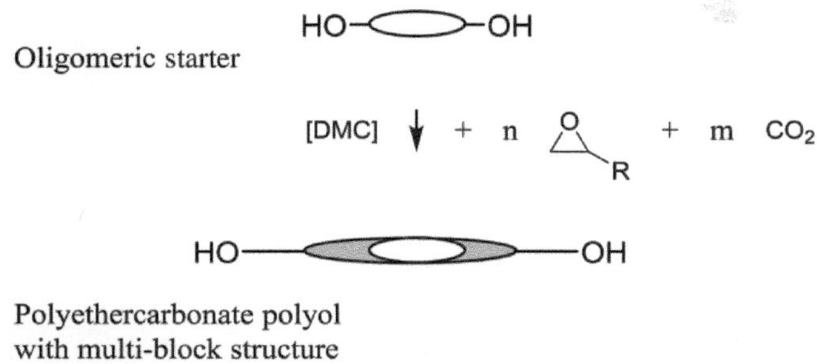

Scheme 6.2 Block structures obtained with polyethercarbonate polyol when using an oligomeric alcohol starter.

The reaction follows an immortal polymerization mechanism and the molecular weight is determined by the ratio of comonomers relative to alcohol starter. Polyols with rather narrow polydispersities are obtained (Fig. 6.1). By adjusting the CO_2 pressure during the synthesis, the CO_2 content can be varied. Note that the CO_2 content in the polymer determines the flexibility of the chain, which is reflected in the viscosity as macroscopic measure. Viscosities are in the range of 1000–4000 Pa.s for bifunctional polyethercarbonate polyols and 2500–8000 Pa.s for trifunctional polyethercarbonate polyols obtained by copolymerization of propylene oxide and carbon dioxide [18]. Interestingly, the glass transition temperature is determined by the ether units and resembles more the glass transition of polyether than of the analogous alternating polycarbonates (Fig. 6.2). Polyethercarbonate polyols have been used for the preparation of elastomeric coatings [20], transparent films [21], and introduced successfully for the production of polyurethanes [18].

The use of cobalt or chromium complexes or selected Lewis acidic compounds as catalysts leads to alternating polycarbonates [12,22–24]. Some catalysts are compatible with the use of an alcohol starter to control molecular weight and functionality of the polycarbonate polyols [25,26]. With regard to torsional flexibility, more rigid bis(hydroxyquinoline) ligands provide a catalytic system for the copolymerization of epoxides and CO_2 and epoxides with superior stability against hydrolysis and oxidative degradation compared to salen ligand based catalysts [26]. Polycarbonate polyols have been used for the preparation of polyurethane hardcoats [27], adhesives [28,29], shape memory materials [30], and flexible foams [31]. The use of polycarbonate polyols in polyurethane foams leads to an increase in typical mechanical properties, such as wet compression set, compression modulus and stress, tensile strength, Young's modulus, and tear

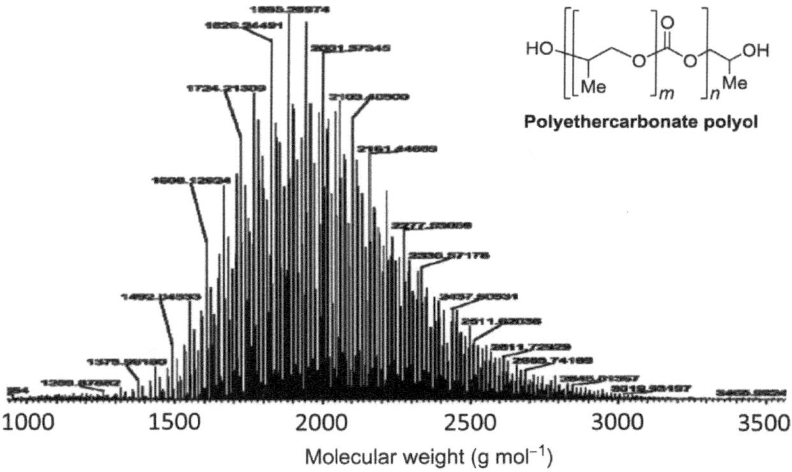

Figure 6.1 Mass spectrum of a typical polyethercarbonate diol with M_n 1885 g mol^{-1} measured with matrix assisted time of flight mass spectroscopy.

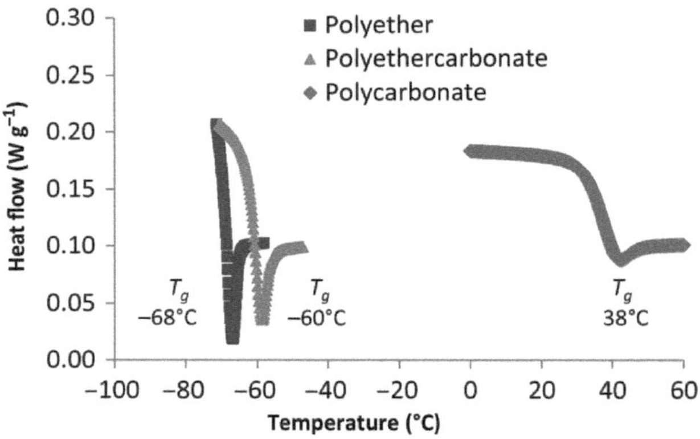

Figure 6.2 The glass transition temperature of polyethercarbonates (here, 7 wt% CO_2 content, triangles) is in between polyethers (squares) and alternating polycarbonates (diamonds); all polyols synthesized from propylene oxide and optionally carbon dioxide.

resistance [31]. At temperatures below 0°C, an increasing concentration of polycarbonate polyol leads to increasing storage modulus [31].

The content of renewable feedstock can also be increased by the introduction of bio-based starters, such as castor oil [32]. Reaction of epoxidized soybean oil and castor oil fatty acid provides polyols which have been used in the preparation of IPDI-based polyurethanes. The

Table 6.1 Properties of polyurethanes made from polyols, which are obtained by reaction of epoxidized soybean oil and castor oil [33]

Sample	Polyol	Polyurethane			
	Molecular weight (g mol^{-1})	Tensile strength (MPa)	Young's modulus (MPa)	Elongation at break (%)	Toughness (MPa)
1	2830	11.6	5.8	232	13.9
2	1670	4.9	2.3	314	7.7
3	589	1.6	0.9	356	2.5

properties of the resulting polyurethanes are compared in Table 6.1 [33]. Generally, higher crosslinking density imparts higher tensile strength, but also lowers the elongation at break. A high content of flexible fatty acid chains together with high cross-linking density in the polyurethane materials provides the highest toughness (Sample 1). The use of a soybean oil-derived polyol leads to highly ordered hard domain structures, reduction of the size of hard domain and the fraction of soft domains, and results in a broad distribution of inter-domain spacing. Connected with the changes in polymer phase morphology, compression and shear modulus increase in the foams [34]. By near-infrared diffuse reflectance spectroscopy changes in compression modulus and strength, as well as thermal conductivity are readily followed [35]. Polyurethane flexible foams can be filled with styrene acrylonitrile copolymers leading to an increase in compression modulus, which is related to slightly reduced foam cell size at otherwise equal phase morphology [34].

6.3 PHASE SEGREGATED POLYURETHANES

Segmented thermoplastic polyurethanes are characterized by substantial strength, toughness, and elasticity. They are characterized by phase segregated domains whose formation is triggered by the presence of soft and hard segments along the polymer chains [36,37]. In the solid state, the hard and soft segments of the copolymer aggregate into separate micro-domains, which results in an elevated rubbery plateau modulus and other enhanced physical properties (see Fig. 6.3). The soft segments, being in a viscous or rubbery state, provide elastomeric character to the polymer. The hard segments, being in a glassy or semicrystalline state, provide dimensional stability to the polymer by acting as a thermally reversible and multifunctional crosslinker as well as reinforcing filler.

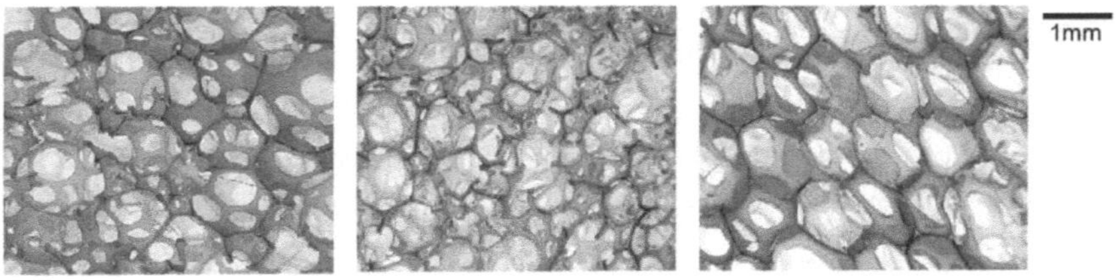

Figure 6.3 Scanning electron microscopy images of polyurethane foams obtained with 30% soybean oil-derived polyol (left), 30% SAN (middle) in comparison to a control sample (right) [34]. *Images reprinted from L. Zhang, H.K. Jeon, J. Malsam, R. Herrington, C.W. Macosko, Substituting soybean oil-based polyol into polyurethane flexible foams. Polymer 48 (22) (2007) 6656–6667, Copyright (2007), with permission from Elsevier.*

Segmented polyurethanes are obtained using an oligomeric diol as the soft segment and a short chain diol in combination with diisocyanates as a hard segment. The nature of the isocyanate building block has significant impact on the properties of the material. Thus, phase mixing and crystallinity is determined by the relation of, for example, the more rigid isophorone diisocyanate (IPDI) and hexamethylene diisocyanate (HDI) [38]. The glass transition temperature decreases and the crystallinity increases as the relative content of HDI increases. The properties of the polyurethane consequently vary between rubbery materials to harder, more crystalline materials with good elastomeric behavior [38]. Also for typical polyurethane foam systems, the formation of lamellae-like polyurea structures with 5–10 nm diameter has been reported [39]. The surfactant in the formulation helps in maintaining the cell structure of the foam until the phase segregation leads to additional stabilization of the network structure.

Also, the mass transfer properties of membranes are determined by the content of hydrophilic segments and the melting temperature of the soft segment [40]. Mixed blocks loosen the inter-chain interactions [41]. Similarly, the hydrogen bonding interactions between the urethane groups are weakened by the introduction of hydrophilic groups and strengthened by the introduction of acidic groups [42]. Consequently, the permeability for water vapor increases with increasing content of hydrophilic polyethylene glycol in the polymer backbone [40]. Above the melting temperature of the soft segments, the permeability is further enhanced. With decreasing size of the crystalline domains [43] and increasing flexibility of the chains in the soft segments the fractional free volume increases, which enhances the permeability for water vapor [44].

Hydroxy-terminated polyalkylenes have been introduced to enhance the resistance of polyurethanes to solvents and oxidative degradation [45,46]. Polyalkylenes are highly inert polyols and characterized by a particularly weak propensity to form hydrogen bonding interactions. The hydroxy-terminated polyalkylenes are obtained readily as low-melting waxy solids by hydrogenation of OH-functionalized polybutadienes. Polyurethane copolymers and block copolymers prepared with hydroxy-terminated polyalkylenes show phase segregation. In case of copolymers obtained by reaction of the hydroxy-terminated polyalkylene polyol with a diisocyanate, trans-1,4-diisocyanatocyclohexane provides a particularly pronounced phase segregation [45]. The resulting copolymers are translucent to opaque due to the nearly matching refractive index of the different phases and are characterized by high solvent resistance, hydrolysis stability, and oxidative stability.

Maleic anhydride grafted polyethylenes can be reacted with polyols to create fine phase dispersions in polyurethanes [47]. The dispersion can be prepared in two different ways: (1) the maleic anhydride grafted polyethylene is mixed first with the polyol to make the co-polymer polyol (Scheme 6.3). Subsequently, the hard segment components are added to finish the polyurethane formation. (2) Alternatively, the maleic anhydride grafted polyethylene can be blended with the polyurethane and heated to temperatures above the urethane reversion temperature to enable the formation of ester bonds between the anhydride and the polyol. In both cases, the polyethylene phase is dispersed in the polyurethane phase at very small length scales. Optically transparent blends are obtained with up to 20% soft segments [47]. Such dispersions are characterized by enhanced tensile strength as well as increased solvent and thermal resistance.

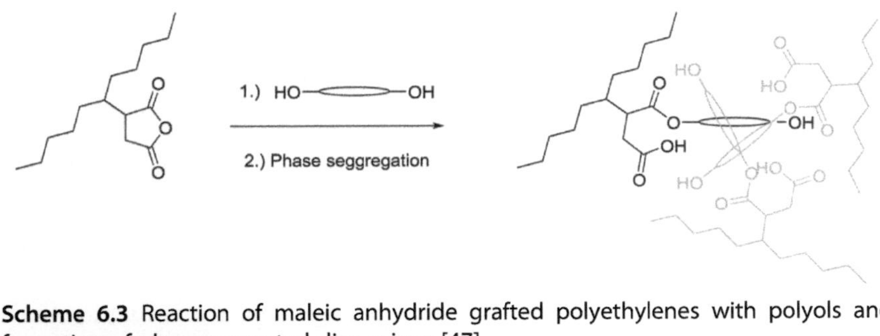

Scheme 6.3 Reaction of maleic anhydride grafted polyethylenes with polyols and formation of phase separated dispersions [47].

Polydimethylsiloxane-based segmented polyurethanes can have a complex three-phase structure consisting of a siloxane phase, a hard segment phase, and an ether rich phase [48,49], in particular when mixtures of silicon-based and polyether polyols are used. Such materials can be obtained with mixed poly(dimethylsiloxane)/poly(hexamethylene oxide) or poly(dimethylsiloxane)/polytetramethylene oxide macrodiols [48]. The fraction of carbonyl groups involved in strong hydrogen bonding interactions, which is associated with the formation of hard domains, increases with increasing poly(dimethylsiloxane)/poly(hexamethylene oxide) ratio in the copolymers. Such three-phase segmented polyurethanes are characterized by excellent mechanical properties, in terms of tensile strength and elongation [49]. The polydimethylsiloxane-segments tend to migrate to the surface of the polymer. When exposed to a flame, the polydimethylsiloxane is oxidized into a silicate-like material, leading to enhanced char yields [49].

Incorporation of fluorinated polyether polyols has essentially no noticeable effect on phase segregation [50]. As blends of polyether polyols and fluorinated polyether polyols are immiscible, three phases are generally present in mixed segmented polymers. Noteworthy, the incorporation of fluorinated polyether polyols results in dramatically enhanced fluorine concentration at the surface [50,51]. This results in a low surface energy and enhanced hydrophobicity of the polymer surface. Such fluorinated polymers are interesting for elastomer applications where a low surface friction is required. The friction coefficient of the fluorinated polyurethanes decreases significantly as the surface fluorine concentration increases. Whereas solution-polymerized polyurethanes with a more homogeneous microstructure are characterized by conventional stable sliding, melt-polymerized polyurethanes with relatively large microscale domain structure undergo a stick-slip motion [51].

6.4 POLYURETHANES WITH INTERPENETRATING POLYMER NETWORKS

Materials characterized by interpenetrating polymer networks can display broader glass transition temperatures compared to materials made of a single polymer, due to the mutual network entanglement. In case of polyurethanes this technology has been utilized for preparing high-performance damping materials [52] and pervaporation membranes [53]. The formation of interpenetrating polymer networks is also an approach to maintain the biocompatibility of polyurethanes, while improving the resistance to biodegradation, stress induced degradation, and surface cracking.

Conventionally, interpenetrating networks are generated by blending of polyurethane and polystyrene [54,55]. Usually, one of the polymer components is used with a linear structure. In this case the relative degree of intermixing depends on the mobility of the linear polymer. Thus, interpenetrating networks with linear polyurethane component shows a higher degree of phase segregation compared to networks in which the less mobile polystyrene component is linear [56]. The morphology generated in the material is determined by the onset point of phase segregation, the rate of phase segregation, and the time of physical interlocking [56]. Phase segregation commences when the molecular weight of the component polymers exceeds a certain critical value so that the Gibb's free energy of mixing ΔG_m becomes positive for the mixture. The mobility of the polymer chains determines on how fast the phases separate. The network is then physically interlocked when both component polymers reach the gel point. Beyond this point, the phase domain size cannot increase any further.

Interpenetrating polymer network of polyurethane and poly(vinyl ester) result in improved properties compared to the use of styrene as conventional comonomer [52,57]. Polymerization with ethyl acrylate resulted in a higher degree of carbonyl hydrogen bonding resulting in better compatibility of the two networks [58]. Thus, a nanometer scale phase range below 20 nm and a dual continuous microstructure is obtained, while the domain size of comparable samples obtained with styrene is 200−500 nm [58]. The finer microstructure resulted in superior damping properties. Interpenetrating polymer networks with polyurethane as the ductile phase and poly(methyl methacrylate) as the stiff phase are transparent [59,60]. Also polyacrylamide is suitable to form interpenetrating polymer networks with polyurethanes [61].

Polyurethanes characterized by interpenetrating polymer networks are frequently used in combination with a suitable filler [62]. The incorporation of graphite particles improves the damping properties of interpenetrating polymer networks significantly [63]. Graphite-filled polyurethane/poly(methyl methacrylate-butyl methacrylate) interpenetrating polymer networks show increased damping over a broadened frequency range and wider temperature range with the highest damping observed at low frequencies (1 Hz) [63]. Upon the addition of graphite particles, the hardness and the tensile strength of the materials were also improved. Silica nanoparticles introduced into polyurethane/poly(ethyl methacrylate) interpenetrating polymer network materials enhance the damping ability of the material and result in improved material strength [64]. Similarly, micro hollow glass beads as fillers in polyurethane/epoxy resin interpenetrating polymer network materials can significantly improve the damping properties [65]. In parallel, the thermal stability and the tensile strength are improved, while the impact resistance of the composites is impaired. Filled with oxide nanoparticles, composites comprising an semi-interpenetrating polymer network of polyurethane and poly(2-hydroxyethyl methacrylate) have pronounced dynamic heterogeneity suppressing the dynamics and increase of the elasticity modulus with temperature [66]. In the composite, an unusual crosslinked structure is formed, whereby two types of covalent bonds are formed in building the network architecture: (1) residual isocyanate groups at the polyurethane network junctions react with the poly(2-hydroxyethyl methacrylate) hydroxyl groups; and (2) covalent bonds are also formed between the polymer matrix and the surface functionalized oxide nanoparticles. Already at a relatively low content of the oxide nanoparticles (0.25 wt%), this structure reduces the segmental dynamics and leads to a two-fold increase in the modulus [66].

For applications as pervaporation membrane used in the separation of water-ethanol mixtures, interpenetrating network materials based on polyurethane and poly (4-vinyl pyridine) are suitable [53]. A hydrophobic polyurethane made from hydroxyl terminated polybutadiene and 4,4′-dicyclohexylmethane diisocyanate (H_{12}MDI) crosslinked with benzoyl peroxide is employed for making the membrane. Already a small amount of poly (4-vinyl pyridine) is adequate as second phase. Thus, for separation of a 90 wt% water-ethanol mixture, membranes with a content of 8 wt% poly (4-vinyl pyridine) have a permeability of 1184 g m^{-2} h and a separation factor of four [53].

6.5 POLYURETHANES WITH SHAPE MEMORY EFFECT

By applying an external stimulus, some polyurethane materials are able to return to their original shape after the material has been deformed [67–69]. This ability is called the shape memory effect. Polyurethane materials which show a shape memory effect are characterized by a two phase structure. The first phase remembers the initial shape of the material. Here, the three-dimensional molecular architecture of polyurethane materials is advantageous to conserve a certain shape. By chemical cross-linking the original shape of the polyurethane material is fixed. Also the formation of micro-crystals can be utilized to inscribe the original shape. Chain entanglement gives additional dimensional stability. The second phase shows a reversible transition between a soft and a rigid state. In shape memory polyurethanes, the weaker inter-chain interactions during formation of crystalline aggregates and glassy states are utilized for constructing the temporarily fixed arrangement of the polymer network. These interactions are thermally reversible and temperature has commonly been applied as external stimulus to trigger the return of the material to the original shape. Nonetheless, other stimuli such as light can also been applied. The temperature behavior is related to the melting temperatures (T_m) of the crystalline aggregates and the glass transition (T_g) between glassy state and the molten, rubber-like state. Segmented polyurethanes with shape memory effect need to have crystalline domains which show a high degree of crystallinity at room temperature and stable rigid domains, which provide physical strong interactions between the chains. The melting temperature of the soft segment crystals decides the response temperature of the shape memory effect. The return speed is then related primarily to the stability of the rigid domains under stress.

Also certain segmented polyurethanes are known for their shape memory effect. To exhibit the shape memory effect the formation of defined network architectures with two phases is highly important. Typically, polyurethanes with shape memory effect are synthesized from diisocyanates, such as 4,4′-methylene(bisphenyl)diisocyanate (MDI), a diol, such as polytetramethylene glycol (M_n 2900 g mol^{-1}) or polyethylene glycol (M_n 3400 g mol^{-1}) and a chain extender, such as 1,4-butane diol [67]. With increasing content of hard segments, a higher shape recovery is generally observed. Due to the more planar shape of 1,4-phenyldiisocyanate (PDI) compared to the bent shape of MDI, more densely packed hard segments are achieved. In consequence, PDI-based polyurethanes show better shape recovery in comparison to MDI-based polyurethanes [70]. Reinforced with functionalized multiwall carbon nanotubes (MWNT) the segmented polyurethanes display an enhanced memory effect [67]. The soft segments absorb external stress by unfolding the entanglement of their molecular chains. The hard segments and the MWNT are required for the material to return to the original shape.

6.6 CONCLUSIONS AND OUTLOOK

Tailoring the inner morphology of polyurethane materials is regularly the key for beneficial properties in the target application. Generally, two different approaches are exploited. Phase segregation can be triggered by utilizing molecular building blocks that lead to formation of polymer phases, which are incompatible with each other. This is the basis of thermoplastic polyurethanes characterized by substantial strength, toughness, and elasticity. Phase segregation leads to the presence of soft and hard segments in the materials. An alternative approach is the formation of interpenetrating networks of two different polymeric materials. Such interpenetrating networks are the basis for memory materials, which return to their original shape upon applying an external trigger. Usually, the shape reversal to the original shape is triggered by heating the material above a certain critical temperature. For the preparation of polyurethanes with their exciting properties, the introduction of new polyols partially derived from sustainable raw materials, such as carbon dioxide or bio-based raw materials provides new chances in developing materials for the next generation of high-performance applications.

REFERENCES

[1] M.F. Sonnenschein, W. Koonce, Polyurethanes, in: H.F. Mark (Ed.), Encyclopedia of Polymer Science and Technology, John Wiley and Sons Inc, New York, 2014, pp. 204–266.
[2] N.B. Shelke, R.K. Nagarale, S.G. Kumbar, Polyurethanes, in: S.G. Kumbar, C.T. Laurencin, M. Deng (Eds.), Natural and Synthetic Biomedical Polymers, Elsevier, Oxford, UK, 2014, pp. 123–144.
[3] L. Maisonneuve, O. Lamarzelle, E. Rix, E. Grau, H. Cramail, Isocyanate-free routes to polyurethanes and poly(hydroxy urethane)s, Chem. Rev. 115 (2015) 12407–12439.
[4] J. Datta, M. Wloch, Progress in non-isocyanate polyurethanes synthesized from cyclic carbonate intermediates and di-or polyamines in the context of structure-properties relationship and from an environmental point of view, Polym. Bull 73 (2016) 1459–1496.
[5] G. Rokicki, P.G. Parzuchowski, M. Mazurek, Non-isocyanate polyurethanes: synthesis, properties, and applications, Polym. Adv. Tech. 26 (2015) 707–761.
[6] H. Blattmann, M. Fleischer, M. Baehr, R. Muelhaupt, Macromol. Rapid Commun 35 (2014) 1238–1254.
[7] M. Hölscher, C. Gürtler, W. Keim, T.E. Müller, M. Peters, W. Leitner, et al., Carbon dioxide as a carbon resource-recent trends and perspectives, Naturforschung 67 (2012) 961–975.
[8] P. Markewitz, W. Kuckshinrichs, W. Leitner, J. Linssen, P. Zapp, R. Bongartz, et al., Worldwide innovations in the development of carbon capture technologies and the utilization of CO_2, Energy Environ. Sci. 5 (2012) 7281–7305.

[9] S. Narang, R. Mehta, S.N. Upadhyay, Copolymerization of propylene oxide and carbon dioxide, Curr. Org. Chem. 19 (2015) 2344–2357.
[10] S. Paul, Y. Zhu, C. Romain, R. Brooks, P.K. Saini, C.K. Williams, Ring-opening copolymerization (ROCOP): synthesis and properties of polyesters and polycarbonates, Chem. Commun. 51 (2015) 6459–6479.
[11] M. Taherimehr, P.P. Pescarmona, Green polycarbonates prepared by the copolymerization of CO_2 with epoxides, J. Appl. Polym. Sci. 131 (2014)41141/1-41141/17
[12] M.R. Kember, A. Buchard, C.K. Williams, Catalysts for CO_2/epoxide copolymerization, Chem. Commun. 47 (2011) 141–163.
[13] A. Noreen, K.M. Zia, M. Zuber, S. Tabasum, A.F. Zahoor, Bio-based polyurethanes: An efficient and environment friendly coating system: a review, Prog. Org. Coat 91 (2016) 25–32.
[14] N.J. Sangeetha, A.M. Retna, Y.J. Joy, A. Sophia, A review on advanced methods of polyurethane synthesis based on natural resources, J. Chem. Pharm. Sci. 7 (2014) 242–249.
[15] S. Elmas, M.A. Subhani, W. Leitner, T.E. Müller, Anion effect controlling the selectivity in the zinc-catalysed copolymerisation of CO_2 and cyclohexene oxide, Beilstein J. Org. Chem. (2015) 11.
[16] M.A. Subhani, C. Gürtler, W. Leitner, T.E. Müller, Nanoparticulate TiO_2-supported double metal cyanide catalyst for the copolymerization of CO_2 with propylene oxide, Eur. J. Inorg. Chem. 13–14 (2016) 1944–1949.
[17] Y. Dienes, W. Leitner, M.G.J. Müller, W.K. Offermans, T. Reier, A. Reinholdt, et al., Hybrid sol-gel double metal cyanide catalysts for the copolymerisation of styrene oxide and CO_2, Green Chem. 14 (2012) 1168–1177.
[18] J. Langanke, A. Wolf, J. Hofmann, K. Böhm, M.A. Subhani, T.E. Müller, et al., Carbon dioxide (CO_2) as sustainable feedstock for polyurethane production, Green Chem. 16 (2014) 1865.
[19] X. Sun, S. Chen, X. Zhang, G. Qi, Double metal cyanide complex catalyst and its catalysis for epoxides-involved polymerization, Huaxue Jinzhan 24 (2012) 1776–1784.
[20] M.A. Subhani, B. Köhler, C. Gürtler, W. Leitner, T.E. Müller, Light-mediated curing of CO_2-based unsaturated polyethercarbonates via thiol-ene click chemistry, Polym. Chem. 7 (2016) 4121–4126.
[21] M.A. Subhani, B. Köhler, C. Gürtler, W. Leitner, T.E. Müller, Transparent films from CO_2-based polyunsaturated poly(ethercarbonate)s: a novel synthesis strategy and fast curing, Angew. Chem. Int. Ed. 55 (2016) 5591–5596.
[22] H. Sugimoto, S. Inoue, Alternating copolymerization of carbon dioxide and epoxide - recent advances, Kobunshi Ronbunshu 62 (2005) 131–146.
[23] G.W. Coates, D.R. Moore, Discrete metal-based catalysts for the copolymerization of CO_2 and epoxides: discovery, reactivity, optimization, and mechanism, Angew. Chem. Int. Ed. 43 (2004) 6618–6639.
[24] D.J. Darensbourg, M.W. Holtcamp, Catalysts for the reactions of epoxides and carbon dioxide, Coord. Chem. Rev. 153 (1996) 155–174.
[25] A.M. Chapman, C. Keyworth, M.R. Kember, A.J.J. Lennox, C.K. Williams, Adding value to power station captured CO_2: tolerant Zn and Mg homogeneous catalysts for polycarbonate polyol production, ACS Catal. 5 (2015) 1581–1588.
[26] S. Elmas, M.A. Subhani, M. Harrer, W. Leitner, J. Sundermeyer, T.E. Müller, Highly active Cr(III) catalysts for the reaction of CO_2 with epoxides, Catal. Sci. Technol. 4 (2014) 1652–1657.
[27] J. Anderson, M. Quinn, T. Samoila, C. Simoneau, G. Stokes, S. Wang, Unique aliphatic polyurethane hardcoats from CO_2-based polyols, Paint Coat. Ind. (2014) 1–5.

[28] K. Oguro, Development and uses of new polyester (polycarbonate)-ether copolyols, Setchaku no Gijutsu 26 (2007) 37–42.
[29] B. Liu, L. Chen, Synthesis, property and application of ω, ω'-hydroxyl-ended aliphatic polycarbonate polyols, Huaxue Tongbao 65 (2002)w49/41-w49/47.
[30] Shape memory polyurethane foam having resistance to hydrolysis for down-hole tool applications, Anon, IP.com J., 2011, 11, 30.
[31] M. DeBolt, A. Kiziltas, D. Mielewski, S. Waddington, M.J. Nagridge, Flexible polyurethane foams formulated with polyols derived from waste carbon dioxide, J. Appl. Polym. Sci. 133 (45) (2016).
[32] A. Shirke, B. Dholakiya, K. Kuperkar, Novel applications of castor oil based polyurethanes: a short review, Polym. Sci. B 57 (2015) 292–297.
[33] C. Zhang, Y. Xia, R. Chen, S. Huh, P.A. Johnston, M.R. Kessler, Soy-castor oil based polyols prepared using a solvent-free and catalyst-free method and polyurethanes therefrom, Green Chem. 15 (2013) 1477–1484.
[34] L. Zhang, H.K. Jeon, J. Malsam, R. Herrington, C.W. Macosko, Substituting soybean oil-based polyol into polyurethane flexible foams, Polymer 48 (2007) 6656–6667.
[35] C.E. Miller, B.E. Eichinger, Analysis of rigid polyurethane foams by near-infrared diffuse reflectance spectroscopy, Appl. Spectrosc. 44 (1990) 887–894.
[36] J.A. Miller, S.B. Lin, K.K.S. Hwang, K.S. Wu, P.E. Gibson, S.L. Cooper, Properties of polyether-polyurethane block copolymers: effects of hard segment length distribution, Macromolecules 18 (1985) 32–44.
[37] I. Yilgor, E. Yilgor, G.L. Wilkes, Critical parameters in designing segmented polyurethanes and their effect on morphology and properties: a comprehensive review, Polymer 58 (2015) A1–A36.
[38] C.M. Gomez, D. Gutierrez, M. Asensio, V. Costa, A. Nohales, Transparent thermoplastic polyurethanes based on aliphatic diisocyanates and polycarbonate diol, J. Elastom. Plast. (2016) 1–19.
[39] B.D. Kaushiva, Structure-property relationships of flexible polyurethane foams, Ph. D. thesis, Virginia Polytechnic Institute and State University, 1999.
[40] S. Mondal, J.L. Hu, Structural characterization and mass transfer properties of nonporous-segmented polyurethane membrane: influence of the hydrophilic segment content and soft segment melting temperature, J. Membrane Sci. 276 (2006) 16–22.
[41] S. Mondal, J. Hu, Structural characterization and mass transfer properties of polyurethane block copolymer: influence of mixed soft segment block and crystal melting temperature, Polym. Int. 55 (2006) 1013–1020.
[42] S. Mondal, J.L. Hu, Structural characterization and mass transfer properties of nonporous segmented polyurethane membrane: influence of hydrophilic and carboxylic group, J. Membrane Sci. 274 (2006) 219–226.
[43] Z.F. Wang, B. Wanga, X.M. Ding, M. Zhang, L.M. Liu, N. Qi, et al., Effect of temperature and structure on the free volume and water vapor permeability in hydrophilic polyurethanes, J. Membrane Sci. 241 (2004) 355–361.
[44] S. Mondal, J.L. Hu, Z. Yong, Free volume and water vapor permeability of dense segmented polyurethane membrane, J. Membrane Sci. 280 (2006) 427–432.
[45] C.A. Byrne, C.R. Desper, N.S. Schneider, B. Chu, Y.J. Li, Polyethylene-based Polyurethane Copolymers and Block Copolymers, Army Research Laboratory, Watertown, 1994.
[46] M.B. Rossman, Synthesis and Characterization of Polybutadiene-Containing Polyurethanes, Master thesis, University of Massachusetts, 1981.
[47] M.F. Sonnenschein, J.M. Virgili, D.D. Adrian, B.L. Wendt, Reactively dispersed olefins in polyurethanes: a new class of polymer/polymer composite, M.F. Sonnenschein, Polymer 76 (2015) 80–88.

[48] T. Choi, J. Weksler, A. Padsalgikar, J. Runt, Influence of soft segment composition on phase-separated microstructure of polydimethylsiloxane-based segmented polyurethane copolymers, Polymer 50 (2009) 2320–2327.

[49] F. Wang, Polydimethylsiloxane modification of segmented thermoplastic polyurethanes and polyureas, Ph.D. thesis, Virginia Polytechnic Institute and State University, 1998.

[50] X.-H. Yu, A.Z. Okkema, S.L. Cooper, Synthesis and physical properties of poly (fluoroalkylether) urethanes, J. Appl. Polym. Sci. 41 (1990) 1777–1795.

[51] Y.-S. Kim, J.-S. Lee, Q. Ji, J.E. McGrath, Surface properties of fluorinated oxetane polyol modified polyurethane block copolymers, Polymer 43 (2002) 7161–7170.

[52] C. Qin, D. Tang, J. Cai, J. Zhang, W. Cai, X. Sun, Study on Polyurethan/vinyl ester resin IPN damping materials, J. Mater. Sci. Technol. 19 (2003) 56–58.

[53] H. Shih-Liang, L. Juin-Yih, HTPB-H12MDI based polyurethane IPN membranes for pervaporation, J. Membrane Sci. 115 (1996) 1–10.

[54] V. Nevissas, J.M. Widmaier, G.C. Meyer, Effect of crosslink density and internetwork grafting on the transparency of polyurethane/polystyrene interpenetrating polymer networks, J. Appl. Polym. Sci. 36 (1988) 1467–1473.

[55] L.E. Fuentes, S. Perez, S.I. Martinez, A.R. Garcia, Interpenetrating polymer networks from modified castor oil - based polyurethane and polystyrene: miscibility and mechanical properties as a function of the composition, Revista Ion 24 (2011) 45–50.

[56] D.S. Lee, S.C. Kim, Polyurethane interpenetrating polymer networks (IPN's) synthesized under high pressure. 3. morphology and Tg behavior of polyurethane-polystyrene semi-IPN's and linear blends, Macromolecules 17 (1984) 2222–2227.

[57] C.-L. Qin, W.-M. Cai, J. Cai, D.-Y. Tang, J.-S. Zhang, M. Qin, Damping properties and morphology of polyurethane/vinyl ester resin interpenetrating polymer network, Mat. Chem. Phys. 85 (2004) 402–409.

[58] C. Qin, Z. Jin, X. Bai, H. Jiang, W. Cai, Compatibility of polyurethane/(vinyl ester resin)(ethyl acrylate) interpenetrating polymer network, Polym. J. 39 (2007) 1365–1372.

[59] K.C. Jajam, S.A. Bird, M.L. Auad, H.V. Tippur, ensile, fracture and impact behavior of transparent interpenetrating polymer networks with polyurethane-poly(methyl methacrylate), Polym. Test. 32 (2013) 889–900.

[60] X. Kong, S.S. Narine, Sequential interpenetrating polymer networks produced from vegetable oil based polyurethane and poly(methyl methacrylate), Biomacromolecules 9 (2008) 2221–2229.

[61] S.H. Baek, B.K. Kim, Synthesis of polyacrylamide/polyurethane hydrogels by latex IPN and AB crosslinked polymers, Colloids Surf. A 220 (2003) 191–198.

[62] S.K. Srivastava, M. Kotal, Recent advances on preparation, properties and applications of polyurethane nanocomposites, Adv. Compos. (2013) 33–92.

[63] M. Liu, G. Song, J. Yi, Y. Xu, Damping analysis of polyurethane/polyacrylate interpenetrating polymer network composites filled with graphite particles, Polym. Compos. 34 (2013) 288–292.

[64] T. Trakulsujaritchok, D.J. Hourston, Damping characteristics and mechanical properties of silica filled PUR/PEMA simultaneous interpenetrating polymer networks, Eur. Polym. J. 42 (2006) 2968–2976.

[65] T. Wang, S. Chen, Q. Wang, X. Pei, Damping analysis of polyurethane/epoxy graft interpenetrating polymer network composites filled with short carbon fiber and micro hollow glass bead, Mater. Des. 31 (2010) 3810–3815.

[66] V.A. Bershtein, V.M. Gun'ko, L.V. Karabanova, T.E. Sukhanova, P.N. Yakushev, L. M. Egorova, et al., Shape memory studies of functionalized MWNT-reinforced polyurethane copolymers, RSC Adv. 3 (2013) 14560.

[67] S. Mondal, J.L. Hu, Shape memory studies of functionalized MWNT-reinforced polyurethane copolymers, Iran. Polym. J. 15 (2006) 135–142.
[68] N. Ahmed, A. Kausar, B. Muhammad, Advances in shape memory polyurethanes and composites: a review, Polym. Plast. Tech. Eng. 54 (2015) 1410–1423.
[69] S. Farzaneh, J. Fitoussi, A. Lucas, M. Bocquet, A. Tcharkhtchi, Shape memory effect and properties memory effect of polyurethane, J. Appl. Polym. Sci. 128 (2013) 3240–3249.
[70] J.H. Yang, B.C. Chun, Y.-C. Chung, J.H. Cho, Comparison of thermal/mechanical properties and shape memory effect of polyurethane block-copolymers with planar or bent shape of hard segment, Polymer 44 (2003) 3251–3258.

CHAPTER 7

Nonisocyanate Polyurethanes

Marcin Włoch and Janusz Datta
Gdańsk University of Technology, Gdańsk, Poland

Contents

7.1 INTRODUCTION

The history of polyurethanes (PU) began in the 1930s, when Otto Bayer described the polyaddition of isocyanates and polyols. The diversity of substrates, which can be used in the synthesis of PU, resulted in obtaining polyurethanes with diversified structure and properties, and as a consequence, polyurethanes has found its way into many industrial applications. Moreover, this polymer can be obtained in the form of elastomers, flexible and rigid foams, fibers and coatings [1–3].

Polyurethanes are synthesized by the polyaddition of aliphatic (e.g., hexamethylene diisocyanates, HDI; isophorone diisocyanate, IPDI) or aromatic (e.g., 2,4- and 2,6-toluene diisocyanate, TDI; 4,4′-methylene diphenyl diisocyanate, MDI) diisocyanates, polyols (based on polyesters or polyethers, e.g., PTHF [poly(tetrahydrofuran)], PPG [poly(propylene glycol)], or PEG [poly(ethylene glycol)]) and low molecular weight chain extenders (e.g., 1,2-etanediol; 1,4-butanediol) [1,2]. Synthesis can be

Polyurethane Polymers: Blends and Interpenetrating Polymer Networks
DOI: http://dx.doi.org/10.1016/B978-0-12-804039-3.00007-5

realized by the one- or two-step method. The two-step method is more beneficial, due to a greater possibility of controlling the polyurethane structure, i.e., domain structure (soft and hard domains), that controls theunusual properties of polyurethanes. The two-step method (Fig. 7.1), also known as the prepolymer method, starts with the reaction between an excess of diisocyanates and polyols to obtain a urethane—prepolymer terminated with isocyanates groups. The second step involves the reaction of this semi-product (urethane—prepolymer) with a low molecular weight chain extender.

The disadvantages of isocyanates (Fig. 7.2.) are phosgene-based synthesis, high moisture sensitivity and human toxicity; these have resulted in investigations connected with the synthesis of polyurethanes without using diisocyanates [4]. Moreover, some disadvantages of isocyanate-based PUs are connected with the elimination of PUs at the end of their life, due to formation of toxic substances during burning and/or removal to landfill, of the waste products. When the waste materials are incinerated, isocyantes and amines are formed, and when the waste goes to landfill, amines are formed and released to the environment.

Moisture sensitivity of the diisocyanates should be considered in two ways. It is highly important in the case of elastomeric PU synthesis and during the storing of diisocyanates. For example, a reaction of 4,4′-diphenylmethane diisocyanate (MDI) with water results in a porous structure (due to generation of CO_2) and a yellow color (due to formation of primary amine groups presented in the structure of 4,4′-diphenylmethane diamine (MDA)) of elastomeric materials. Moreover, the formation of urea bonds has some effect on the mechanical properties of polyurethane materials (e.g., increasing stiffness). On the other hand, the reaction of isocyanate groups with water is the main reaction during the synthesis of polyurethane foams. The porous structure is formed due to the generation of carbon dioxide.

Nonisocyanate polyurethanes (NIPU) can be obtained by polycondensation or via the polyaddition method (Fig. 7.3) [5,6]. A comparison of both methods are presented in Table 7.1. From an industrial point of view, more attention is paid to a NIPU obtained by polyaddition of multifunctional cyclic carbonates and primary diamines or polyamines, which can be realized as a solvent- and catalyst-free reaction without the formation of any byproduct. Nonisocyanate polyhydroxyurethanes (HNIPUs) obtained by polyaddition of cyclic carbonates and diamines or polyamines are better characterized in the literature (especially in the terms of

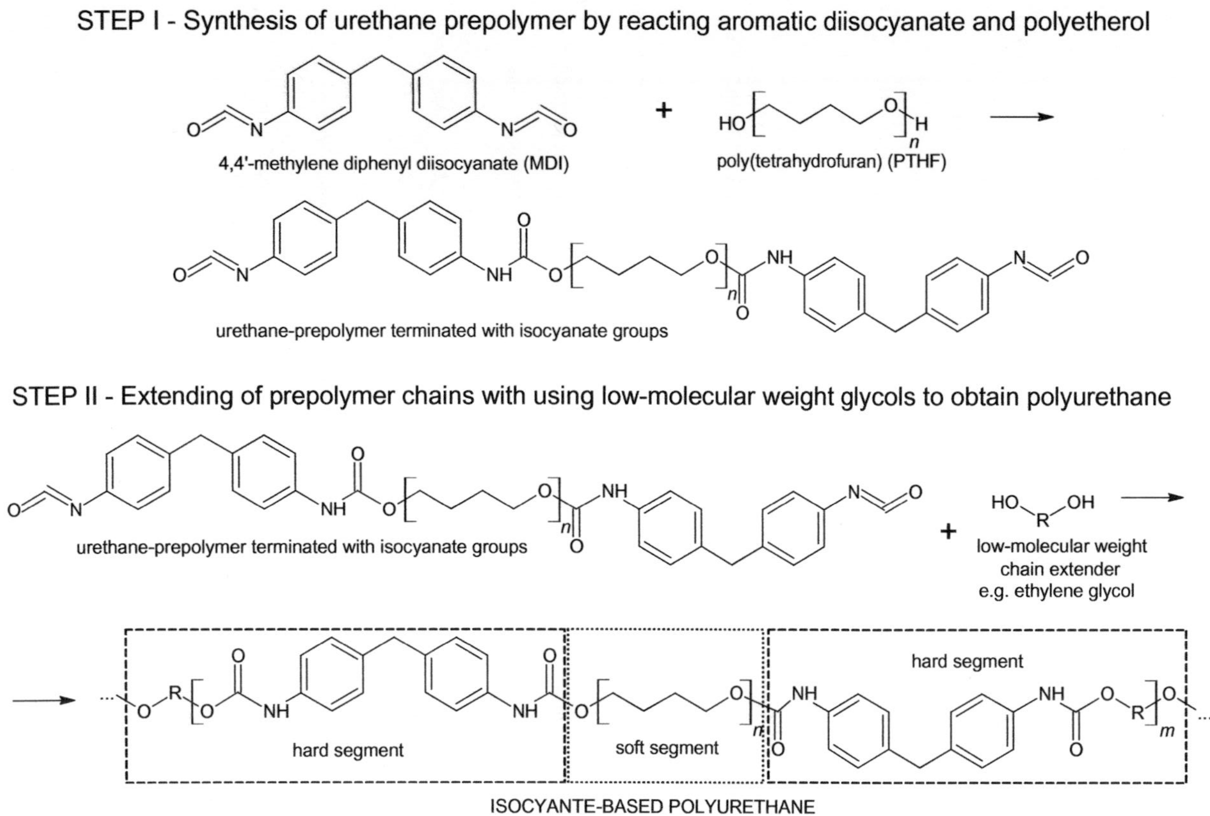

Figure 7.1 Two-step synthesis of polyurethanes by reacting aromatic diisocyanate, polyetherdiol and low molecular weight chain extender.

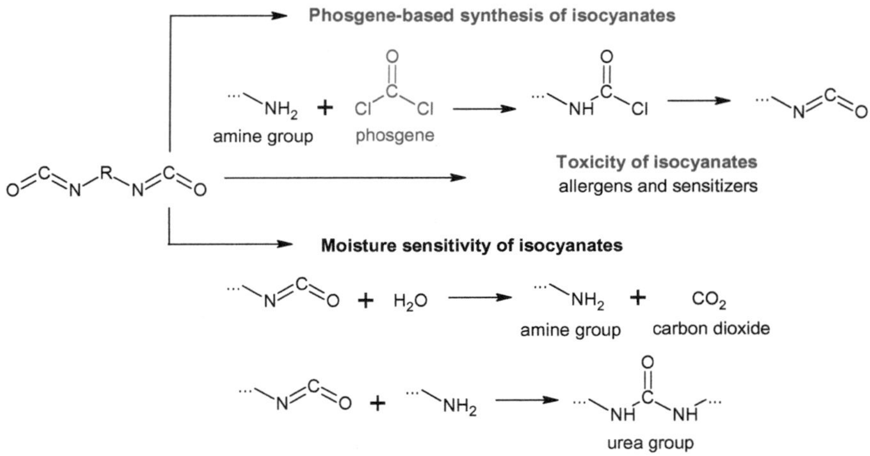

Figure 7.2 Disadvantages of diisocyanates.

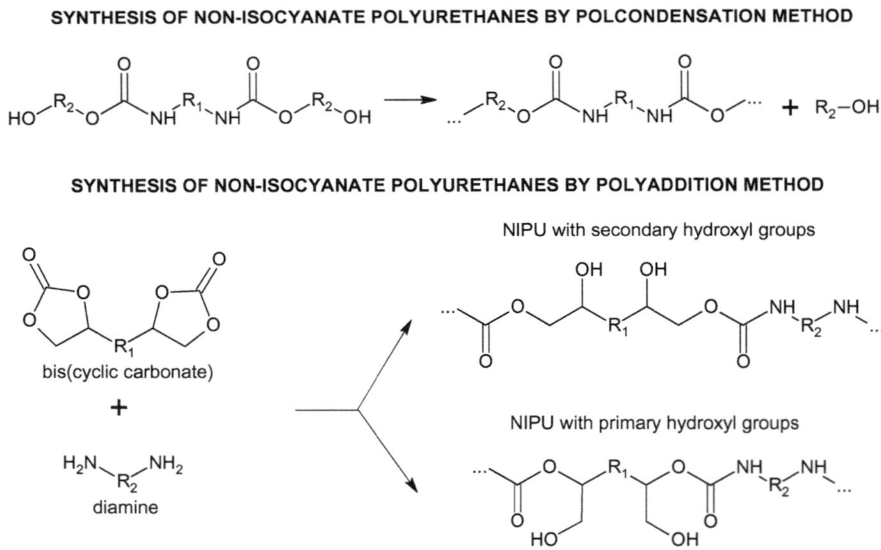

Figure 7.3 Synthesis of nonisocyanate polyurethanes by polycondensation and polyaddition.

mechanical properties) and its synthesis are more economic from the industrial point of view—in the terms of the absence of solvents (needed to synthesize NIPUs and purification of obtained polymers), catalysts and byproducts.

Table 7.1 Comparison of nonisocyanate polyurethanes synthesized by polycondensation and polyaddition

Analyzed factor	Nipus synthesized by polycondensation	Nipus synthesized by polyaddition
Reaction medium	Solvent	–
Temperature of reaction	Higher in the case of NIPU synthesized by polycondensation	
Time of reaction	Higher in the case of NIPU synthesized by polycondensation	
Catalyst of reaction	Only in the case of NIPU synthesized by polycondensation	
Substrates	Several different components e.g., dihydroxyurethanes	di- or poly-cyclic carbonate intermediates and di- or poly-amines
Byproducts	e.g., ethylene glycol or phenol	–
Hydroxyl groups in the structure of final polymer	Depending on substrates used during the synthesis	Always present primary and secondary hydroxyl groups
Purification of final polymers	With using solvent	–
Molecular weight of final products	Higher in the case of NIPU synthesized by polycondensation (if polyaddition is realized by solvent-free method)	

The polyaddition is performed generally at medium temperatures (ca. 60–80°C), and this condition is similar in the case of traditional isocyanate-based PUs (synthesis of ureathe prepolymer and chain extending). The polycondensation is realized at high temperatures (ca. 150–220°C), due to necessity of removing the byproduct from the reaction mixture. The applications of NIPU are listed in Table 7.2, and are very similar to applications of isocynate-based polyurethanes.

During the abovementioned reaction, primary and secondary hydroxyl groups are formed, so these types of NIPUs are also known as polyhydroxyurethanes. The presence of the hydroxyl groups in the polymer chain permit the formation of intra molecular hydrogen bonds, resulting in increased mechanical properties and the chemical resistance of the final polymer. Moreover, the presence of hydroxyl groups in the structure of final polymer permits chemical functionalization.

Table 7.2 Proposed applications of nonisocyanate polyurethanes

Applications	Refs.
Chemical-resistant coatings	[7,8]
Adhesives	[9–11]
Imitation leathers	[12,13]
Foams	[14,15]
Modificators of epoxy resins	[16]
Modificators of fillers (carbonated silane coupling agents)	[17,18]
Gas barrier	[19]

Modern polymer science and technology focuses on the development of green polymers. This is related to green chemistry trends, proposing natural and renewable resources. For NIPUs, some of cyclic carbonate intermediates are synthesized using as starting molecules such as vegetable oils [20], glycerol [21], pentaerythritol [21], limonene [22], vanilline [23], isosorbide [24], or cardanol [25]. The trends of green chemistry also occupied special place in polyurethanes synthesized by conventional method connected with using isocyantes. In accordance to the literature, there are known bio-based diisocyanates [26,27] and bio-based polyols [28–32].

The main aim of this chapter is to present a general overview of the NIPU synthesis method (i.e., polycondensation and polyaddition), and to describe structure and properties of selected intermediates and final polymeric products. The focus is on five-membered ring cyclic carbonate intermediates, which were obtained by cycloaddition of carbon dioxide to selected glycidyl ethers and thiol-ene coupling of unsaturated cyclic carbonates monomers with thiols. We also discuss the synthesis and properties of bio-based NIPU, obtained from, e.g., carbonated vegetable oils.

7.2 SYNTHESIS, STRUCTURE, AND PROPERTIES OF NONISOCYANATE POLYURETHANES OBTAINED BY THE POLYCONDENSATION METHOD

7.2.1 Possibilities of Polycondensation Process Realization

The first synthesis of the NIPU by polycondensation method was realized by Morgan in 1963 [33] who discussed a process based on the reaction of alkylene bis(chloroformate)s with diamines (Fig. 7.4).

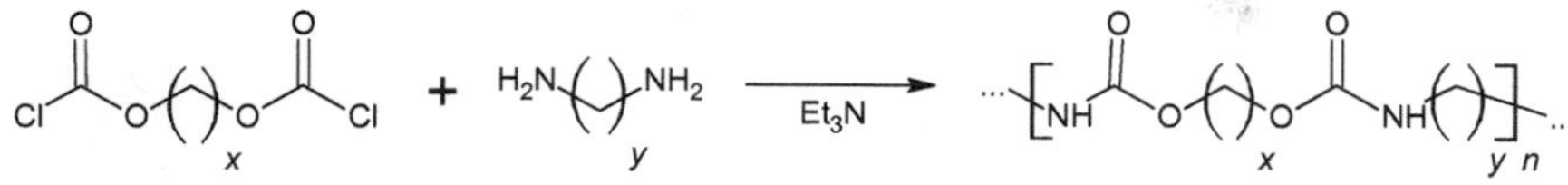

Figure 7.4 Synthesis of nonisocyanate polyurethanes by polycondensation of bis (chloroformate)s and diamines.

Polycondensation resulting in NIPUs can be realized by several different methods, using a variety of components. However, the two main types of the methods are based on the reaction of carbamates with diols, and chloroformates with diamines. The most important methods (Fig. 7.5) include:

- self-polycondensation of dihydroxydicarbamates (obtained by the reaction of cyclic carbonate with diamine) and polycondensation of dihydroxydicarbamates with diols [34,35],
- cationic ring-opening polymerization of cyclic urethane (obtained by ring-closing of dihydroxycarbamate, which was synthesized from aminoalcohol and ethylene carbonate [36],
- polycondensation of dimethyl dicarbamates and diols [37],
- polycondensation of phenoxycarbonyloxymethyl ethylene carbonate (from glycerol carbonate and phenyl chloroformate) with diamines [38].

The NIPU obtained by polycondensation methods have a linear structure, and generally hydroxyl groups are absent in the structure of final polymers. One of exceptions is the polycondensation of phenoxycarbonyloxymethyl ethylene carbonate with diamines (showed at the bottom of Fig. 7.5).

NIPU can be also obtained by the reaction of polycarbamate and aldehyde or polyaldehyde (Fig. 7.6) [39]. Both components can be reacted at ambient temperature (generally less than 80°C) to form nonisocyanate polyurethane coating, with good chemical resistance, adhesion, mechanical properties and weatherability. The polycarbamates can be obtained by the reaction of polyol (e.g., acrylic polyol) with urea in the presence of catalyst and at suitable temperature. As polyaldehyde components, 1,3-cyclohexanedicarboxyaldehyde and 1,4-cyclohexanedicarboxyaldehyde are used.

7.2.2 Properties of Nonisocyanate Polyurethanes Obtained by the Polycondensation Method

The properties of NIPUs obtained by polycondensation are affected by synthesis conditions (i.e., temperature, time, presence and type of catalyst,

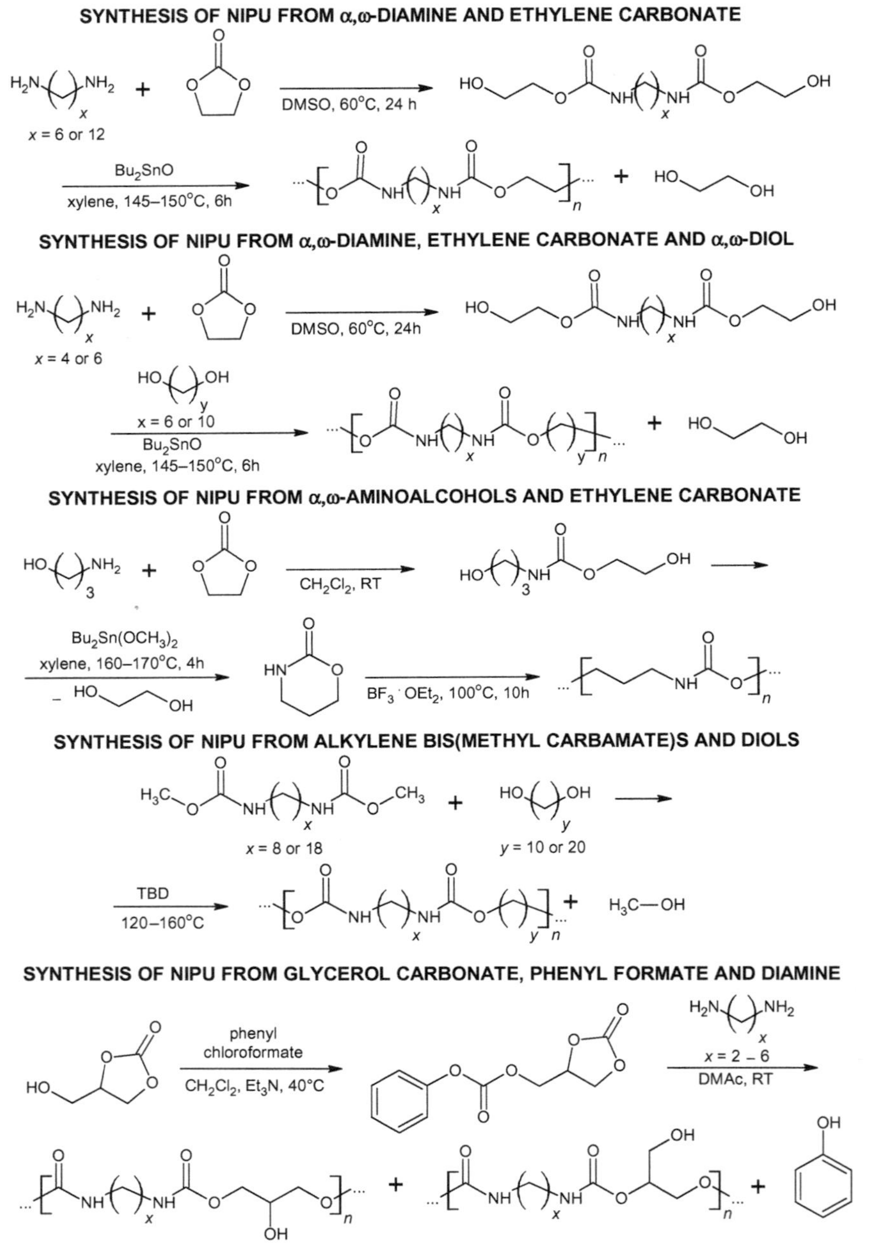

Figure 7.5 Selected methods of nonisocyanate polyurethanes by the polycondensation method.

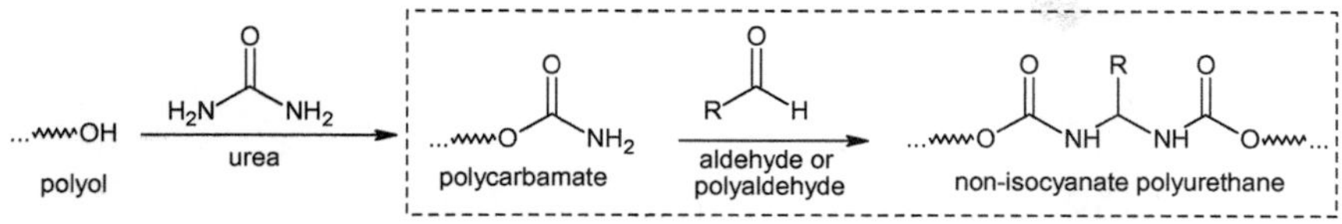

Figure 7.6 Synthesis of NIPU by the reaction of polycarbamate and aldehyde or polyaldehyde.

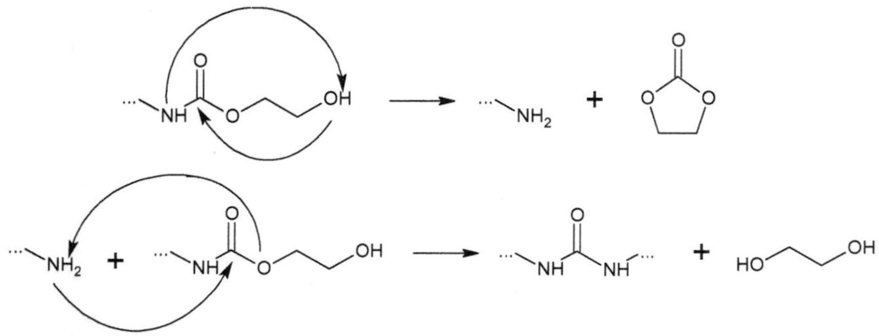

Figure 7.7 Formation of urea bonds during synthesis of NIPUs by the polycondensation method.

and type of solvent) and the structure of substrates used to prepare intermediates (which determine the structure of the final polymers). It is very important to apply suitable temperature and time of reaction, because when the optimal conditions are exceeded, the average molecular weight (M_n) can decrease, but the polydispersity index (M_w/M_n) increases, resultant from side reactions connected with the formation of urea bonds (Fig. 7.7). Improvement of yield is connected to a reduction of reaction time and an increase in average molecular weight obtained by using suitable catalysts, and also effective methods of removing byproducts [35].

Generally, NIPU that are obtained by polycondensation method possess lower M_n than NIPUs synthesized by solvent-free polyaddition of cyclic carbonates and diamines. Moreover, in the case of NIPUs produced by polycondensation, properties described in the literature include only glass transition temperature (determined by differential scanning calorimetry) and thermal stability (determined by thermogravimetric analysis). These properties are mostly affected by the type of components that are used, especially diamines, and similar trends are observed in NIPUs obtained by the polyaddition method.

7.3 SYNTHESIS, STRUCTURE, AND PROPERTIES OF NONISOCYANATE POLYURETHANES BY THE POLYADDITION METHOD

7.3.1 Di- and Poly-cyclic Carbonate Intermediates

7.3.1.1 Types of Cyclic Carbonates and Chemistry of Its Reaction with Amine Curing Agents

The most described in the literature NIPUs are synthesized from five-membered cyclic carbonates (5CC). The mechanism of the reaction between 5CC and diamines or polyamines [40] is presented in Fig. 7.8.

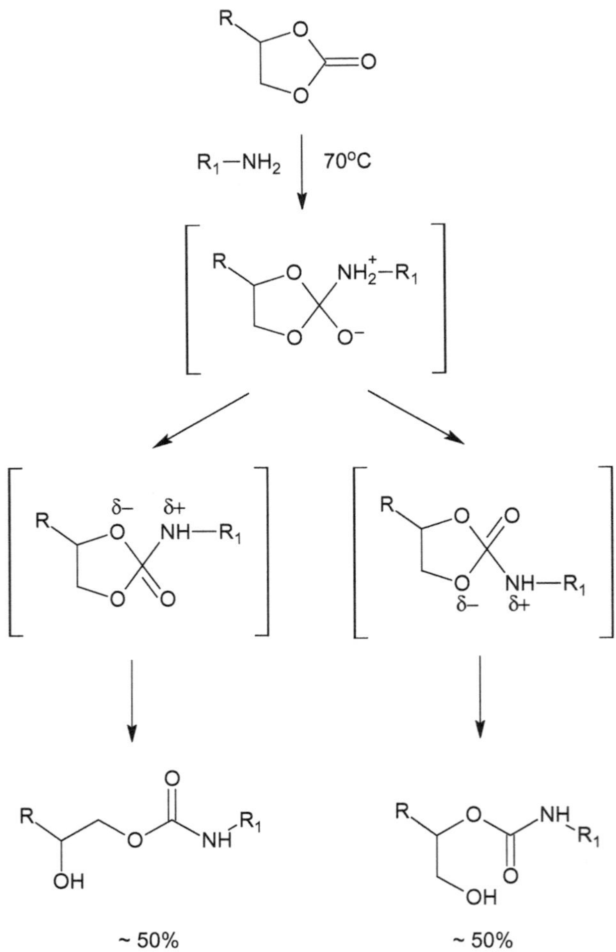

Figure 7.8 Mechanism of five-membered cyclic carbonate with amine.

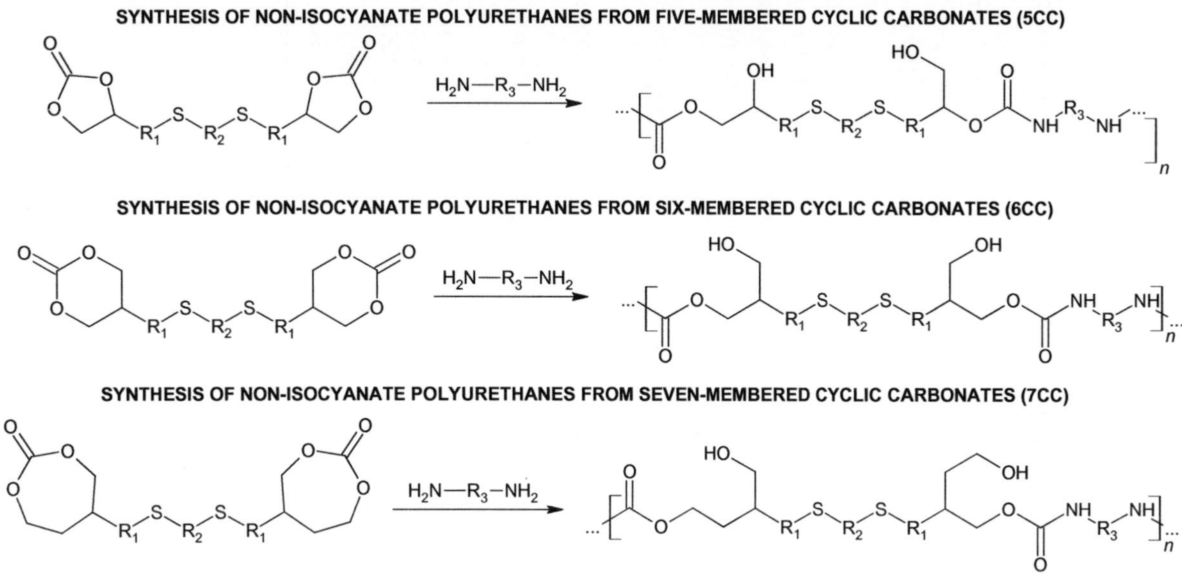

Figure 7.9 Synthesis of nonisocyanate polyurethanes with using six-membered and seven-membered cyclic carbonates and diamines.

It is also possible to synthesize the NIPU from six-membered (B6CC) [41–43] and seven-membered cyclic (B7CC) [44] carbonates and diamines (Fig. 7.9). B6CC and B7CC are more reactive than five-membered cyclic carbonates (B5CC), resulting from larger ring strain than in case of five-membered cyclic carbonates [40,42]. The reactivity of these cyclic carbonates with aliphatic and aromatic amines are compared in Table 7.3. Unfortunately, these semiproducts are synthesized using phosgene-based compounds, e.g., ethyl chloroformate and triphosgene.

The five-membered cyclic carbonates react quite slowly with primary amine groups at room temperature, but the higher effectiveness is observed around 70°C (maximum of exotermic peak in DSC study presented by Camara et al. [45]). The reaction of cyclic carbonates with secondary amines groups (presented, e.g., along the chain of triethylenetetramine) is possible only at temperature up to 200°C, but using high temperatures can be connected with thermal degradation of the obtained polymer (see Fig. 7.10).

The lower reactivity of five-membered cyclic carbonates means that the use of medium temperature (i.e., 80°C) and a catalyst is required to obtain high conversion of cyclic carbonate groups to hydroxyurethane groups.

Table 7.3 Rate constants for reactions of cyclic carbonates with n-hexylamine and benzylamine [41,42,44]

Amine	**Rate constant of reaction cyclic carbonate with amine k [dm³/mol]**		
	Five-membered cyclic carbonate (5CC)	**Six-membered cyclic carbonate (6CC)**	**Seven-membered cyclic carbonate (7CC)**
n-hexylamine	0.02	1.19	48.5
Benzylamine	0.01	0.29	11.0

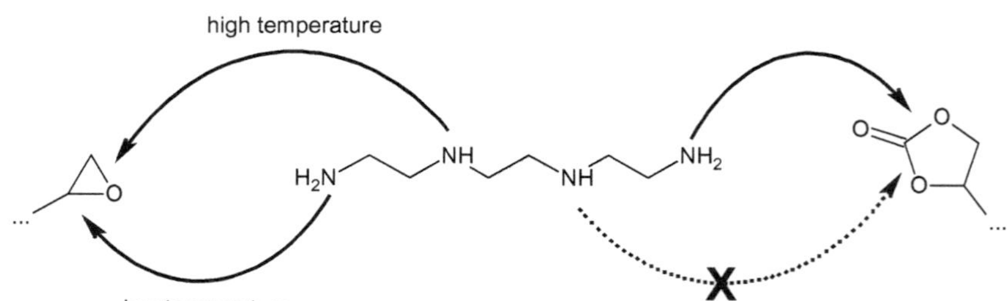

Figure 7.10 Comparison of epoxy and five-membered cyclic carbonate groups with primary and secondary amine groups presented in the structure of triethylenetetramine [45].

The catalyst promotes the nucleophilic attack of the amine on the carbonyl group of the carbonate (**Fig. 7.11**), examples include 1,5,7-triazabicyclo [4.4.0]dec-5-ene (TBD), 1,8-diazabicyclo[5.4.0]undec-7-ene (DBU), and N,N′-disubsituted aromatic-aliphatic thioureas [5,46,47].

In the case of NIPU obtained by the polyaddition method, side reactions during the synthesis are also possible [48,49]. As the main products of hydroxyurethanes with primary or secondary hydroxyl groups are formed (**Fig. 7.8**), side reactions result in the formation of many byproducts (Fig. 7.12). As an

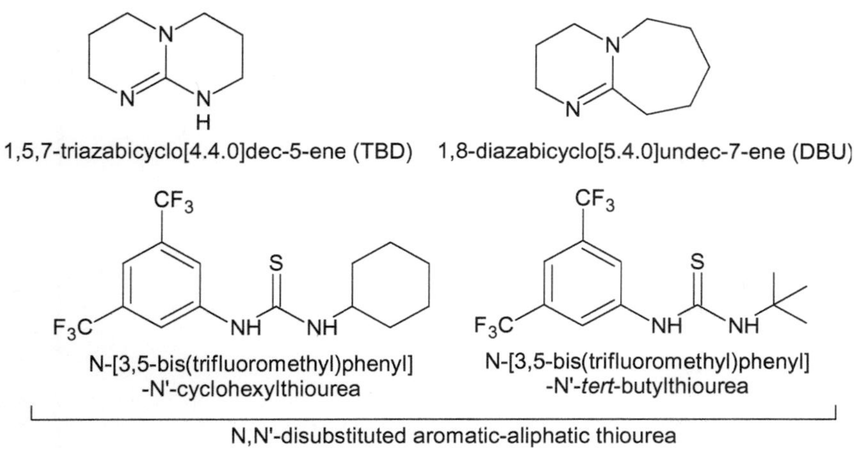

Figure 7.11 Organocatalysts used in the reaction of five-membered cyclic carbonates with primary amine groups.

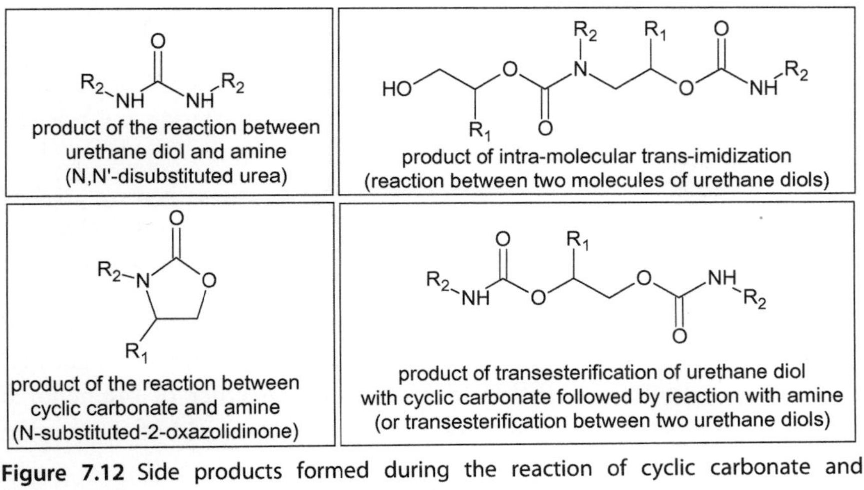

Figure 7.12 Side products formed during the reaction of cyclic carbonate and primary amine.

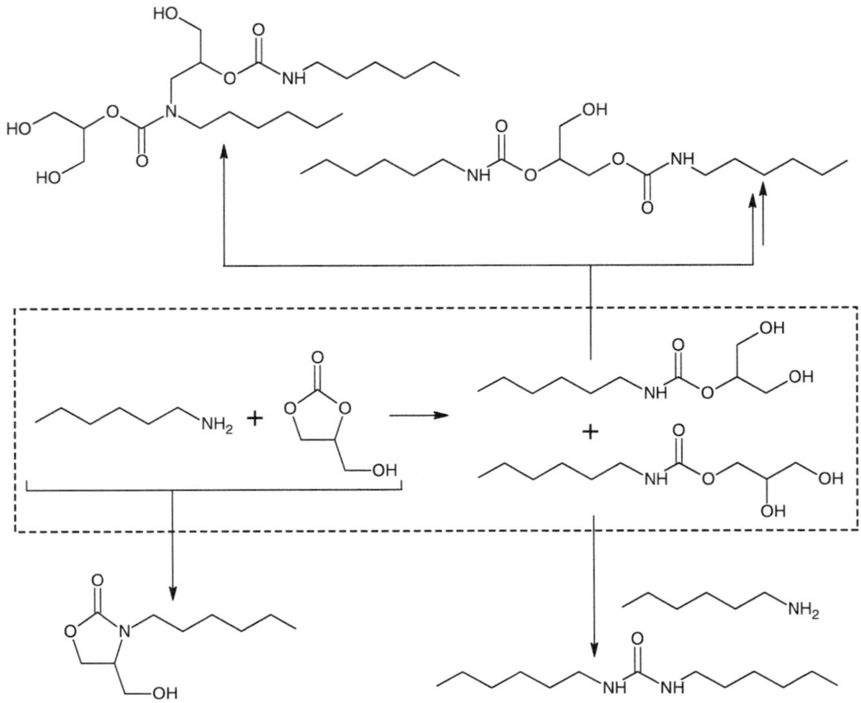

Figure 7.13 Main and side products formed during the reaction between glycerol carbonate and n-hexylamine (investigated by Besse et al. [49]).

example, the reaction between glycerol carbonate and n-hexylamine (with formation of main and side products) is presented in Fig. 7.13. The formation of byproducts is probably the reason for low molecular masses of NIPU as described in the literature. The NIPU with high molecular masses can be probably obtained by low temperature reactions and the presence of a catalyst, which activates the cyclic carbonate and not the amine during the reaction [47,49,50].

7.3.1.2 Synthesis and Structure of Cyclic Carbonate Intermediates

7.3.1.2.1 Cycloaddition of Carbon Dioxide into di- or Poly-glycidyl Ethers

The bis- and poly(cyclic carbonate) intermediates, for the synthesis of NIPU, can be obtained by the reaction of diglicidyl ethers (e.g. diglicydyl ether of bisphenol A (DGEBA) [51] or diglicydyl ether of bisphenol F (DGEBF) [13]) with carbon dioxide (Fig. 7.14). The yield of the process is around 100%.

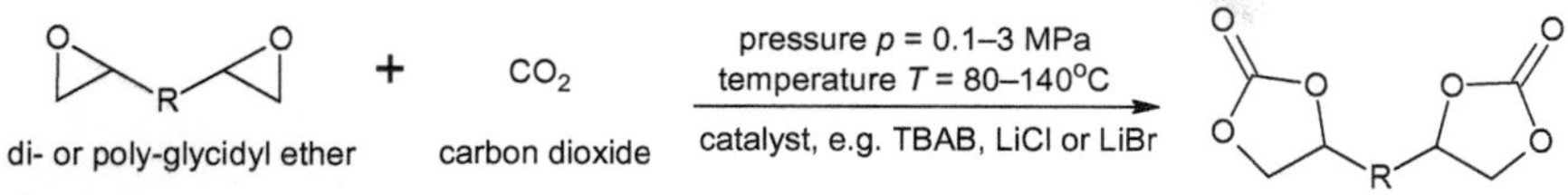

Figure 7.14 Cycloaddition of carbon dioxide into glycidyl ethers.

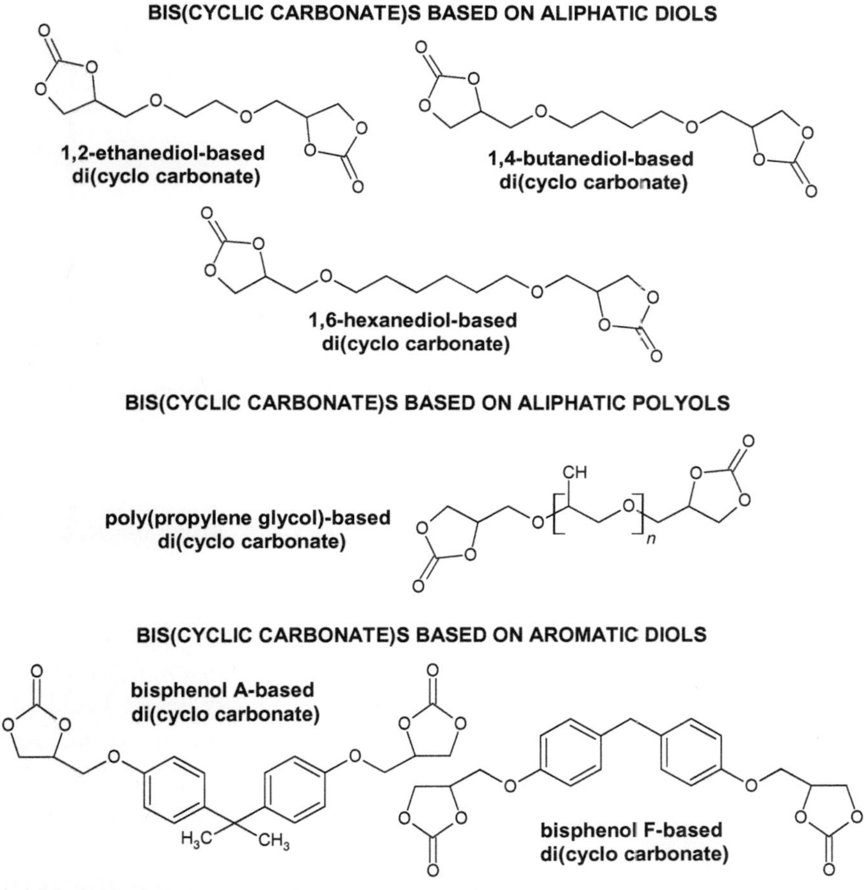

Figure 7.15 Bis(cyclic carbonate)s based on aliphatic and aromatic diols.

Diglicidyl ethers of diols can be used as a substrate for the synthesis of bis(cyclic carbonate)s, e.g., 1,2-ethanediol [52], 1,4-butanediol [45], 1,6-hexanediol [13], or poly(propylene glycol) [18]. The structure of obtained bis(cyclic carbonate)s from these compounds are showed in Fig. 7.15.

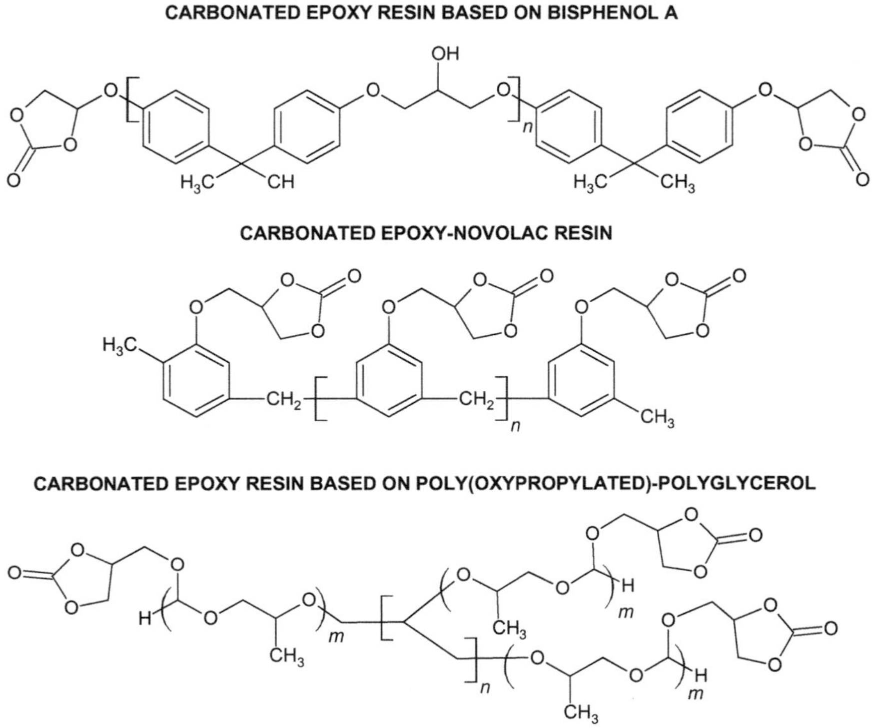

Figure 7.16 Bis- and poly-(cyclic carbonate)s based on commercially available epoxy resins.

This process can be also successfully applied for the modification of commercially-available products (Fig. 7.16) e.g., epoxy resins based on bisphenol A or poly(oxypropylated)-polyglycerol and also epoxy-novolac resins [12,13,53–56].

Some research interests are occupied by intermediates obtained by using bio-based substances as starting molecules, e.g., glycerol [21], resorcinol[45], cardanol [25], pentaerythritol [21,45], trimethylolpropane [21], limonene [22], and isosorbide [24] (Fig. 7.17). Increased interest is focused on cycloaddition of CO_2 into epoxidized vegetable oils, e.g., soybean [57–61], linseed [61,62], cottonseed [63], or vernonia oil [64]. Epoxidized soybean and linseed oils are nowdays produced on large scales by several companies, and used as modifiers of commercial epoxy resins or plasticizers for items such as poly(vinyl chloride) [65]. It is also possible to produce terminal and internal carbonated fatty acid diesters and

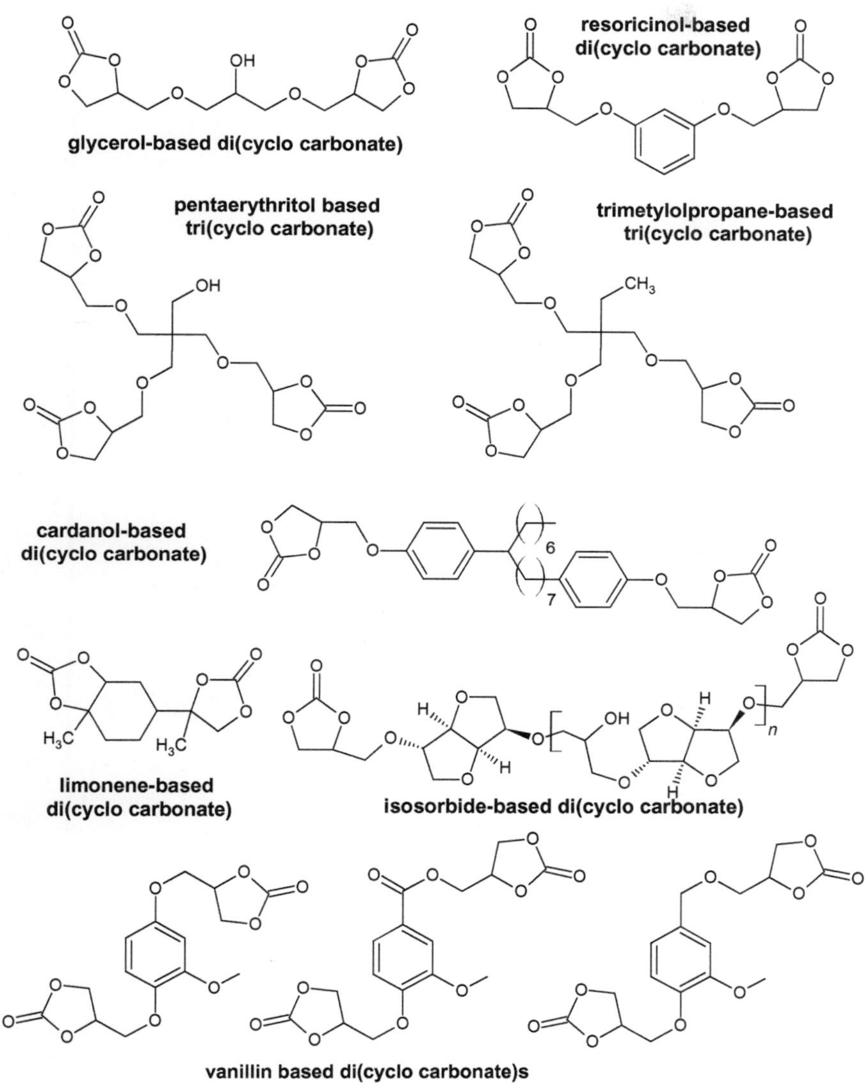

Figure 7.17 Bis- and tris-(cyclic carbonate)s synthesized with using as starting molecules bio-based substances.

diamines, where fatty acids are obtained from vegetable oils, e.g., oleic acid from sunflower oil [66,67]. The cyclic carbonates obtained from vegetable oils and fatty acids derivatives are presented in Fig. 7.18. The application of bio-based and renewable feedstock in the synthesis of polyurethanes is connected with trends of green chemistry.

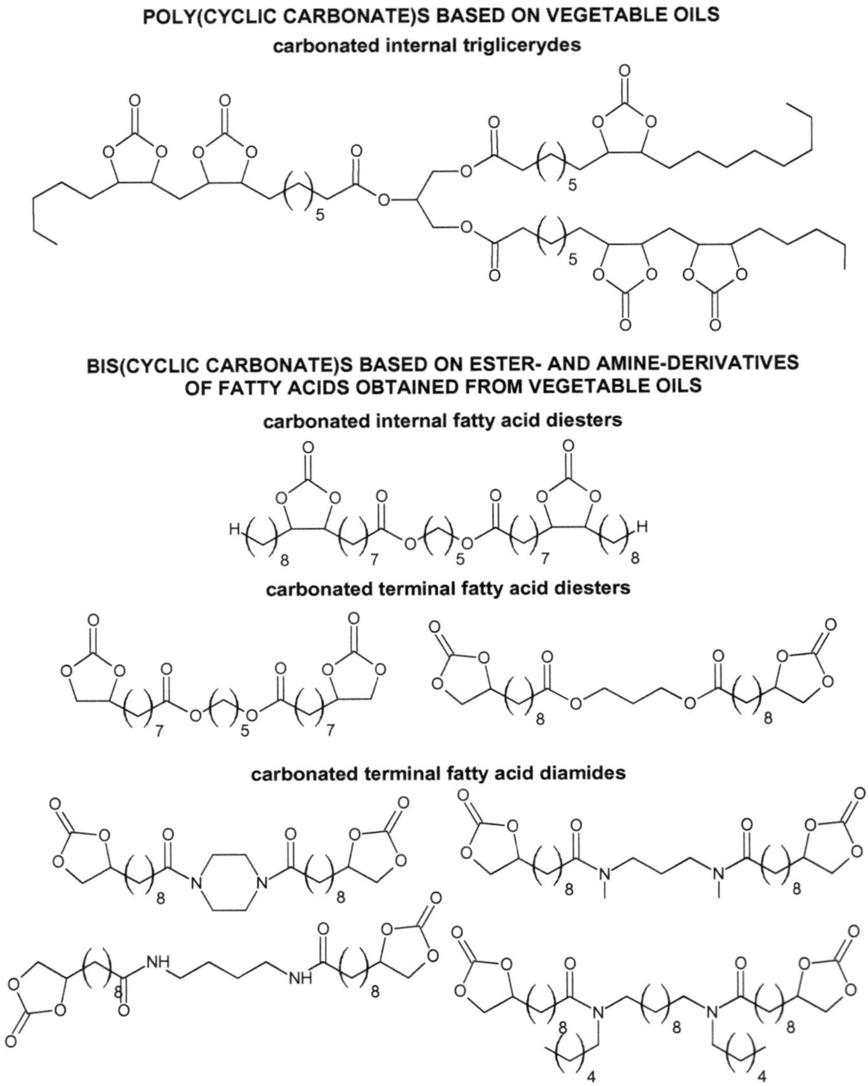

Figure 7.18 Bio-based carbonated vegetable oils and internal or terminal carbonated fatty acid diesters or diamides.

7.3.1.2.2 Thiol-ene Coupling of Unsaturated Derivatives of Cyclic Carbonates and Thiols

The other way of cyclic carbonate intermediates synthesis is connected with thiol-ene coupling, and this process is based on reacting di- or poly-thiol with unsaturated cyclic carbonate monomers (Fig. 7.19).

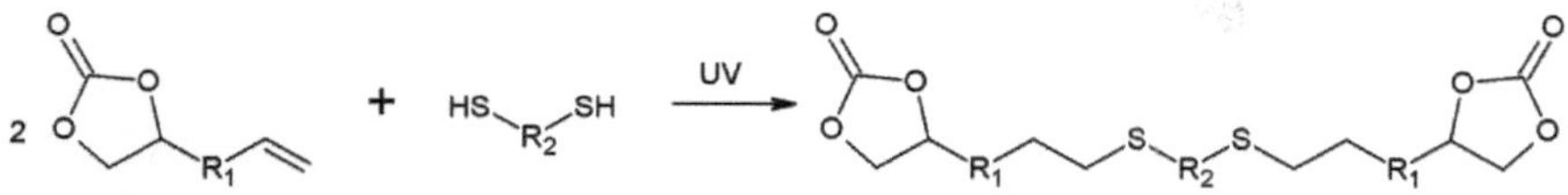

Figure 7.19 Synthesis of cyclic carbonate intermediates by thiol-ene coupling.

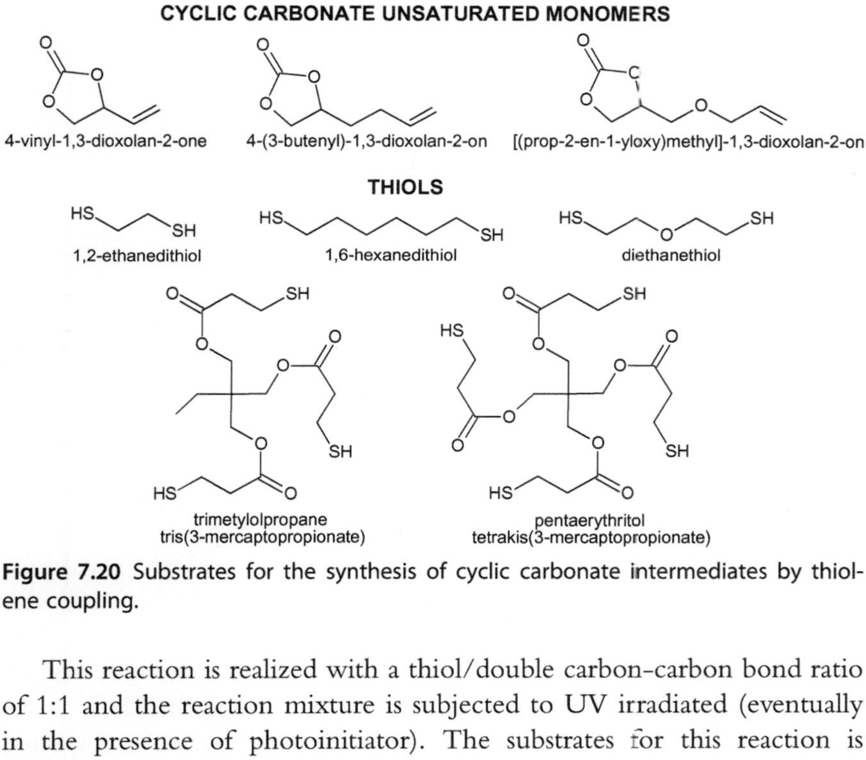

Figure 7.20 Substrates for the synthesis of cyclic carbonate intermediates by thiol-ene coupling.

This reaction is realized with a thiol/double carbon-carbon bond ratio of 1:1 and the reaction mixture is subjected to UV irradiated (eventually in the presence of photoinitiator). The substrates for this reaction is shown in Fig. 7.20, and the products obtained by this method are shown in Fig. 7.21. The thiol-ene coupling, as a method for synthesis of intermediates for the synthesis of NIPUs by polyaddition method, gives the possibility of realization as a solvent-free process, the absence of byproducts, high yields, and short reaction time [42,68,69].

7.3.1.2.3 Other Methods for Synthesis of Cyclic Carbonate Intermediates

Cyclic carbonate intermediates, can be also synthesized by other methods (Fig. 7.22), especially:

- esterification of dicarboxylic acids with glycerol carbonate—e.g., terephthalic acid with glycerol carbonate [70],

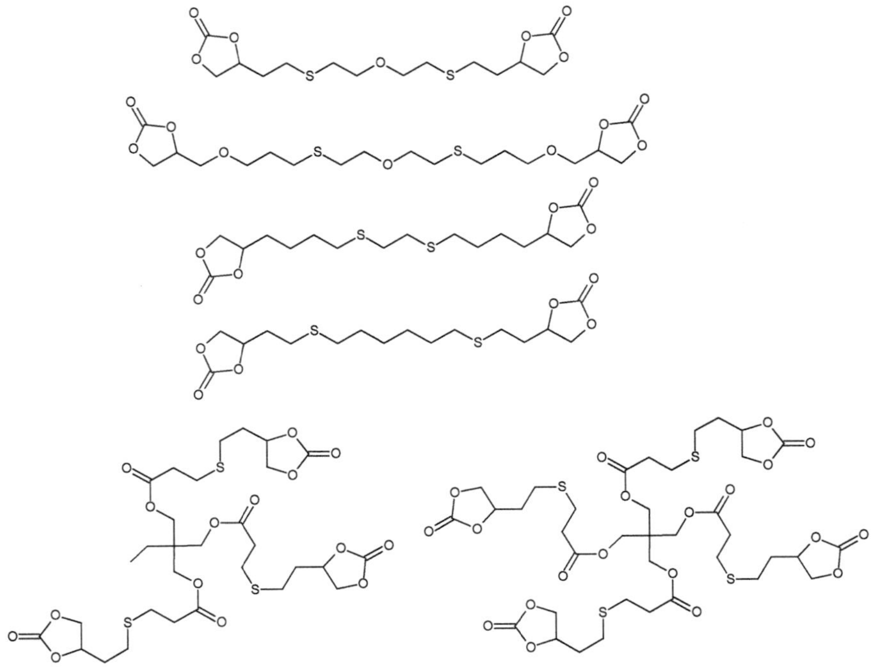

Figure 7.21 Bis- and tris(cyclic carbonate)s obtained by thiol-ene coupling.

- reaction of multifunctional hydroxyl-compounds with dialkyl carbonates—e.g., D-sorbitol or hyperbranched polyglycerol (Fig. 7.23) with using dimethyl carbonate [71,72],
- ring-opening polymerization of trimethylene carbonate with using a hydroxyl compounds (Fig. 7.24) as a chain transfer agent [73],
- tosylation of diols with using glycerol carbonate tosylate – as starting molecules can be used e.g., poly(ethylene glycol) or poly(propylene glycol) (Fig. 7.25)—this method is characterized by good yields and can be realized without using a solvent [50],
- radical polymerization of unsaturated derivatives of glycerol carbonate (Table 7.4) [74–78],
- radical copolymerization of unsaturated derivatives of glycerol carbonate with other unsaturated monomers (Table 7.5) [75,77,79,80],
- cycloaddition of carbon dioxide to polyepoxides [81–83].

7.3.2 Di- and Poly-amine Curing Agents

Diamines and polyamines used in the synthesis of NIPU (by poliaddition, and also polycondensation method) are shown in Fig. 7.26 and Fig. 7.27.

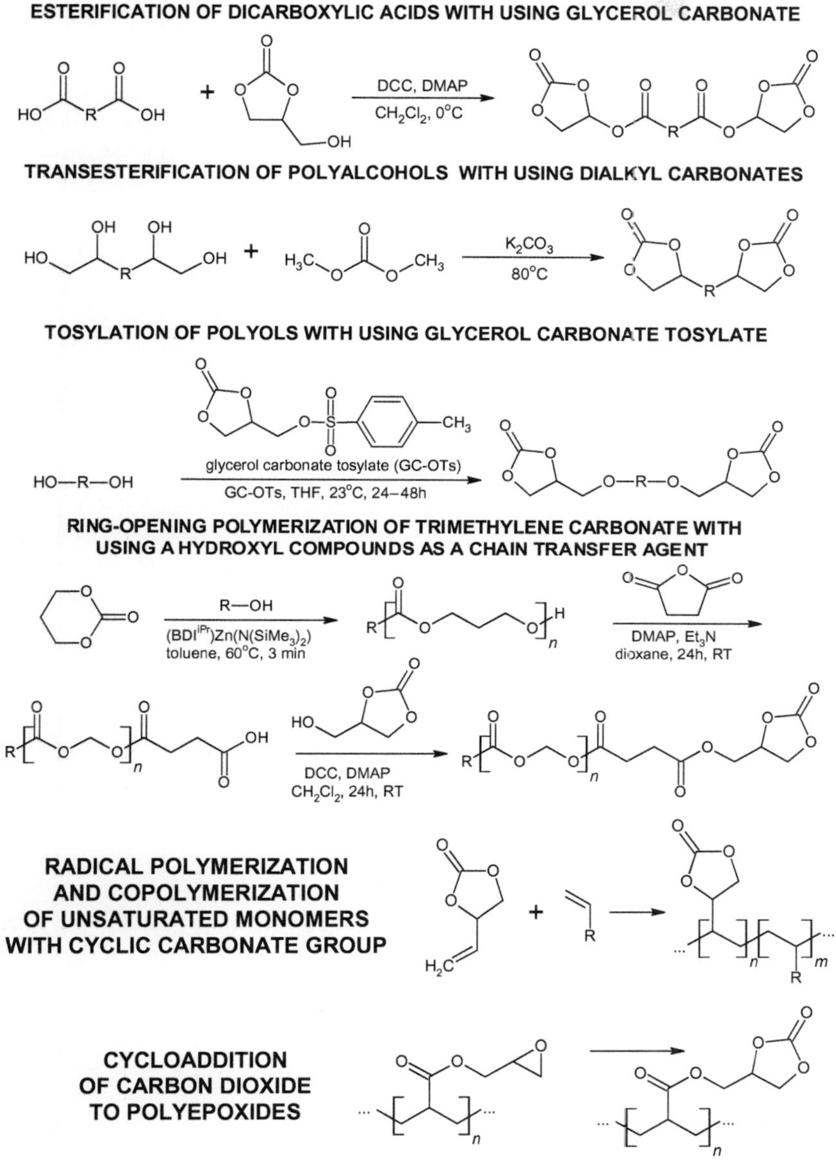

Figure 7.22 Other synthesis methods of cyclic carbonate intermediates.

In the diamines, examples are: aliphatic (e.g., 1,4-butanediamine, 1,6-hexametylenediamine), cycloaliphatic (e.g., isophorone diamine), aromatic diamines (e.g., m-xylene diamine or p-xylene diamine), and linear polyetheramines (e.g., 3,6-dioxaoctane-1,8-diamine, JEFFAMINEs from

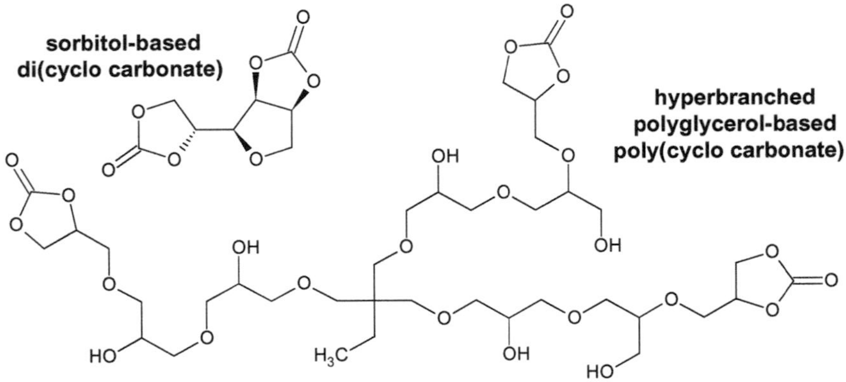

Figure 7.23 Cyclic carbonates obtained by transesterification of polyalcohols with using dimethyl carbonate.

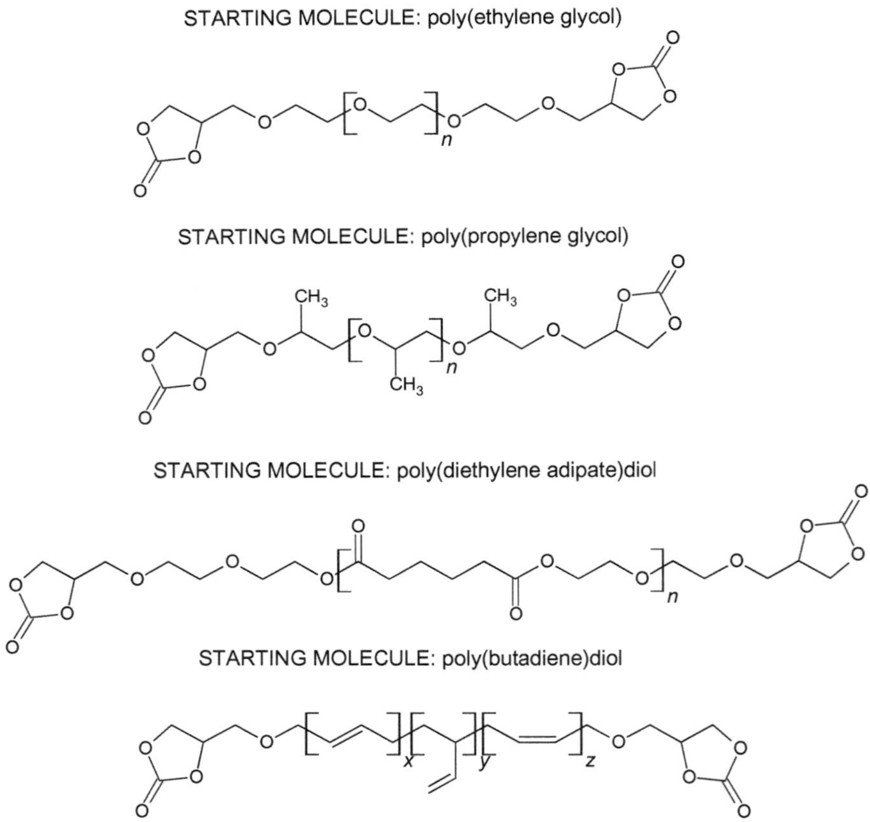

Figure 7.24 Cyclic carbonates obtained by tosylation of polymer diols and semi-products obtained by this way.

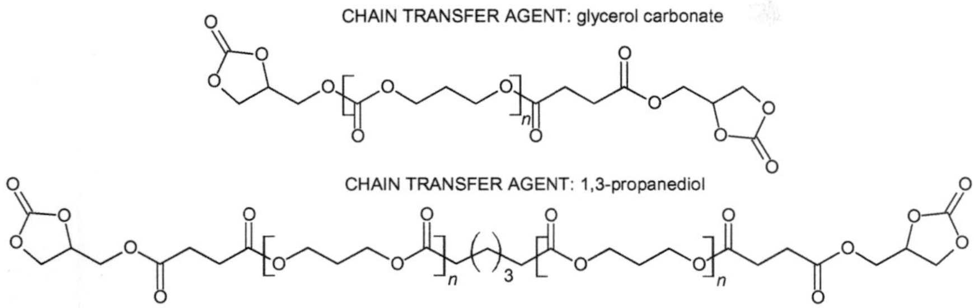

Figure 7.25 Di(cyclic carbonate) telechelic poly(trimethylene carbonate)s.

Table 7.4 Examples of unsaturated cyclic carbonate monomers used in radical polymerization method

Unsaturated cyclic carbonate monomer		**Refs.**
Glycerol carbonate methacrylate (GCM)	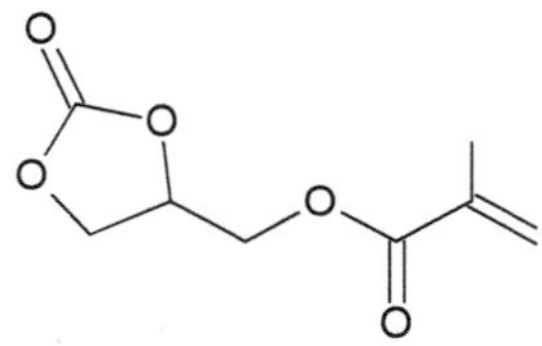	[74,75]
2-Methacrylic acid 2-(2-oxo-1,3-dioxolan-4-ylmethoxy carbonylamino)ethyl ester (CUMA)	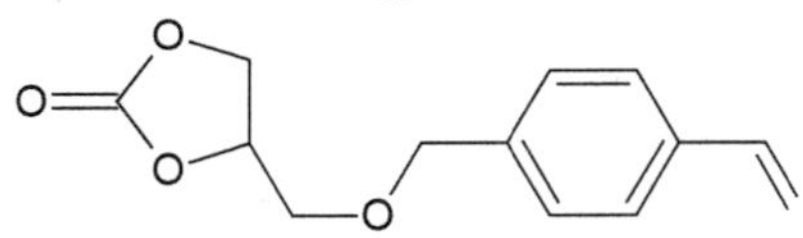	[75]
(2-oxo-1,3-dioxolan-4-yl) methyl vinyl ether (OVE)	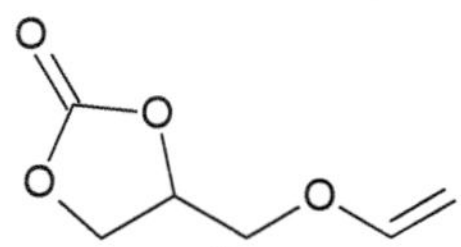	[76]
Norbornene-based cyclocarbonate (NCC)	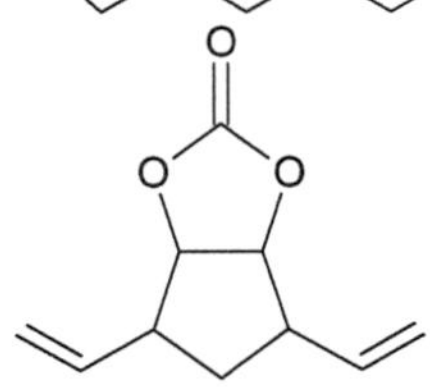	[77]

HUNTSMAN). The group of polyamines consist of: linear polyamines (e.g., diethylenetriamine, triethylenetetramine), branched polyamines (e.g., LUPASOL amines from BASF) and branched amino-amides (e.g., citric acid hexamethylene amino-amide). Some of the amines can be obtained from bio-based components e.g., lysine, lysinol or amine derivative of dimerized fatty acids (e.g., PRIAMINES from CRODA).

Table 7.5 Examples of unsaturated cyclic carbonate monomers and unsaturated comonomers used in radical copolymerization method

Unsaturated cyclic carbonate monomer	**Unsaturated comonomer**	**Refs.**
Vinyl ethylene carbonate (VEC) 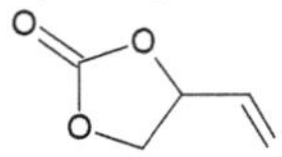	Vinyl acetate 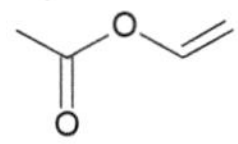	[79]
	Butyl acrylate 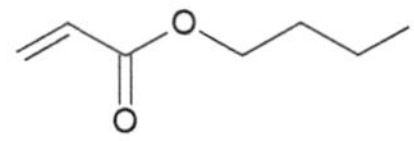	[79]
	Vinyl neodecanoate 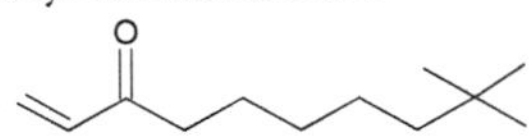	[80]
	Vinyl neononanoate 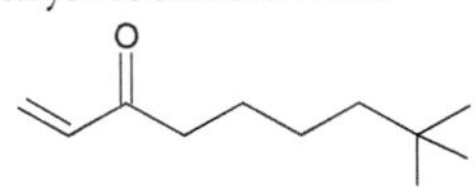	[80]
2-Methacrylic acid 2-(2-oxo- 1,3-dioxolan-4-ylmethoxy carbonylamino)ethyl ester (CUMA) 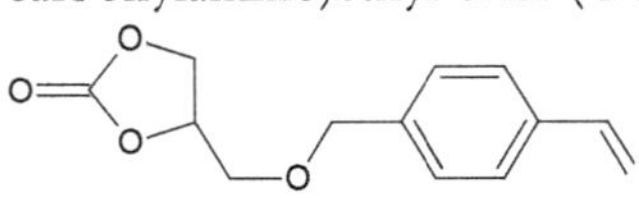	Butyl acrylate 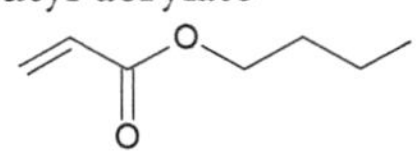	[75]
	Butyl methacrylate 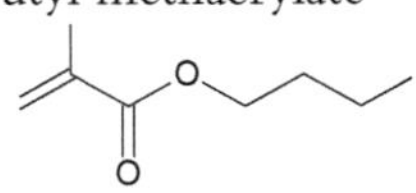	[75]
	Diethylene glycol methyl ether methacrylate 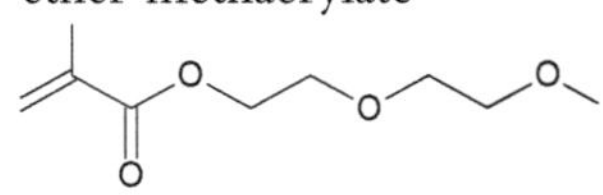	[75]
Norbornene-based cyclocarbonate (NCC) 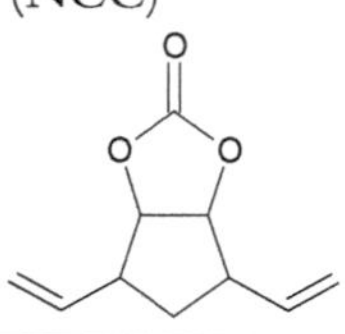	Cyclooctane 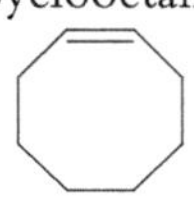	[77]

7.3.3 Structure and Properties Relationship in the Case of Nonisocyanate Polyurethanes by the Polyaddition Method

In the context of a structure–properties relationship, is known that properties of NIPUs are affected by structure and functionality of

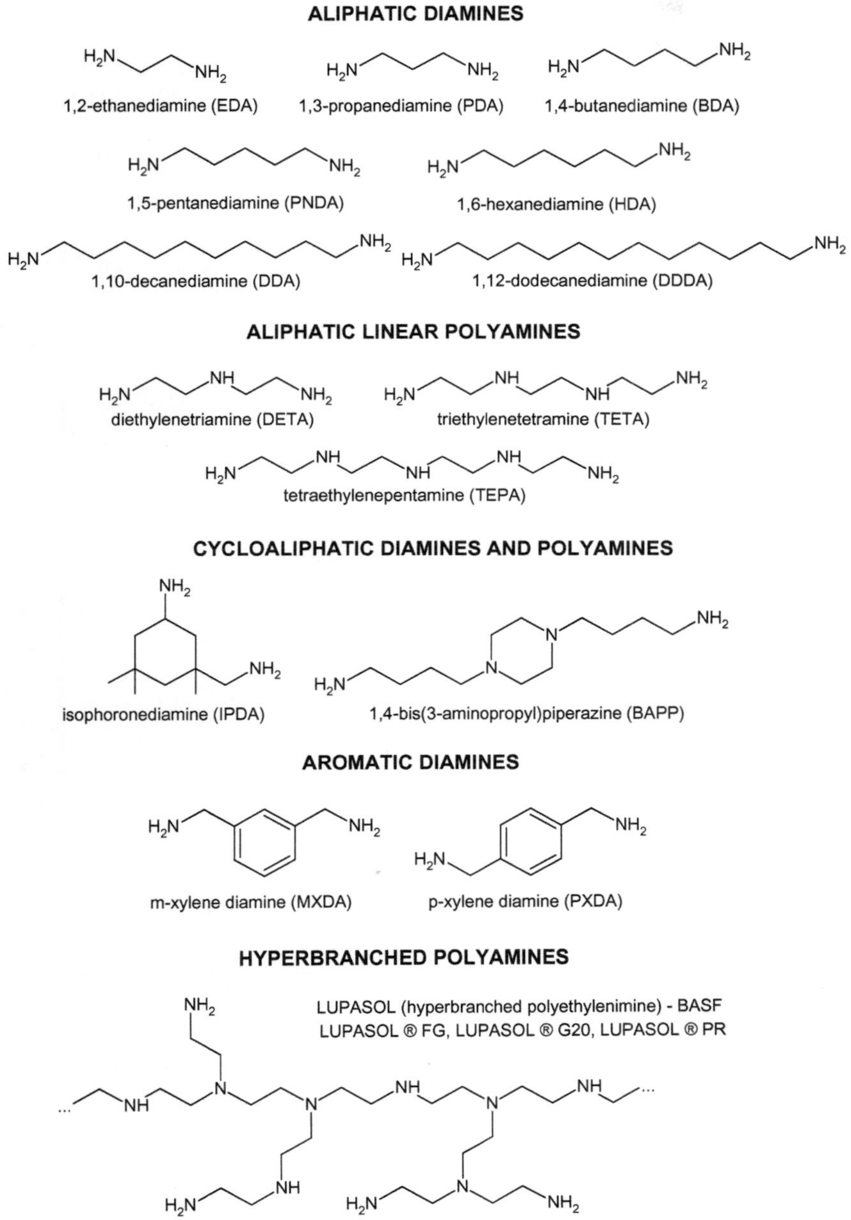

Figure 7.26 Diamines and polyamines used for the synthesis of NIPUs from cyclic carbonate intermediates.

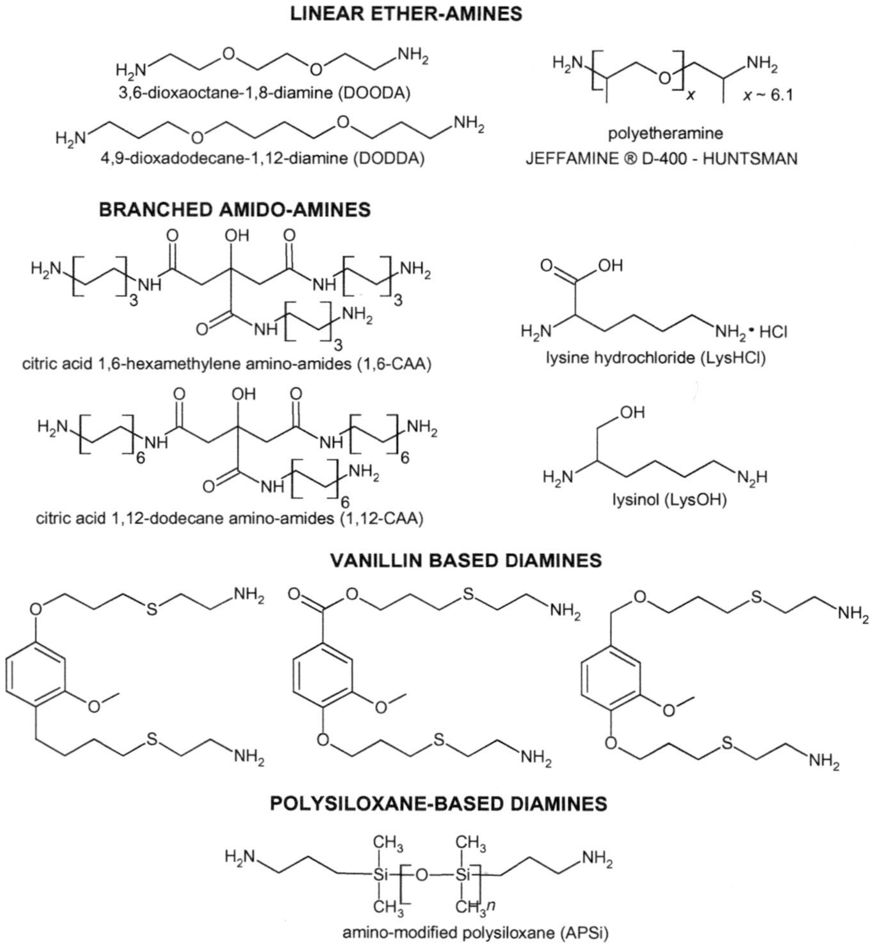

Figure 7.27 Amine curing agents with ether, amide, and polysiloxane moieties and bio-based amines from lysine and vanillin.

cyclic carbonate intermediates and diamines or polyamines. Conclusions from a literature review are presented in Table 7.6. A review of structure–properties relationship comparing mechanical properties (i.e., tensile properties, hardness), thermal properties (i.e., glass transition temperature by DSC method) and thermal stability (temperature at which 5% and 10% weight loss of sample and maximum degradation rate occurred (by TGA method)) is also presented in the literature [84].

Table 7.6 Structure–properties relationship in the case of nonisocyanate polyurethanes obtained by reacting cyclic carbonate intermediates and diamines or polyamines

Factor		Influence on properties
Structure of cyclic carbonate intermediates	Length of intermediate chains	Increasing of chain length (based on polysiloxane or poly(propylene glycol) results in decreasing of glass transition temperature (higher mobility of polymer chains) and increasing of materials flexibility
	Presence of cycloaliphatic and aromatic moieties	Increasing of glass transition temperature (resulted from rigid structure of intermediates)
Functionality of cyclic carbonate intermediates	Increasing of cyclic carbonate groups content (more than two 5CC groups)	Increasing of stiffness and glass transition temperature (caused by reduction of polymer chains mobility due to higher cross-link density)
Structure of amine curing agents	Length of aliphatic chain	Increasing of aliphatic chain of diamines result in decreasing of glass transition temperature (higher mobility of polymer chains) and increasing of thermal stability of final polymer
	Presence of ether bonds in structure of linear diamine	Increasing of thermal stability and decreasing of glass transition temperature (higher flexiblility of nipus structure)
	Presence of cycloaliphatic moieties—e.g., Introduced by IPDA	Decreasing of tensile strength and elongation at break, but increasing of tensile modulus (improvement of stiffness) and glass transition temperature (resulted from rigid structure of ipda) in comparison to linear aliphatic diamines
	Presence of aromatic moieties—e.g., Introduced by PXDA or MXDA	Increasing of glass transition temperature (resulted from rigid structure of PXDA or MXDA)

(Continued)

Table 7.6 (Continued)

Factor		Influence on properties
Functionality of amine curing agent	Increasing of amine groups content (more than two NH_2 groups)—e.g. Mixture of 1,6-HDA with triamines e.g., 1,6-CAA	Increasing of stiffness and glass transition temperature (caused by reduction of polymer chains mobility due to higher cross-link density)
Molar ratio of poly(cyclic carbonate)s to diamines	B5CC:$NH_2$1:0.5	Decreasing of tensile properties and glass transition temperature (incomplete cross-linking)
	B5CC:$NH_2$1:1	The highest tensile strength, hardness and glass transition temperatures (resulted ftom the highest cross-link density)
	B5CC:$NH_2$1:2	Occurring of side reactions e.g. Aminolysis of ester groups and formation of amide groups

7.4 SUMMARY

The first nonisocyanate polyurethane was obtained by polycondensation method approximately 30 years after Otto Bayer's discovery of isocyanate-based polyurethanes in 1930s. However, NIPUs are still not applied in industrial settings. NIPU can be obtained by polycondensation and poly-addition methods, but from an industrial point of view, only the methods based on polyaddition of cyclic carbonates and amines are valuable, resulting in high yields, absence of byproducts, purification of obtained polymer and relatively fast reaction at room temperature.

Yet, in the literature is a growing trend is observed of developing new NIPU, especially with using bio-based substances, related to green chemistry trends. From the literature, very efficient synthesis methods of cyclic carbonate intermediates by cycloaddition carbon dioxide to di- or poly-glycidyl ethers is known. This method is characterized by high yields (ca. 100%), can be realized without using solvent, and can be successful applied to chemical modification of commercially available products such as epoxy resins based on diglycidyl ether of bisphenol A or epoxidized vegetable oils. Obtained intermediates in the next step are reacted with diamines or polyamines to obtain nonisocyanate polyhydroxyurethanes. Other methods for synthesis of cyclic carbonate intermediates are also known (e.g., thiol-ene coupling of unsaturated cyclic carbonates with dithiols, tosylation of diols with using glycerol carbonate tosylate) but some additional investigations should be performed.

NIPU eliminate using toxic and moisture sensitive diisocyanates, which are produced using phosgene. Unfortunately the cyclic carbonate-amine reaction rate is quite slower than isocyanate-alcohol, and using medium temperature and a catalyst is required to obtain high conversion of cyclic carbonate groups to hydroxyurethane groups. Both reactions are exothermic and postcuring is required for completion of the reaction. The mechanical properties of NIPU are quite lower than isocyanate-based polyurethanes, so hybrid systems (especially with epoxy systems) have been developed.

Further studies in the area of NIPU should be focused on investigating the structure—properties relationship in the context of domain structure of polyurethanes (consisted of soft and hard segments), which has very high influence on properties of final polymer, directly connected with application of any developed polymers.

REFERENCES

[1] A. Prociak, G. Rokicki, J. Ryszkowska, Polyurethane Materials, PWN Press, Warsaw, Poland, 2014.

[2] Z. Wirpsza, Polyurethanes – Chemistry, Technology and Applications, WNT Press, Warsaw, Poland, 1993.

[3] H.W. Engels, H.G. Pirkl, R. Albers, R.W. Albach, J. Krause, A. Hoffmann, et al., Polyurethanes: versatile materials and sustainable problem solvers for today's challenges, Angew. Chem. Int. Ed. 52 (36) (2013) 9422–9441.

[4] J. Guan, Y. Song, Y. Lin, X. Yin, M. Zuo, Y. Zhao, et al., Progress in study of non-isocyanate polyurethane, Ind Eng Chem Res 50 (11) (2011) 6517–6527.

[5] G. Rokicki, P.G. Parzuchowski, M. Mazurek, Non-isocyanate polyurethanes: synthesis, properties, and applications, Polym. Adv. Technol. 26 (2015) 707–761.

[6] O.L. Figovsky, A. Leykin, L. Shapovalov, Non-isocyanate poyurethanes – yesterday, today and tomorrow. Int. Sci. J. Altern. Energy Ecol. 191–192:95–108.

[7] O.L. Figovsky, L. Shapovalov, O. Axenov, Advanced coatings based upon non-isocyanate polyurethanes for industrial applications, Surf. Coat. Int. B Coat. Trans. 87 (2004) 83–90.

[8] O.L. Figovsky, L. Shapovalov, F. Buslov, Ultraviolet and thermostable non-isocyanate polyurethane coatings, Surf. Coat. Int. B Coat. Trans. 88 (2005) 67–71.

[9] O. Figovsky, L. Shapovalov, Cyclocarbonate based polymers including non-isocyanate polyurethane adhesives and coatings, in: P. Somasundaran (Ed.), Encyclopedia of Surface and Colloid Science –, Volume 3, Taylor & Francis Group, CRC Press, New York, 2006, pp. 1633–1653.

[10] O. Figovsky, L. Shapovalov, Nonisocyanate polyurethanes for adhesives and coatings. In: First International IEEE Conference on Polymers and Adhesives in Microelectronics and Photonics. Proceedings, IEEE, Germany, pp 257–264; 2001.
[11] O. Figorsky, L. Shapovalov, Hybrid nonisocyanate polyurethane adhesives. In: International Conference on Polymer Bonding 2004. The Bonding of Rubber and Plastics to Various Substrates, Rapra Technology Ltd., pp 99–103; 2004.
[12] K. Hanada, K. Kimura, K. Takahashi, O. Kawakami, M. Uruno, Polysiloxane-modified polyhydroxy polyurethane resin, method for producing same, heat-sensitive recording material using the resin, imitation leather, thermoplastic polyolefin resin skin material, material for weather strip, and weather strip. US Patent 20120231184 A1; 2012.
[13] K. Hanada, K. Kimura, K. Takahashi, O. Kawakami, M. Uruno, Polysiloxane-modified polyhydroxy polyurethane resin, method for producing same, heat-sensitive recording material using the resin, imitation leather, thermoplastic polyolefin resin skin material, material for weather strip, and weather strip. US Patent 20120237701 A1; 2012.
[14] A. Cornille, S. Dworakowska, D. Bogdal, B. Boutevin, S. Caillol, A new way of creating cellular polyurethane materials: NIPU foams, Eur. Polym. J. 66 (2015) 129–138.
[15] O. Figovsky, L. Shapovalov, A. Leykin, O. Birukova, R. Potashnikova, Advances in the field of nonisocyanate polyurethanes based on cyclic carbonates, Chem. Chem. Technol. 7 (2013) 79–87.
[16] G. Rokicki, C. Wojciechowski, Epoxy resin modified by aliphatic cyclic carbonates, J. Appl. Polym. Sci. 41 (1990) 647–659.
[17] M. Kathalewar, A. Sabnis, G. Waghoo, Effect of incorporation of surface treated zinc oxide on non-isocyanate polyurethane based nano-composite coatings, Prog. Org. Coat. 76 (2013) 1215–1229.
[18] Z. Hoşgör, N. Kayaman-Apohan, S. Karataş, Y. Menceloğlu, A. Güngör, Preparation and characterization of phosphine oxide based polyurethane/silica nanocomposite via non-isocyanate route, Prog. Org. Coat. 69 (4) (2010) 366–375.
[19] K. Matsumoto, A. Kokai, T. Endo, Synthesis and properties of novel poly(hydroxyurethane) from difunctional alicyclic carbonate and *m*-xylylenediamine and its possibility as gas barrier materials, Polym. Bull. 73 (3) (2016) 677–686.
[20] D. Miloslavskiy, E. Gotlib, O. Figovsky, D. Pashin, Cyclic carbonates based on vegetable oils, Int. Lett. Chem. Phys. Astron. 8 (2014) 20–29.
[21] M. Fleischer, H. Blattmann, R. Mülhaupt, Glycerol-, pentaerythritol- and trimethylolpropane-based polyurethanes and their cellulose carbonate composites prepared via the non-isocyanate route with catalytic carbon dioxide fixation, Green Chem. 15 (2013) 934–942.
[22] M. Bähr, A. Bitto, R. Mülhaupt, Cyclic limonene dicarbonate as a new monomer for non-isocyanate oligo- and polyurethanes (NIPU) based upon terpenes, Green Chem. 14 (2012) 1447–1454.
[23] M. Fache, E. Darroman, V. Besse, R. Auvergne, S. Caillol, B. Boutevin, Vanillin, a promising biobased building-block for monomer synthesis, Green Chem. 16 (2014) 1987–1998.
[24] V. Besse, R. Auvergne, S. Carlotti, G. Boutevin, B. Otazaghine, S. Caillol, et al., Synthesis of isosorbide based polyurethanes: an isocyanate free method, React. Funct. Polym. 73 (2013) 588–594.
[25] M. Kathalewar, A. Sabnis, D. D'Mello, Isocyanate free polyurethanes from new CNSL based bis-cyclic carbonate and its application in coatings, Eur. Polym. J. 57 (2014) 99–108.

[26] T. Calvo-Correas, A. Santamaria-Echart, A. Saralegi, L. Martin, A. Valea, M.A. Corcuera, et al., Thermally-responsive biopolyurethanes from a biobased diisocyanate, Eur. Polym. J. 70 (2015) 173–185.
[27] L. Hojabri, X. Kong, S.S. Narine, Fatty acid-derived diisocyanate and biobased polyurethane produced from vegetable oil: synthesis, polymerization, and characterization, Biomacromolecules 10 (4) (2009) 884–891.
[28] S. Caillol, M. Desroches, G. Boutevin, C. Loubat, R. Auvergne, B. Boutevin, Synthesis of new polyester polyols from epoxidized vegetable oils and biobased acids, Eur. J. Lipid Sci. Technol. 114 (2012) 1447–1459.
[29] A. Campanella, L.M. Bonnaillie, R.P. Wool, Polyurethane foams from soyoil-based polyols, J. Appl. Polym. Sci. 112 (2009) 2567–2578.
[30] P. Rojek, A. Prociak, Effect of different rapeseed-oil-based polyols on mechanical properties of flexible polyurethane foams, J. Appl. Polym. Sci. 125 (2012) 2936–2945.
[31] J. Datta, E. Głowińska, Effect of hydroxylated soybean oil and bio-based propanediol on the structure and thermal properties of synthesized bio-polyurethanes, Ind. Crops Prod. 61 (2014) 84–91.
[32] X. Luo, S. Hu, X. Zhang, Y. Li, Thermochemical conversion of crude glycerol to biopolyols for the production of polyurethane foams, Bioresour. Technol. 139 (2013) 323–329.
[33] P.W. Morgan, Low-temperature solution polycondensation, J. Polym. Sci. C Polym. Symp. 4 (1963) 1075–1096.
[34] E. Dyer, H. Scott, The preparation of polymeric and cyclic urethans and ureas from ethylene carbonate and amines, J. Am. Chem. Soc. 79 (1957) 672–675.
[35] B. Ochiai, T. Utsuno, Non-isocyanate synthesis and application of telechelic polyurethanes via polycondensation of diurethanes obtained from ethylene carbonate and diamines, J. Polym. Sci. A Polym. Chem. 51 (2013) 525–533.
[36] G. Rokicki, A. Piotrowska, A new route to polyurethanes from ethylene carbonate, diamines and diols, Polymer 43 (2002) 2927–2935.
[37] M. Unverferth, O. Kreye, A. Prohammer, M.A.R. Meier, Renewable non-isocyanate based thermoplastic polyurethanes via polycondensation of dimethyl carbamate monomers with diols, Macromol. Rapid Commun. 34 (2013) 1569–1574.
[38] L. Ubaghs, N. Fricke, H. Keul, H. Hocker, Polyurethanes with pendant hydroxyl groups: synthesis and characterization, Macromol. Rapid Commun. 25 (2004) 517–521.
[39] J.N. Argyropoulos, D. Bhattacharjee, P. Foley, G.E. Spilman, H. Zhang, Ambient temperature curable isocyanate-free compositions for preparing crosslinked polyurethanes, US Patent 2011/0313091 A1; 2011.
[40] H. Tomita, F. Sanda, T. Endo, Model reaction for the synthesis of polyhydroxyurethanes from cyclic carbonates with amines: Substituent effect on the reactivity and selectivity of ring-opening direction in the reaction of five-membered cyclic carbonates with amine, J. Polym. Sci. A Polym. Chem. 39 (2001) 3678–3685.
[41] H. Tomita, F. Sanda, T. Endo, Reactivity comparison of five- and six-membered cyclic carbonates with amines: basic evaluation for synthesis of poly(hydroxyurethane), J. Polym. Sci. A Polym. Chem. 39 (2001) 162–168.
[42] H. Tomita, F. Sanda, T. Endo, Polyaddition behavior of bis(five- and six-membered cyclic carbonate)s with diamine, J. Polym. Sci. A Polym. Chem. 39 (2001) 860–867.
[43] L. Maisonneuve, A.L. Wirotius, C. Alfos, E. Grauab, H. Cramail, Fatty acid-based (bis) 6-membered cyclic carbonates as efficient isocyanate free poly(hydroxyurethane) precursors, Polym. Chem. 5 (2014) 6142–6147.

[44] H. Tomita, F. Sanda, T. Endo, Polyaddition of bis(seven-membered cyclic carbonate) with diamines: a novel and efficient synthetic method for polyhydroxyurethanes, J. Polym. Sci. A Polym. Chem. 39 (2001) 4091–4410.
[45] F. Camara, S. Benyahya, V. Besse, G. Boutevin, R. Auvergne, B. Boutevin, et al., Reactivity of secondary amines for the synthesis of non-isocyanate polyurethanes, Eur. Polym. J. 55 (2014) 17–26.
[46] M. Blain, L. Jean-Gérard, R. Auvergne, D. Benazet, S. Caillol, B. Andrioletti, Rational investigations in the ring opening of cyclic carbonates by amines, Green Chem. 16 (2014) 4286–4291.
[47] R.H. Lambeth, T.J. Henderson, Organocatalytic synthesis of (poly)hydroxyurethanes from cyclic carbonates and amines, Polymer 54 (2013) 5568–5573.
[48] J.H. Clements, Reactive applications of cyclic alkylene carbonates, Ind. Eng. Chem. Res. 42 (2003) 663–674.
[49] V. Besse, F. Camara, F. Méchin, E. Fleury, S. Caillol, J.P. Pascault, et al., How to explain low molar masses in polyhydroxyurethanes (PHUs), Eur. Polym. J. 71 (2015) 1–11.
[50] L. Annunziata, A.K. Diallo, S. Fouquay, G. Michaud, F. Simon, J.M. Brusson, et al., α,ω-Di(glycerol carbonate) telechelic polyesters and polyolefins as precursors to polyhydroxyurethanes: an isocyanate-free approach, Green Chem. 16 (4) (2014) 1947–1956.
[51] B. Ochiai, S. Inoue, T. Endo, One-pot non-isocyanate synthesis of polyurethanes from bisepoxide, carbon dioxide, and diamine, J. Polym. Sci. A Polym. Chem. 43 (2005) 6613–6618.
[52] X. Sheng, G. Ren, Y. Qin, X. Chen, X. Wang, F. Wang, Quantitative synthesis of bis(cyclic carbonate)s by iron catalyst for non-isocyanate polyurethane synthesis, Green Chem. 17 (2015) 373–379.
[53] O. Figovsky, L. Shapovalov, Preparation of oligomeric cyclocarbonates and their use in nonisocyanate or hybrid nonisocyanate polyurethanes, US Patent 7232877 B2; 2007.
[54] O. Figovsky, L. Shapovalov, N. Blank, F. Buslov, Cyclocarbonate groups containing hydroxyamine oligomers from epoxycylclocarbonates. US Patent 6407198 B1; 2002.
[55] L. Rappoport, Water-compatible urethane-containing amine hardener. US Patent 6218480 B1; 2001.
[56] S.L. Herlihy, B. Rowatt, R.S. Davidson, Carbonate containing energy-curable compositions. US Patent 20080286486 A1; 2008.
[57] I. Javni, D.P. Hong, Z.S. Petrović, Soy-based polyurethanes by nonisocyanate route, J. Appl. Polym. Sci. 108 (2008) 3867–3875.
[58] I. Javni, D.P. Hong, Z.S. Petrović, Polyurethanes from soybean oil, aromatic, and cycloaliphatic diamines by nonisocyanate route, J. Appl. Polym. Sci. 12 (2013) 8566–8571.
[59] K.M. Doll, S.Z. Erhan, The improved synthesis of carbonated soybean oil using supercritical carbon dioxide at a reduced reaction time, Green Chem. 7 (2005) 849–854.
[60] Z. Li, Y. Zhao, S. Yan, X. Wang, M. Kang, J. Wang, et al., Catalytic synthesis of carbonated soybean oil, Catal. Lett. 123 (2008) 246–251.
[61] M. Bähr, R. Mülhaupt, Linseed and soybean oil-based polyurethanes prepared via the non-isocyanate route and catalytic carbon dioxide conversion, Green Chem. 14 (2012) 483–489.
[62] A.R. Mahendran, N. Aust, G. Wuzella, U. Müller, A. Kandelbauer, Bio-based non-isocynate urethane derived from plant oil, J. Polym. Environ. 20 (2012) 926–931.
[63] L. Zhang, Y. Luo, Z. Hou, Z. He, W. Eli, Synthesis of carbonated cotton seed oil and its application as lubricating base oil, J. Am. Oil Chem. Soc. 91 (2014) 143–150.

[64] N. Mann, S.K. Mendon, J.W. Rawlins, F. Thames, Synthesis of carbonated vernonia oil, J. Am. Oil Chem. Soc. 85 (2008) 791–796.
[65] J. Datta, M. Włoch, Selected biotrends in development of epoxy resins and their composites, Polym. Bull. 71 (2014) 3035–3049.
[66] A. Boyer, E. Cloutet, T. Tassaing, B. Gadenne, C. Alfos, H. Cramail, Solubility in CO_2 and carbonation studies of epoxidized fatty acid diesters: towards novel precursors for polyurethane synthesis, Green Chem. 12 (2010) 2205–2213.
[67] L. Maisonneuve, A.S. More, S. Flotran, C. Alfos, F. Robert, Y. Landais, et al., Novel green fatty acid-based bis-cyclic carbonates for the synthesis of isocyanate-free poly (hydroxyurethane amide)s, RSC Adv. 4 (2014) 25795–25803.
[68] C.N. Tang, H.B. Nulwala, K. Damodaran, P. Kaur, D.R. Luebke, Tunable poly (hydroxyl urethane) from CO_2-Based intermediates using thiol-ene chemistry, J. Polym. Sci. A Polym. Chem. 49 (2011) 2024–2032.
[69] S. Benyahya, M. Desroches, R. Auvergne, S. Carlotti, S. Caillol, B. Boutevin, Synthesis of glycerin carbonate-based intermediates using thiol-ene chemistry and isocyanate free polyhydroxyurethanes therefrom, Polym. Chem. 2 (2011) 2661–2667.
[70] S. Benyahya, J.P. Habas, R. Auvergne, V. Lapinte, S. Caillol, Structure-property relationships in polyhydroxyurethanes produced from terephthaloyl dicyclocarbonate with various polyamines, Polym. Int. 61 (2012) 1666–1674.
[71] K.M. Tomczyk, P.A. Guñka, P.G. Parzuchowski, J. Zachara, G. Rokicki, Intramolecular etherification of five-membered cyclic carbonates bearing hydroxyalkyl groups, Green Chem. 14 (2012) 1749–1758.
[72] P.G. Parzuchowski, M. Kiźlińska, G. Rokicki, New hyperbranched polyether containing cyclic carbonate groups as a toughening agent for epoxy resin, Polymer 48 (2007) 1857–1865.
[73] M. Helou, J.F. Carpentier, S.M. Guillaume, Poly(carbonate-urethane): an isocyanate-free procedure from α,ω-di(cyclic carbonate) telechelic poly(trimethylene carbonate)s, Green Chem. 13 (2) (2011) 266–271.
[74] S. Jana, H. Yu, A. Parthiban, C.L.L. Chai, Controlled synthesis and functionalization of PEGylated methacrylates bearing cyclic carbonate pendant groups, J. Polym. Sci. A Polym. Chem. 48 (2010) 1622–1632.
[75] F. Camara, S. Caillol, B. Boutevin, Free radical polymerization study of glycerin carbonate methacrylate for the synthesis of cyclic carbonate functionalized polymers, Eur. Polym. J. 61 (2014) 133–144.
[76] T. Nishikubo, A. Kameyama, M. Sasano, Synthesis of functional polymers bearing cyclic carbonate groups from (2-oxo-1,3-dioxolan-4-yl) methyl vinyl ether, J. Polym. Sci. A Polym. Chem. 32 (1994) 301–308.
[77] T. Hino, N. Inoue, T. Endo, Ring-opening metathesis copolymerization behaviors of cyclooctene and norbornene bearing a five- or six-membered ring cyclic carbonate, J. Polym. Sci. A Polym. Chem. 43 (2005) 6599–6604.
[78] B. Ochiai, Y. Ootani, T. Maruyama, T. Endo, Synthesis and properties of polymethacrylate bearing cyclic carbonate through urethane linkage, J. Polym. Sci. A Polym. Chem. 45 (2007) 5781–5789.
[79] D. Webster, A. Crain, Synthesis of cyclic carbonate functional polymers, Functional Polymer, Modern Synthetic Methods and Novel Structures. ACS Symposium Series, Vol. 704, American Chemical Society, 1998, pp. 303–320.
[80] D.C. Webster, A.L. Crain, Synthesis and applications of cyclic carbonate functional polymers in thermosetting coatings, Prog. Org. Coat. 40 (2000) 275–282.
[81] T. Miyata, K. Matsumoto, T. Endo, S. Yonemori, S. Watanabe, Synthesis and radical polymerization of styrene-based monomer having a five-membered cyclic carbonate structure, J. Polym. Sci. A Polym. Chem. 50 (2012) 3046–3051.

[82] A.L. Brocas, G. Cendejas, S. Caillol, A. Deffieux, S. Carlotti, Controlled synthesis of polyepichlorohydrin with pendant cyclic carbonate functions for isocyanate-free polyurethane networks, J. Polym. Sci. A Polym. Chem. 49 (2011) 2677–2684.
[83] D.W. Park, K.H. Kim, S.Y. Park, C.K. Sung, S.W. Park, Direct incorporation of carbon dioxide in poly(glycidyl methacrylate) and its application to the synthesis of hydrogel, React. Kinet. Catal. Lett. 75 (2002) 375.
[84] J. Datta, M. Włoch, Progress in non-isocyanate polyurethanes synthesized from cyclic carbonate intermediates and di- or polyamines in the context of structure–properties relationship and from an environmental point of view, Polym. Bull. 73 (5) (2016) 1459–1496.

CHAPTER 8

Conducting Polyurethane Blends: Recent Advances and Perspectives

Raghvendra K. Mishra[1,2], Jiji Abraham[1], Nandakumar Kalarikkal[1], Karingamanna Jayanarayanan[3], Kuruvilla Joseph[2] and Sabu Thomas[2]
[1]Mahatma Gandhi University, Kottayam, Kerala, India
[2]Indian Institute of Space Sience and Technology, Thiruvananthapuram, Kerala, India
[3]Amrita University, Coimbatore, Tamil Nadu, India

Contents

8.1 INTRODUCTION

8.1.1 Polyurethane Blends

A polymer blend is a mixture of two or more polymers in suitable amounts with only secondary interactions between them. Blending of polymers is an excellent method for improving the different properties [1]. A composite material can be defined as macroscopic

Polyurethane Polymers: Blends and Interpenetrating Polymer Networks
DOI: http://dx.doi.org/10.1016/B978-0-12-804039-3.00008-7

combination of two or more materials (reinforcing elements, fillers, and composite matrix binder which join together by basic adhesion theory) which differ in form or composition on a macro scale. The constituents in composites retain their identities; these can be physically identified and exhibit an interface between one another. A composite material is created from a powder (or reinforcement) and an appropriate compatibilizers and their use has increased over the past 25 years. In the past 15 years, research has focused on increasing the commercialization of polymer blends systems [2] for two purposes: to reduce the domain size of the different polymers in the blend and thus improve its morphology; and to enhance the adhesion between the domain boundaries by providing chemical bonding across them [3]. To obtain the miscibility and compatible of polymer blends Gibb's free energy plays a significant role. For a miscible polymer blend, Gibb's free energy should be negative and mixing should be exothermic, leading to free energy of mixing only being negative if the enthalpy of mixing (ΔH_{mix}) is negative [4], which is not normally the case if there is no attraction between the polymers in the blend. As an example, in hydrogen bonding or dipole–dipole bonding, the entropy (ΔS_{mix}) refers to negligible value, and the enthalpy of mixing (ΔH_{mix}) is positive which lead to free energy of mixing being positive [4]. There are three different types of blends: miscible (exhibit one glass transition temperature, T_g), partially miscible (one parts of the two polymers are dissolved in each other and the blend exhibits two T_g that are shifted towards each other) and immiscible blends (the interphase region is limited between the blended polymers and coarse phase morphology), this type of blends exhibits two T_g values [4].

$$\Delta G = \Delta H_{\text{mix}} - T\Delta S_{\text{mix}} \tag{8.1}$$

8.1.2 Examples of Polyurethane Blends

The polyurethane prepolymer was produced by a reaction of polyester polyol and 2,4-toluene diisocyanate and then end-capped with phenol. Soluble polyamide was prepared by two-step synthesis from 2,2′-bis (3,4-dicarboxyphenyl) hexafluoro propane dianhydride and 3,3′ diamino-4, 4′-dihydroxy biphenyl. Cast films were obtained by mixing solutions of PU and PI. Dynamic analysis showed the polyamide content increases the T_g of the films when it shifts to high temperatures [5]. Thermal decomposition and combustion reaction of polyether–polyurethane and polyester–polyurethane in air and nitrogen atmosphere were investigated by thermogravimetry analysis to measure the thermal decomposition mechanism. Thermo grams of PU from differential scanning calorimetry (DSC) exhibits two glass transition temperatures corresponding to the hard and soft segments of the polyurethane. The thermal stability of polyurethane is dependent on the molecular weight of polyurethane. Polyester–polyurethane was found to be more thermally stable than polyether–polyurethane [6]. In another study, researchers synthesized biodegradable elastomeric epoxy modified polyurethanes based on caprolectone and poly (ethylene glycol). 1,6-hexamethylene diisocyanate was used for the preparation of isocyanate terminated PU, which were subsequently blocked with glycidol to prepare epoxy polyurethane. Optimum degradation rates and mechanical properties were obtained by this method [7].

An interpenetrating polymer network (IPN) is defined as a combination of two polymer networks where at least one polymer is synthesized or crosslinked in the presence of other and has a network structure. IPN is another way to combine two different polymers with improved mechanical properties and crosslink density. The IPN is an important parameter for adhesive strength [8,9]. This property enhancement is attributed to increased crosslinked density or due to permanent physical change entanglement in the IPNs [10]. The IPN consisting of 2-hydroxy ethyl methacrylate terminated polyurethanes (HPU) and PUs with different ether type polyols have been studied and it was found that the compatibility of polymers in IPN formation depends on molecular weight of prepolymers. The hard to soft segment ratio of materials has a profound effect on the water absorption of the IPNs. The maximum tensile strength occurred at 25 wt% HPU content for all the IPN systems [11]. The excellent damping properties of epoxy resin/polyurethane (EP/PU) semi-interpenetrating polymer networks (sIPNs) have been studied using DMA. The researchers found that the same composition (70/30) shows better tensile strength, elongation at break, and cavitation resistance [12].

The effect of the isocyanate/glycol ratio, glycol type, and glycol mixture on impact strength, dimensional stability, and thermal properties have examined by using different molecular weight polyethylene glycol and polypropylene glycol and it was found that impact strength increased with an increase in NCO/OH ratio [13].

Fibers of polyurethane can be prepared through wet and dry spinning. Fibers have moderate water absorption, low shrinkage in hot water, modest lowering of tensile properties under wet conditions, high tensile and work recoveries, and good stability to ultraviolet light which are suited to clothing textile uses. Polyurethanes from piperazines and 2,2-bis (4-hydroxy-3,5-dichlorophenyl) propane bischloroformate had unusually high thermal transitions and good solubility combined with ease of orientation and crystallization in fiber form. The fibers from the 2-methylpiperazine and homo piperazine urethanes can be converted to fabrics and these fabrics exhibit a silk like handle; good wash wrinkle recovery, resistance to soil and bleach, stability to ironing at 200°C and very low flammability [14].

Epoxy resin/polyurethane hybrid networks can be synthesized by frontal polymerization (FP) in which the polymerization take place through the reaction vessel. Frontal polymerization is initiated when a heat source contacts a solution of monomer and thermal initiator at one end of the tubular reactor. Once initiated, no further energy was required for polymerization to occur. Finally samples were characterized with a Fourier transform infrared spectrometer, thermogravimetric analysis, and a scanning electron microscope. EP/PU hybrid networks with same properties can also be synthesized by batch polymerization, but the FP method requires significantly less time and lower energy input [15]. Epoxy resin modified using a polyurethane prepolymer based on hydroxyl-terminated polyester has been reported. The researchers found that an isocyanate-terminated polyurethane reacted with epoxy resin led to significant improvement in fracture toughness and the use of a chain extender with the polyurethane prepolymer caused a seven-fold increase in the impact strength of these systems [16]. In another study, epoxy resin modified using phenolic hydroxyl-terminated (HTPU) and aromatic amine terminated (ATPU) polyurethanes was investigated and the results showed that epoxy modified with HTPU possesses improved fracture toughness than the ATPU [17]. Epoxy resin/polyurethane networks by a FP using polypropylene glycol, toluene diisocyanate, and 1,4-butanediol with diglycidyl ether of bisphenol A (DGEBA) and stannous caprylate (as a catalyst) has been reported. The properties of the blends were same as those synthesized by batch polymerization, however less time and lower energy input was required for this type of polymerization process [15]. The effect of polyurethane on mechanical properties are demonstrated in which an epoxy resin was modified with aliphatic polyurethane (PUR) synthesized from poly (ethylene glycol) and 4, 4′-diisocyanato dicyclohexyl methane (HMDI) in the absence of a solvent. This amine-cured epoxy composition containing 5 phr PUR exhibited enhanced thermal stability, flexural strength, storage modulus and adhesion strength [18]. In another study, the glass transition temperature, contact angle and mechanical properties of epoxy/polyurethane (EP/PUR) systems were evaluated. The results show that the interfacial and mechanical properties reached to a maximum values at 40 phr PUR as result of hydrogen bonding between hydroxyl groups of EP and isocyanate groups of PUR [19]. A series of semi-interpenetrating polymer networks based on polyurethane and epoxy resin were obtainable and characterized by various techniques. From the damping characterization it is found that heat treatment conditions significantly alter the viscoelastic properties of the blend [20]. An improvement in fracture energy was noticed when epoxy was modified using polyurethane prepolymers synthesized from polyether diol and MDI using different types of coupling agents [21]. Organic polyphosphazenes are used in polymer blends and IPNs to impart flame retardancy [22] due to the alternate arrangement of phosphorous and nitrogen moiety in the backbone. Even though a large variety of polyphosphazenes are reported with different side groups [23–25] their use is only limited to military applications due to their high cost.

Synthesis of hydroxylated polyester (HP) based polyurethane polyols containing internal carboxyl group with different diisocyanates were reported [26]. These polyurethane polyols were partly acetoacetylated with ethyl acetoacetate to incorporate ß-ketoester in the polyurethane polyol backbone and the synthesized polyols were characterized by Fourier transform infrared spectroscopy, nuclear magnetic resonance and DSC. Polyurethane polyols and their acetoacetylated cousins were used to develop PUDs, the PUDs were crosslinked with hexamethoxy methyl melamine and their film properties were studied by dynamic mechanical and thermal analyzers and thermogravimetric analysis.

The moisture permeable segmented polyurethane films with different content of polytetramethyleneglycol (PTMG) and polyethylene glycol (PEG) as soft segment were synthesized [27]. Recently, polymer fibers with submicron diameter prepared by electrospinning has received special interest [28,29]. In this technique, an electric field is induced by a high voltage power supply, which is then applied to a polymer solution or melt placed in a container that has a millimeter size nozzle, causing it to be ejected from the capillary tip of the nozzle in the form of a liquid jet. Electrospinning has the advantage of allowing for the simple processing and easy preparation of nanofibers, compared with previous spinning techniques [30]. Moreover, electrospun fibers have a high surface area and porosity, which is suitable for variety of applications such as membranes, filters, artificial blood vessels, reinforced composite materials, and drug delivery systems [31,32]. For example, electrospun PSF nonwovens are suitable for use in hemodialysis membranes [32,33]. However, electrospun nonwovens have low mechanical properties, due to the orientation within the fibers as there is no drawing involved in their processing [34,35]. The research conducted so far on the electrospinning process can be classified into various categories, including studies of the processing parameters, the spinning conditions, the morphology of the electrospun fibers, and the instability zone between the capillary tip and the collector [30–36].

8.2 POLYURETHANE CONDUCTING POLYMERS

The intrinsically conducting polymers (ICP) are special class of polymer because of their unique electrical, optical, and chemical properties. Conductivity of these materials can be varied from semiconducting to metallic range by doping. Potential applications for conducting polymers include diodes, battery anodes or cathodes, semiconductors, energy storage and conversion devices, electroluminescence devices, nonlinear optical materials, EMI shielding, radar absorbers oscillators, amplifiers, frequency converters, and sensors [37–45]. Conducting polymers are superior to traditional dielectric materials because of their interesting properties in electrical and microwave frequencies. Nonbiological materials exhibiting the dielectric properties of biological tissue at microwave frequencies have been used extensively to evaluate hyperthermia applicators, assess microwave imaging systems, determine electromagnetic absorption patterns and as phantoms. Recently conducting polymers are being used as doping in insulating polymers to improve their electrical properties. In order to increase the performance and conductivity of polyurethane polymer, various conducting polymers include poypyrrole, polythiopene, polyaniline, polyactelene, and polypheylenevinylene have been used for blending with polyurethane.

Conducting polymers have received a lot of attention since the discovery of polyacetylene in the 1970s by Hideki Shirakawa, Alan Heeger, and Alan MacDiarmid. In 2000, these gentlemen were awarded the Nobel Prize in Chemistry for their discovery of electrically conducting polymers. The process of making a polymer to be conductive is known as 'doping'. For conducting polymers, the doping process increases the conductivity of the polymer from 10^{-10} to 10^{3} S cm^{-1} and oxidation of the polymer leads to p-type doping, while the use of a reducing agent leads

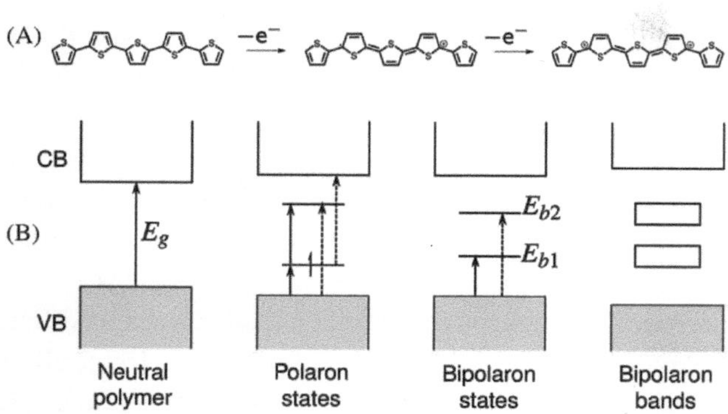

Figure 8.1 The natural polymer, band gap of bipolarons, and polarons, A) oxidation state of polymer, B) band gap with respect to the oxidation state (doping) [47].

to n-type doping [46]. For most of the conducting polymers such as polypyrrole, polythiophene, and polyaniline, the doping causes the energy distortion of the polymer energy band gap by the formation of polarons and bipolarons as shown in Fig. 8.1 [47], where undoped polymer is designated as natural polymer, and first oxidation (slightly doped), second oxidation and further oxidation state of conducting polymer is defined as polarons, bipolarons, and bipolarons band, respectively.

The total conductivity of a conducting polymer is a sum of its intra chain (intramolecular, includes the number of defect sites and the extent of conjugation) and inter chain (intermolecular, depends on conjugated crosslinks between polymer chains and the degree of oriented micro/macro crystal domains) conductivity. Most conducting polymers are rigid due to the strong aromatic interaction between the polymer chains. Several efforts have been made to increase the processability of these polymers including using special solvents, soft templates, self-doping and the use of solubilizing groups [48]. Conducting polymers have a vast range of applications such as batteries, supercapacitors, light emitting diodes (LEDs), sensors, transistors, photovoltaics, and electromagnetic interference shielding materials [48].

8.2.1 Poyurethene/Polyaniline Blends

Polyaniline is a unique polymer and since its discovery, numerous studies of it and its blends have been carried out, starting in the early 1980s. Advantages over its competitors, additional to its conductivity include factors such as a lower cost of aniline and easy synthesis, carried out through either chemical oxidation, aqueous or organic solvent solution processes. Polyaniline has a greater air stability compared to the much more conducting but air sensitive polyacetylene [49]. The doping process of polyaniline is quite simple and unique because the doping/dedoping of polyaniline is based on the acid/base reaction whereas other conducting polymers follow the redox reaction. Doped polyaniline with different acids can attain conductivity more than 100 S cm^{-1} while the conductivity of dedoped neutral form drops in the range of insulators.

A polymer and conventional polymer blended with PANI can exhibit good mechanical properties and interesting electrical properties. However, formation of highly heterogeneous two-phase (immiscible) morphology limits the application of these blends. The grafting of PANI segments with blocks of another polymer has been recently reported, in which polyethylene glycol or polyacrylic acid is grafted onto a PANI backbone. The use of a thermoplastic elastomer and conducting polymer blends is more attractive due to the excellent mechanical and processability properties without vulcanization [50]. Moreover, the interaction between NH in PANI and NHCOO in PU may improve the compatibility of the system to obtain the desired property [51]. The PANI/PU blend also refers to the phase mixing due to the strong tendency of PANI and PU to form hydrogen bonding [52]. One research work explained the interpenetrating network of PU/PMMA system, this report mentions that the hydrogen bonding formation between NH of PANI and NHCOO group of PU/PMMA polymer network had some effect on the physicomechanical, electrical, and thermal properties of the conducting IPN. It was also reported that the interaction between the carboxyl groups in PU and imine groups in PANI could induce miscibility of PANI/PU blends [50,53]. It was recognized that the incorporation of (PAni)CSA complex into PU/PMMA IPNs increased the tensile strength from 1.16 to 1.38 MPa and electrical properties of PAni filled IPNs were increased with an increase in PAni concentration. From this it may be concluded that the PAni acts both as a conducting and reinforcing agent in PU/PMMA blends [54].

PANI doped with methane sulphonic acid (MSA) has recently been reported to produce three-dimensional variable range hopping (VRH) conduction, which is not found with HCl-doped PANI [55]. The conduction mechanism in polyaniline and polypyrrole has been illustrated by the hopping of electrons along the polymeric chain [56–58]. Various techniques were used to measure the charge transport in polyaniline protonated fully with MSA and PANI (MSA)-PU blend including temperature dependence of thermoelectric power and magnetic susceptibility and electron spin resonance at room temperature [55]. The blend follows a one-dimensional variable range hopping type of conduction and the electric field dependence of its conductivity exhibits the Poole-Frenkel effect.

The temperature-dependent magnetic susceptibility measurements indicated the presence of Pauli and Curie spins in both protonated polyanilines PANI (MSA)-PU blends [55]. The percentage of Lorentzian and Gaussian spins was estimated from electron spin resonance measurements. In case of PANI-MSA, a larger number of spins was noted to be delocalized. In another study [59], DC conductivity was described by the VRH model whereby the large value of VRH exponent observed (>0.5) indicated that hopping transport occurs between the superlocalized states of polymer with the frequency (ω) dependence of conductivity satisfying the ωs power law. The onset frequency ωs is proportional to the conductivity σs, while the permittivity increased sharply at low frequency and high temperature. One group measured conductivity using the collinear four-point probe method [60]. The polymer film was fastened to wire contacts using conductive silver paint. A constant current was applied to the outer electrodes and the resulting voltage across the inner elecrodes was recorded. In a more recent work, the authors measured resistance (R) of the material using a four-probe measurement instrument but this time the conductivity was obtained from the formula $s = L/R\ A$, where L is thickness and A is the cross-section area [61].

Variation in conductivity of series of blends of conductive PAni/PU, PAni/polystyrene-isoprene-copolymer (PAni/SIS), and carbon black (CB)/PU composite have been detailed [62]. Among them the PAni–PU blend exhibited a conductivity of 10^{-5} S cm^{-1} at low percolation threshold (11.5 wt%) compared with immiscible blend of PAni/SIS (16.2 wt%) and CB/PU (25 wt%). The electrical conductivity of PAni/PU (11.5/88.5, v/v) decreased (to 2×10^{-12} S cm^{-1}) along with aging and the change in morphology was also found as time increased. There has been an attempt to produced conducting membranes of PAni–PU for fuel cells and electro dialysis application [63] with 10^{-5} S cm^{-1} conductivities range. There was an increase in the PU glass transition temperature with the addition of PAni and SPAni because of their interaction with the PU chains. The electrical conductivity behavior of conducting PAni-DBSA/thermoplastic PU (TPU) free-standing films has been studied [64]. The commercial TPU was immiscible with prepared PAni-DBSA and it was found that electrical conductivity was low, about 1×10^{-6} S cm^{-1}, due to phase separation in the blends. Conductivity values and percolation threshold of PAni-DBSA/TPU blends depends on the methods used to prepare blends, compatibility, fine dispersion, and conducting pathways of PAni-DBSA. When the concentration of PAni-DBSA is increased twice, the conductivity was enhanced to 10,000 fold and reached 1.5×10^{-2} S cm^{-1}, this was owing to the formation of interconnecting networks. For the simultaneous improvement of the toughness and conductivity of the PU–PAni blend, a suitable method has been demonstrated [65,66]. Two series of toughened semiconductive polyaniline (PAni)/polyurethane (PU)–epoxy (PAni/PU–EPOXY) nanoblends were prepared using a conducting polymer PAni and PU prepolymer-modified-diglycidyl ether of bisphenol A (DGEBA) epoxy. After the careful evaluation of thermal, morphological, mechanical, and electrical properties, it was found that impact strength was enhanced 100% in PU (PPG 2000)-modified composites. In addition, the thermal stability of this composite was superior to that of neat epoxy resin, regardless of PU content at 27.5 wt%. The conductivity of the blend increased to the range 10^{-9} to 10^{-3} S cm^{-1} on addition of PAni in the frequency range 1 kHz–13 MHz. The electrical properties of polyurethane elastomer/ polyaniline (PU/PAni–HCl) blend films under tensile deformation have been reported [67]. Both surface-modified and volume-modified blends of PU and PAni–HCl were prepared for this purpose. Surface modification of PU film was done by swelling the parent film in aniline followed by oxidation to form PAni–HCl and volume-modified PU was prepared by mixing the polymer components in a joint solution and then solution casting. For surface-modified samples, nonlinear current–voltage characteristics were observed, whereas linear characteristics were observed for volume-modified samples. Deformation of the polymer composites caused moderately reversible decrease of their conductivity due to the deformation of a fractal percolation network. Nanostructured PAni blended castor oil–PU coatings were prepared by a nanotechnological approach in which nanostructured MO–PAni (MO = methyl orange, an organic template for nanostructures) were dispersed in castor oil polyurethane (COPU)-based composites [68]. The effect of loading of nanostructured MO–PAni in COPU on the spectral, physicochemical, and morphological properties has been analyzed and the fine dispersion of nanostructured PAni at lower loadings as well as its intermolecular hydrogen bonding with COPU remarkably increase the overall performance of the nanocomposite. The conductivity of pristine MO–PANI was found to be 10^{-3} S cm^{-1} while that of 0.5-MO–PAni/COPU, 1.0-MO–PAni/COPU, and 2.0-

MO–PAni/COPU were found to be 5.7×10^{-4}, 4.3×10^{-4}, and 3.5×10^{-4} S cm^{-1}, respectively. It was also reported that COPU /PAni blends have wide range application potential as corrosion protection. Electroactive polymers (EAPs) are an important candidate for actuation, due to their large >10% strains characteristics [69] and they have been widely used for energy conversion or energy harvesting [70–74]. Polyurethane is a electroactive polymer which can be used for actuator and transducer applications due to large deformations, high specific energy, and shorter response time [75]. It has been reported that incorporation of nanofillers such as nanocopper or carbon black can remarkably increase the strains induced by an applied electric field [72,75,76]. Numerous studies proposed that the electrostrictive characteristics of a polymer depend on dielectric permittivity and modulus property [76]. The dielectric permittivity is an important parameter because it directly influences the achievable electrical field induced strains in actuator applications. Thus, high dielectric constant is a basic requirement for new electrostrictive polymers for achieving large electric field induced strains. The incorporated metal or other highly conductive fillers in to polymer matrix is the most popular technique that can enhance the interfacial polarization of materials. In contrast, the microstructure of the phase distribution between fillers and polymer matrix is also important parameter for the dielectric composites. The major drawback of the metal or other conductive fillers based conducting polymer composites is the formation of agglomerates of filler when the filler loading is increased. Therefore, conductive polymer based polymer blends are of interest for improved electrical properties and microstructure. Polyaniline is a conductive polymer, which has been extensively used in conducting blend preparation for electrical engineering applications due to its relatively high conductivity, ease of polymerization, environmental stability, and cheap cost [77–79]. In conductive fillers based polymer composites, the dielectric permittivity and conductivity of the two-phase composites were enhanced due to free charges which give rise to Maxwell Wagner interfacial polarization [72]. In addition, as the filler loading increases, the dielectric constant also increases due to polarization. However, a change in the chemical or phase composition in the polymer chain also affects the electric properties which are due to the chargecarrier mobility [80]. The loss tangent depends on the intrinsic conductivity of fillers in the composites. The electrical conductivity depends on the number of charge carries in the materials [72,80,81].

The loss tangent resultant composites increased not only the conductivity effect, but also the interface interactions between hard and soft segments of matrix.

8.2.2 Polyurethene/Polypyrrole Blends

Polypyrrole (PPy) has been extensively studied because of its very high conductivity in the doped state and ease of chemical or electrochemical polymerization, which makes polypyrrole also interesting for many applications, e.g., sensor, biosensor, modified electrodes, actuators, and electronic devices [82,83]. However, their poor mechanical properties and poor stability of PPy at ambient conditions limit their utilization for various commercial applications. PPy exhibits insolubility in most organic solvents and is infusible because of its decomposition is lower than melting temperature. Polypyrrole cannot be shaped by existing plastic processing techniques such as melt processing and solution casting may not usually be applied to this material [84]. Most research on polypyrrole has been carried out in an effort to improve its stability, mechanical properties, and its processability.

Work attempting to make the protonic acid modified PPy soluble in nonpolar or weakly polar organic solvents [85,86] has been undertaken. In order to overcome these problems, blending of PPy with other insulating polymers approach has been reported, which involves the blending of PPy with polymers, e.g., poly(vinyl methyl ketone), or poly(vinyl alcohol). However, the immiscible nature of PPy refers to phase separation process when blended with other polymers and restricts the formation of conducting blends [87]. Recently, improvement in the compatibility of these blends and electrical and mechanical properties has been obtained by doping of functionalized protonic acid between PANI chains [64,88].

However, another method reported using dodecylbenzenesulfonic acid (DBSA) to functionalize a protonic acid with long alkyl chains and increase the space between parallel polymer backbones. This allowed the organic to quickly diffuse in between the polymer chains, and change the electrical conductivity and morphology of the blends [89,90]. In addition, DBSA is made up of hydrophilic and hydrophobic parts; the hydrophobic parts are adsorbed on the conducting polymers, forming the ionic form of the dodecylbenzenesulfonic acid pyrrole salt (DBSPy) [91]. DBSA trapped inside PPy undergoes crosslinking [92] thereby increasing its compatibility with insulating polymers as well as solubility in organic solvents. These composites are produced by solution casting [93,94], melt blending [95,96], or by coprecipitation methods. An in situ method has also been employed, where pyrrole is subjected to the emulsion polymerization with an organic host polymer [97]. In order to get a better combination of mechanical properties and processability without any vulcanization, the formation of heterogeneous conductive blends of thermoplastic elastomers has been developed. PU is one of the most versatile engineering block copolymers existing in different forms, such as foams, thermoplastic elastomers, composites, fibers, and reaction molding plastics [98]. It consists of alternating hard (aromatic diisocyanate extended with a short chain diol) and soft (polyesters or polyethers types) segments [92, 99–101]. Investigations on conductivity and thermal stabilities of thermoplastic polyurethane (TPU)/dodecylbenzenesulfonic acid doped polypyrrole (PPy.DBSA) nanoblends prepared by solution intercalation (SB) and in situ (IS) methods have been carried out and it was observed that the electrical conductivity (σdc) of nanoblends is influenced by the interaction between PPy.DBSA and TPU. The maximum value of σdc has been found at 30 wt% PPy.DBSA for the SB (0.26 S cm^{-1}) nanoblend because of the presence of the hexagonal network. The percolation threshold is obtained at 2.5 wt% of PPy.DBSA for IS and SB nanoblends. Dependence of conductivity on temperature for different nanoblends follows one-dimensional variable range hopping (VRH) model. A significant improvement in thermal stability has been observed at 50 wt% loss for SB nano blend containing 30 wt% PPy.DBSA [102].

Several attempts have been applied to improve the poor mechanical properties of conductive polymers through blending or preparation of their composites with other conventional polymers and fillers e.g., PPy-poly(methyl methacrylate) [103,104], PPy-poly(vinyl alcohol) [105], and PPypoly(tetrafluoroethylene) [106–108]. The characteristics of methanol and supercritical CO_2 process based conducting polypyrrole /PU foams were analyzed by several research groups [109–111]. They found that the conductivity of the conducting foam was in the range of 10^{-7} to 10^{-2} S cm^{-1}.

8.2.3 Polyurethane/Polythiophene Blends

Polythiophenes are potential candidates to be used as electrical conductors, nonlinear optical devices, polymer LEDs, electrochromic or smart windows, photoresists, antistatic coatings, sensors, batteries, electromagnetic shielding materials, artificial noses and muscles, solar cells, electrodes, microwave-absorbing materials, new types of memory devices, nanoswitches, optical modulators and valves, imaging materials, polymer electronic interconnects, nanoelectronic and optical devices, and transistors. In one study, polyurethane/polythiophene (PU/PT) conducting blends were produced by using electrochemical method [112] and were characterized by cyclic voltammetry (CV). The conductivity was in the range of 7.2×10^{-2} to 1.4×10^{-5} S cm^{-1}. Magnetic susceptibility experiments were also conducted on the composites. Negative magnetic susceptibility values indicated diamagnetism and positive values are due to paramagnetism. Diamagnetic values suggested that the conducting mechanisms of polythiophenes are of bipolaron nature and paramagnetic values demonstrated that the conducting mechanism is of polaron nature. PU/PTh blends were synthesized by solution blending and in situ polymerization and the researchers found that with an increase of PTh content, the shape recovery was increased [113].

8.3 IONIC CONDUCTIVITY IN POLYURETHANE BLENDS

By incorporating the highly ionic conducting component in TPU, its ionic conductivity can be improved and this improvement depends on the compatibilization and the interface between different phases. The interfacial adhesion between different phases can affect the ion transfer at the interface. The blend phase morphology and the interfacial adhesion has a significant effect on the mechanical property, ionic conductivity and other properties of polymer electrolytes. Polyether modified polysiloxanes can be prepared by either grafting or inserting the alkylene oxide containing segments into silicone [114]. A series of polyether modified polysiloxanes (PEMPS) electrolytes with diverse structures such as block copolymer, mono-comb or double-comb copolymer were synthesized and reported to have high ionic conductivity up to 10^{-5} S cm^{-1} at room temperature [115,116]. Their superior ionic conductivity is associated with their flexible and amorphous nature. This promotes high ionic conductivity, but gives poor dimensional stability. PEMPS electrolytes cannot function as separators in cell applications [117]. The ionic conductivity of TPU/ polyether modified polysiloxane (PEMPS) electrolytes is temperature-dependent and the addition of PEMPS to TPU increased the ionic conductivity of TPU/PEMPS electrolytes. The room temperature ionic conductivity of TPU/PEMPS electrolytes with LiTFSI can go up to 2.49×10^{-5} S cm^{-1}, which is in the application range. Due to the multiphase morphology of TPU/PEMPS electrolytes, the contribution of PEMPS to the ionic conductivity improvement of TPU/PEMPS electrolytes includes: (1) PEMPS phase itself provides high ionic conductivity; (2) the interaction of lithium salts with PEMPS and soft segment/hard segment of TPU induced the phase compatibilization of TPU and PEMPS and enhanced the interfacial interaction, which made the Li^+ transfer from one phase to the other phase more easily instead of circumventing the interface boundary [118].

One of the interesting applications of TPU−PPy composites is on electrodes for a bending-electrostrictive polyurethane actuator, as reported by Watanabe et al. [107]. The wrinkled electrode was prepared through in situ deposition of polypyrrole onto the polyurethane elastomer film that was being uniaxially drawn. Then the film was released from the drawing to make the electrode wrinkle.

In one early report, conductive composite foams were prepared by exposing iodine-loaded PU foams to pyrrole vapor. The dopant for the polypyrrole (PPy) was primarily I^{3-}, which formed a charge-transfer complex (PPy$-I_2$) with the amine group of the PPy. Concentration of the PPy$-I_2$ complex has a positive effect on the conductivity of the composite foams, and the conductivity also depends on the distribution of the PPy$-I_2$ complex in the PU matrix and the concentration ratio of PPy and iodine [84].

The effect of conducting polypyrrole (PPy) on the morphology and ionic conductivity of thermoplastic polyurethane (TPU) doped with $LiClO_4$ has been described and compared with those of pure TPU system [119]. The TPU−PPy composites were produced by chemical polymerization of pyrrole inside TPU films. From TGA, the thermal stability of the composite was found to be higher than pure TPU. However, a decrease in T_g of the soft segment was observed, which was attributed to the interaction of PPy−NH groups with either carbonyl or ether oxygens of TPU, leading to phase separation of the hard and soft segments. The composite retained the typical cauliflower-like morphology of PPy. The ionic conductivity was observed higher than pure TPU due to the of coordination of Li^+ ions with pyrrole nitrogens to enhance mobility of ClO_4^- ions and hence increase the ionic conductivity.

8.4 APPLICATIONS OF CONDUCTING POLYURETHANE BLENDS

The potential usage of conducting polymers and its blends include as electromagnetic shielding plastic batteries, sensors, conductive surfaces, magnetic recording, and solar cells.

8.4.1 Shape Memory of Polyurethane

Shape memory polymers (SMPs) (see Fig. 8.2) are a smart class of materials, which have the capability of responding to external stimulus [121]. It can attain the original shape at specific conditions and the shape memory effect (SME) depends upon a large value of recoverable strain greater than 100% and recovery rate not less than 80% [122]. The thermal and elastic properties are dependent on the molecular architecture of the polymers. However, enthalpy, dielectric properties, permeability, and elastic properties are the function of the property changes occurring at T_{tran} [123]. Technical applications of shape memory polymers include medical devices, medical implants, sensors, transducers, and actuators [124]. At the same time, they have disadvantages of high manufacturing cost, toxicity, limited recovery, and complicated surgical problems [125]. So new shape memory materials as polymers [126,127], ceramics [128], and hydrogels [129] have been produced. Polyurethane has been used as insulation for various electronics and electrical devices because these polymers are less toxic and fulfill all the basic requirements for a shape memory polymer. Different alternative ratios of hydroxyl and isocyanates have been analyzed to obtain the desired product [130].

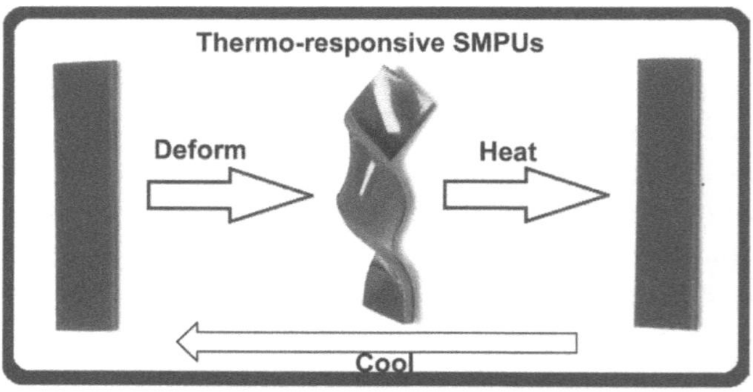

Figure 8.2 Shape memory of polyurethane [120].

Some polyurethanes can show both one-way shape memory (OWSM) or two-way shape memory behavior (TWSM). In the case of one-way shape memory, polyurethane is heated above its transition temperature and deforms into a temporary shape. By lowering the temperature (below T_{tran}) it will fix into the shape because the cooling has caused the shape to store the strain at below T_{tran}. Upon heating above T_{tran}, the soft segment regains its original configuration by releasing the deformation. When shape memory polyurethanes (SMPUs) are again cooled, they might undergo thermal contraction with very little change [131,132]. On the other hand, with two-way shape memory behavior they can deform into one configuration and then attain different configuration with subsequent cooling. Liquid crystal elastomers demonstrate TWSM behavior due to the variation in crosslinking patterns at different temperature [133–135]. The physical and mechanical properties of polyurethanes depend upon the type and amount of chain extender [136,137]. Polyurethanes are normally used as adhesives and coating materials and the thermal, mechanical, and viscoelastic properties depend upon the crosslinking pattern and concentration/number of double bonds [138]. Due to a high cyclic recoverable strain, a rubbery modulus and toughness properties, chemical crosslinking is more advantageous than physical crosslinking in SMPUs [139,140]. Polyurethanes that show SME have the ability to present different mechanical behaviors related to its hard and soft segment morphology [141,142]. The hard segment morphology is responsible for permanent shape and soft morphology allows the passage from permanent to temporary shape. Soft segments are usually exploited for molecular switching when heat is applied and is responsible for SME [143].

Electroactive shape memory blends using PU-block copolymer and conducting polypyrrole by chemical oxidative polymerization have been investigated and in one study a voltage-triggered SME was observed. The synthesized PU had a transition temperature near 460°C and the presence of polypyrrole enhance the electrical conductivity to the order of 10^{-2} S cm^{-1} (at 6–20 wt% polypyrrole). This conductivity was sufficient to show electroactive shape recovery by heating above the transition temperature of 40–450°C due to the melting of the polycaprolactone soft segment domain. Thus, a good shape recovery of 85%–90% was found in the shape recovery test with bending mode when an electric charge of 40 V was applied [144].

8.4.2 Electromagnetic Interference Shielding

Electromagnetic interference (EMI) shielding enclosures are essential because digital circuits generate substantial noise levels throughout a wide frequency spectrum and can serve as a source of EMI. EMI can exist in a wide frequency range from 50 Hz to about 1010−1011 Hz (typical of radar transmission) [145] of the total electromagnetic spectrum. The most common electromagnetic disturbance occurs at about 104 Hz, usually generated from communications, radio, and television. Various metals can be used for EMI shielding, however, conducting polymer composite formulations [146] can serve as both a structural material and a electromagnetic interference shielding materials. They are lightweight, corrosion resistant and cost-effective, and their performance can be varied by adjusting the filler loading and thickness of materials to satisfy requirements dictated by the specific device demands. The shielding effectiveness (SE) of a material is the ratio of the incident field strength (V/m) to that fraction, which is transmitted, expressed in a decibel scale as 20 $\log_{10}$(Ei/Et) (dB). Total SE is the sum of three parts: reflection, absorption, and rereflection. SE measuring details are dealt with elsewhere in the literature [146,147].

Many authors found that PU−PAni systems are the most suitable conducting polymers for EMI shielding [49−51]. Initially microwave-absorbing materials (RAMs) were manufactured with ferrites, carbonyl iron, and CB [148]. But problems related to the cost, ease of application, dispersion in a support matrix and weight, limited the use of RAMs produced using such materials. Conducting polymers have been used as an alternative to these traditional materials in the production of RAMs [149,150]. The microwave absorption, microwave reflection, EMI shielding, and mechanical properties of PAni−PU composite were analyzed and the shielding efficiency of the composite increased with the thickness of the sample, hence this material is ideal for shielding at 2.23 GHz and at 8.82 GHz. The variations of reflection coefficient with frequency at S band and X band show similar behaviors. Reflection of microwaves from the PAni−PU films is very dependent on the frequency. The position, intensity and number of dips are dependent on the thickness of the sample [148,151].

8.4.3 Corrosion Protection

PU is a versatile polymeric material that offers a wide range of applications such as adhesives, elastomers, coatings for textiles/paper, footwear, furniture/ foams, packaging material, and for automotive finishes [152,153]. It is also a promising biomaterial for use in humans [153]. The biocompatible products of PU range from nasogastric catheters, peritoneal dialysis, infusion pumps to implanted pacemaker parts [153]. PU coatings are applied on different surfaces, to improve their wear resistance, appearance, and corrosion resistance. Different types of PU coatings are used in construction, where floors, steel structures, and concrete supports are spray coated to make them more durable against environmental deterioration.

8.4.4 Sensors

A new compressible conducting material ($\sigma = 1.4 \times 10^{-3}$ S cm^{-1}) has been fabricated by coating PU /conducting polypyrrole (PPy) foam for application of pressure sensing [154]. It has been used for breathing monitoring, in which the conductivity of the materials are affected by change in pressure during breathing. The conductance of the conducting foam is found to linearly change as a function of applied force [167]. Another group developed a micropyramid PDMS array in which the PDMS micropyramids were coated with a PEDOT:PSS-PUD (PU dispersion) blend, which functioned as a piezoresistive electrode. By the application of pressure, each pyramid tends to spread laterally and this increases the contact interface area (ACI), contact perimeter (WPE), and the thickness of current path (DPE), thereby improving the current conduction. This sensor worked well even at a high elongation of 40%, and showed a very high sensitivity of 10.32 kPa at that elongation [155].

8.4.5 Stretchable electronics

Stretchable electronics can include the stretchable logic gates, memories, and stretchable display units. Stretchable electronics are attached with basic elements, such as transistors, LEDs, and dielectrics. The important procedure for implementing stretchable electronics is a combination of rigid, active components, and stretchable interconnections, which play an important role in accommodating external strains [156–158]. Amongst materials, conductors, semiconductors, and dielectrics are most relevant stretchable materials. The blending of conductive polymer and elastomeric materials can produce conducting materials for stretchable electronics devices [159–166]. In one study, elastomeric PU–PPY composite foams was prepared by in situ polymerization of pyrrole in preformed PU foams, and conductivity was evaluated 10^{-5} S cm^{-1} and an elongation at break of 160% were obtained from a 6 wt% PPY-containing composite [109]. The conductivity of the PU composite foams can be improved by introducing conductive nanostructures. In addition, PU-PEDOT blends were synthesized from liquid mixtures of EDOT and different contents of PU in tetrahydrofuran (THF) without engaging porous elastomers [167]. The researchers reported conductivity range in 10^{-5} S cm^{-1} at a 200% strain for the blends. Moreover, in another study of PDMS-PEDOT:PSS blends, it found that the miscibility can be enhanced by introducing a copolymer, and demonstrated a conductivity up to 2 S cm^{-1} and a fracture strain of 75% [168].

CONCLUSION

It has been demonstrated that PU-conductive blends are promising materials for many technological uses because of their chemical versatility, stability, processability, and low cost. PU-conducting blends can be produced either by chemical or electrochemical techniques. This chapter discussed various conducting PU blends and their applications related to EMI, sensors, and shape memory field. The reports on corrosion protection studies by PU−PAni are limited and this will be emerging and focus area for many new future developments.

REFERENCES

[1] S.K. Malhotra,, K. Goda, M.S. Sreekala, Part One Introduction to Polymer Composites; 2012.

[2] S.A. Samsudin, A. Hassan, M. Mokhtar, S.M.S. Jamaluddin, Effect of SEBS on the mechanical properties and miscibility of polystyrene rich polystyrene/polypropylene blends, Progr. RubberRubber Plast. Recycl. Technol. 21 (4) (2005) 261−276.

[3] L.D. Giudice, R.E. Cohen, G. Attalla, F. Bertinotti, Compatibilizing effect of a diblock copolymer of isotactic polystyrene and isotactic polypropylene in blends of the corresponding homopolymers, J. Appl. Polym. Sci. 30 (11) (1985) 4305−4318.

[4] C. Koning, M. Van Duin, C. Pagnoulle, R. Jerome, Strategies for compatibilization of polymer blends, Progr. Polym. Sci. 23 (4) (1998) 707−757.

[5] T. Takeichi, K. Suefuji, K. Inoue, Preparation of high-temperature polyurethane by alloying with reactive polyamide, J. Polym. Sci. Part A Polym. Chem. 40 (20) (2002) 3497−3503.

[6] L. Shufen, J. Zhi, Y. Kaijun, Y. Shuqin, W.K. Chow, Studies on the thermal behavior of polyurethanes, Polym. Plast. Technol. Eng. 45 (1) (2006) 95−108.

[7] H. Yeganeh, M.M. Lakouraj, S. Jamshidi, Synthesis and properties of biodegradable elastomeric epoxy modified polyurethanes based on poly (ε-caprolactone) and poly (ethylene glycol), Eur. Polym. J. 41 (10) (2005) 2370−2379.

[8] Y.C. Chou, L.J. Lee, Mechanical properties of polyurethane-unsaturated polyester interpenetrating polymer networks, Polym. Eng. Sci. 35 (12) (1995) 976–988.
[9] S.A.M. Ali, D.J. Hourston, K. Manzoor, D.F. Williams, The stability of poly (ether urethane)/polydimethylsiloxane interpenetrating polymer networks, J. Appl. Polym. Sci. 55 (5) (1995) 733–740.
[10] D.S. Lee, W.K. Kang, J.H. An, S.C. Kim, Gas transport in polyurethane-polystyrene interpenetrating polymer network membranes. II. Effect of crosslinked state and annealing, J. Membr. Sci. 75 (1) (1992) 15–27.
[11] T.T. Hsieh, K.H. Hsieh, G.P. Simon, C. Tiu, Interpenetrating polymer networks of 2-hydroxyethyl methacrylate terminated polyurethanes and polyurethanes, Polym. 40 (11) (1999) 3153–3163.
[12] Y. Li, S. Mao, Study on the properties and application of epoxy resin/polyurethane semi-interpenetrating polymer networks, J. Appl. Polym. Sci. 61 (12) (1996) 2059–2063.
[13] H.D. Rozman, G.S. Tay, A. Abusamah, The effect of glycol type, glycol mixture, and isocyanate/glycol ratio on flexural properties of oil palm empty fruit bunch-polyurethane composites, J. Wood Chem. Technol. 23 (3–4) (2003) 249–260.
[14] P.W. Morgan, Fibers from polyurethanes based on halogen-substituted bisphenols, J. Appl. Polym. Sci. 40 (11–12) (1990) 2037–2047.
[15] S. Chen, Y. Tian, L. Chen, T. Hu, Epoxy resin/polyurethane hybrid networks synthesized by frontal polymerization, Chem. Mater. 18 (8) (2006) 2159–2163.
[16] H. Harani, S. Fellahi, M. Bakar, Toughening of epoxy resin using synthesized polyurethane prepolymer based on hydroxyl-terminated polyesters, J. Appl. Polym. Sci. 70 (13) (1998) 2603–2618.
[17] H.H. Wang, J.C. Chen, Modification and compatibility of epoxy resin with hydroxyl-terminated or amine-terminated polyurethanes, Polym. Eng. Sci. 35 (18) (1995) 1468–1475.
[18] S. Ong, J. Ismail, M.A. Bakar, I.A. Rahman, C.S. Sipaut, C.K. Chee, Polyurethane-modified epoxy resin: solventless preparation and properties, J. Appl. Polym. Sci. 111 (6) (2009) 3094–3103.
[19] S.J. Park, J.S. Jin, Energetic studies on epoxy–polyurethane interpenetrating polymer networks, J. Appl. Polym. Sci. 82 (3) (2001) 775–780.
[20] M. Cristea, S. Ibanescu, C.N. Cascaval, D. Rosu, Dynamic mechanical analysis of polyurethane-epoxy interpenetrating polymer networks, High Perform. Polym. 21 (5) (2009) 608–623.
[21] H.H. Wang, J.C. Chen, Toughening of epoxy resin by reacting with functional terminated-polyurethanes, J. Appl. Polym. Sci. 57 (6) (1995) 671–677.
[22] C.W. Allen, The use of phosphazenes as fire resistant materials, J. Fire Sci. 11 (4) (1993) 320–328.
[23] J.E. Mark, H.R. Allcock, R. West, Inorganic Polymers, Polymer Science and Engineering Series; 1992.
[24] H.R. Allcock, Inorganic—organic polymers, Adv. Mater. 6 (2) (1994) 106–115.
[25] Doi, M. (1983). Sci., Polym. Lett. Ed. 19, 265 1981; M. Doi. J. Polym. Sci. Polym. Phys. Ed, 21, 667.
[26] R. Narayan, D.K. Chattopadhyay, B. Sreedhar, K.V.S.N. Raju, N.N. Mallikarjuna, T.M. Aminabhavi, Synthesis and characterization of crosslinked polyurethane dispersions based on hydroxylated polyesters, J. Appl. Polym. Sci. 99 (1) (2006) 368–380.
[27] M. Shibaya, Y. Suzuki, M. Doro, H. Ishihara, N. Yoshihara, M. Enomoto, Effect of soft segment component on moisture-permeable polyurethane films, J. Polym. Sci. Part B Polym. Phys. 44 (3) (2006) 573–583.

[28] G.I. Taylor, The criterion for turbulence in curved pipes, Proceedings of the Royal Society of London A: Mathematical, Physical and Engineering Sciences, Vol. 124, The Royal Society, London, 1929, pp. 243–249, No. 794.
[29] G. Taylor, Disintegration of water drops in an electric field, Proceedings of the Royal Society of London A: Mathematical, Physical and Engineering Sciences, Vol. 280, The Royal Society, London, 1964, July, pp. 383–397, No. 1382.
[30] M.S. Khil, H.Y. Kim, Y.S. Kang, H.J. Bang, D.R. Lee, J.K. Doo, Preparation of electrospun oxidized cellulose mats and theirin vitro degradation behavior, Macromol. Res. 13 (1) (2005) 62–67.
[31] S.R. Bhattarai, N. Bhattarai, H.K. Yi, P.H. Hwang, D.I. Cha, H.Y. Kim, Novel biodegradable electrospun membrane: scaffold for tissue engineering, Biomater. 25 (13) (2004) 2595–2602.
[32] P. Gibson, H. Schreuder-Gibson, D. Rivin, Transport properties of porous membranes based on electrospun nanofibers, Colloid Surf. A Physicochem. Eng. Asp. 187 (2001) 469–481.
[33] Y.J. Ryu, H.Y. Kim, K.H. Lee, H.C. Park, D.R. Lee, Eur. Polym. J. 39 (2003) 1883.
[34] K.H. Lee, H.Y. Kim, Y.M. La, D.R. Lee, N.H. Sung, Influence of a mixing solvent with tetrahydrofuran and N, N-dimethylformamide on electrospun poly (vinyl chloride) nonwoven mats, J. Polym. Sci. PartPart B Polym. Phys. 40 (19) (2002) 2259–2268.
[35] K.H. Lee, H.Y. Kim, H.J. Bang, Y.H. Jung, S.G. Lee, The change of bead morphology formed on electrospun polystyrene fibers, Polym. 44 (14) (2003) 4029–4034.
[36] M.S. Khil, D.I. Cha, H.Y. Kim, I.S. Kim, N. Bhattarai, Electrospun nanofibrous polyurethane membrane as wound dressing, J. Biomed. Mater. Res. Part B Appl. Biomater. 67 (2) (2003) 675–679.
[37] V. Saxena, B.D. Malhotra, Prospects of conducting polymers in molecular electronics, Curr. Appl. Phys. 3 (2) (2003) 293–305.
[38] F.C. Chen, Y. Yang, M.E. Thompson, J. Kido, High-performance polymer light-emitting diodes doped with a red phosphorescent iridium complex, Appl. Phys. Lett. 80 (13) (2002) 2308–2310.
[39] T. Ahuja, I.A. Mir, D. Kumar, Biomolecular immobilization on conducting polymers for biosensing applications, Biomater. 28 (5) (2007) 791–805.
[40] A.L. Sharma, V. Saxena, S. Annapoorni, B.D. Malhotra, Synthesis and characterization of a copolymer: Poly (aniline-co-fluoroaniline), J. Appl. Polym. Sci. 81 (6) (2001) 1460–1466.
[41] A.L. Sharma, M. Gerard, R. Singhal, B.D. Malhotra, S. Annapoorni, Synthesis and characterization of fluoro-substituted polyaniline, Appl. Biochem. Biotechnol. 96 (1-3) (2001) 155–165.
[42] A.L. Sharma, R. Singhal, A. Kumar, K.K. Pande, B.D. Malhotra, Immobilization of glucose oxidase onto electrochemically prepared poly (aniline-co-fluoroaniline) films, J. Appl. Polym. Sci. 91 (6) (2004) 3999–4006.
[43] R.H. Friend, Conjugated polymers. New materials for optoelectronic devices, Pure Appl. Chem. 73 (3) (2001) 425–430.
[44] M. Angelopoulos, Conducting polymers in microelectronics, IBM J. Res. Dev. 45 (1) (2001) 57–75.
[45] W.R. Salaneck, D.T. Clark, E.J. Samuelsen, Science and Application of Conducting Polymers, lOP Publishing, Bristol, 1991, p. 168.
[46] J. Argyropoulos, P. Popa, G. Spilman, D. Bhattacharjee, W. Koonce, J Coat. Technol. Res. 6 (2009) 501–508.
[47] L. Beverina, G.A. Pagani, M. Sassi, Multichromophoric electrochromic polymers: colour tuning of conjugated polymers through the side chain functionalization approach, Chem. Commun. 50 (41) (2014) 5413–5430.

[48] P. Chandrasekhar, Conducting Polymers, Fundamentals and Applications: A Practical Approach, Springer Science & Business Media, New York, 2013.
[49] V.G. Kulkarni, L.D. Campbell, W.R. Mathew, Thermal stability of polyaniline, Synth. Met. 30 (3) (1989) 321–325.
[50] T.L. Wang, C.H. Yang, Y.T. Shieh, A.C. Yeh, Synthesis and properties of conducting organic/inorganic polyurethane hybrids, Eur. Polym. J. 45 (2) (2009) 387–397.
[51] N.A. Rangel-Vázquez, R. Salgado-Delgado, E. García-Hernández, A.M. Mendoza-Martínez, Characterization of copolymer based in polyurethane and polyaniline (PU/PANI), J. Mex. Chem. Soc. 53 (4) (2009) 248–252.
[52] K.S. Ho, K.H. Hsieh, S.K. Huang, T.H. Hsieh, Polyurethane-based conducting polymer blends: I. Effect of chain extender, Synth. Met. 107 (1) (1999) 65–73.
[53] J. Luo, X. Wang, J. Li, X. Zhao, F. Wang, Conductive hybrid film from polyaniline and polyurethane–silica, Polym. 48 (15) (2007) 4368–4374.
[54] T. Jeevananda, Synthesis and characterization of polyaniline filled PU/PMMA interpenetrating polymer networks, Eur. Polym. J. 39 (3) (2003) 569–578.
[55] B. Sanjai, A. Raghunathan, T.S. Natarajan, G. Rangarajan, S. Ṭhomas, P.V. Prabhakaran, et al., Charge transport and magnetic properties in polyaniline doped with methane sulphonic acidand polyaniline-polyurethane blend, Phys. Rev. B 55 (16) (1997) 10734.
[56] M. Chipara, D. Hui, P.V. Notingher, M.D. Chipara, K.T. Lau, J. Sankar, et al., On polyethylene–polyaniline composites, Compos. Part B Eng. 34 (7) (2003) 637–645.
[57] B. Sixou, N. Mermilliod, J.P. Travers, Aging effects on the transport properties in conducting polymer polypyrrole, Phys. Rev. B 53 (8) (1996) 4509.
[58] N.F. Mott, E.A. Davis, Electronic Processes in Non-crystalline Solids, Clarendon, Oxford, 1979, p. 465.
[59] P. Dutta, S. Biswas, M. Ghosh, S.K. De, S. Chatterjee, The dc and ac conductivity of polyaniline–polyvinyl alcohol blends, Synth. Met. 122 (2) (2001) 455–461.
[60] M. Liu, R.V. Gregory, A blend of polyurethane and a urethane substituted polythiophene in the β-position, Synth. Met. 69 (1) (1995) 349–350.
[61] Y.Z. Wang, Y.C. Hsu, R.R. Wu, H.M. Kao, Synthesis and structure properties of polyurethane based conducting copolymer I. 13 C NMR analysis, Synth. Met. 132 (2) (2003) 151–160.
[62] H. Yoshikawa, T. Hino, N. Kuramoto, Effect of temperature and moisture on electrical conductivity in polyaniline/polyurethane (PANI/PU) blends, Synth. Met. 156 (18) (2006) 1187–1193.
[63] F.D.R. Amado, L.F. Rodrigues, M.M.C. Forte, C.A. Ferreira, Properties evaluation of the membranes synthesized with castor oil polyurethane and polyaniline, Polym. Eng. Sci. 46 (10) (2006) 1485–1489.
[64] D.S. Vicentini, G.M. Barra, J.R. Bertolino, A.T. Pires, Polyaniline/thermoplastic polyurethane blends: Preparation and evaluation of electrical conductivity, Eur. Polym. J. 43 (10) (2007) 4565–4572.
[65] W.C. Chiou, D.Y. Yang, J.L. Han, S.N. Lee, Synthesis and characterization of composites of polyaniline and polyurethane-modified epoxy, Polym. Int. 55 (11) (2006) 1222–1229.
[66] W.C. Chiou, J.L. Han, S.N. Lee, Synthesis and studies of the physical properties of polyaniline and polyurethane-modified epoxy composites, Polym. Eng. Sci. 48 (2) (2008) 345–354.
[67] E. Hrehorova, V.N. Bliznyuk, A.A. Pud, V.V. Shevchenko, K.Y. Fatyeyeva, Electrical properties and fractal behavior of polyurethane elastomer/polyaniline composites under mechanical deformation, Polymer 48 (15) (2007) 4429–4437.

[68] J. Huang, L. Zhang, Effects of NCO/OH molar ratio on structure and properties of graft-interpenetrating polymer networks from polyurethane and nitrolignin, Polym. 43 (8) (2002) 2287–2294.
[69] D. Guyomar, P.J. Cottinet, L. Lebrun, C. Putson, K. Yuse, M. Kanda, et al., The compressive electrical field electrostrictive coefficient M33 of electroactive polymer composites and its saturation versus electrical field, polymer thickness, frequency, and fillers, Polym. Adv. Technol. 23 (6) (2012) 946–950.
[70] D. Guyomar, L. Lebrun, C. Putson, P.J. Cottinet, B. Guiffard, S. Muensit, Electrostrictive energy conversion in polyurethane nanocomposites, J. Appl. Phys. 106 (1) (2009) 014910.
[71] P.J. Cottinet, D. Guyomar, B. Guiffard, C. Putson, L. Lebrun, Modeling and experimentation on an electrostrictive polymer composite for energy harvesting, Ultrason. Ferroelectrics Frequency Control IEEE Trans. 57 (4) (2010) 774–784.
[72] C. Putson, D. Jaaoh, N. Muensit, Large electromechanical strain at low electric field of modified polyurethane composites for flexible actuators, Mater. Lett. 172 (2016) 27–31.
[73] A. Eddiai, M. Meddad, K. Sbiaai, Y. Boughaleb, A. Hajjaji, D. Guyomar, A new technique for maximizing the energy harvested using electrostrictive polymer composite, Opt. Mater. 36 (1) (2013) 13–17.
[74] P. Sukwisute, N. Muensit, S. Soontaranon, S. Rugmai, Micropower energy harvesting using poly (vinylidene fluoride hexafluoropropylene), Appl. Phys. Lett. 103 (6) (2013) 063905.
[75] K. Wongtimnoi, B. Guiffard, A. Bogner-Van de Moortele, L. Seveyrat, C. Gauthier, J.Y. Cavaillé, Improvement of electrostrictive properties of a polyether-based polyurethane elastomer filled with conductive carbon black, Compos. Sci. Technol. 71 (6) (2011) 885–892.
[76] B. Guiffard, L. Seveyrat, G. Sebald, D. Guyomar, Enhanced electric field-induced strain in non-percolative carbon nanopowder/polyurethane composites, J. Phys. D Appl. Phys. 39 (14) (2006) 3053.
[77] C.H. Chen, C.F. Mao, S.F. Su, Y.Y. Fahn, Preparation and characterization of conductive poly (vinyl alcohol)/polyaniline doped by dodecyl benzene sulfonic acid (PVA/PANDB) blend films, J. Appl. Polym. Sci. 103 (5) (2007) 3415–3422.
[78] A.G. MacDiarmid, A.J. Epstein, Polyanilines: a novel class of conducting polymers, Faraday Discuss. Chem. Soc. 88 (1989) 317–332.
[79] G.M.O. Barra, R.R. Matins, K.A. Kafer, R. Paniago, C.T. Vasques, A.T.N. Pires, Thermoplastic elastomer/polyaniline blends: evaluation of mechanical and electromechanical properties, Polym. Test. 27 (7) (2008) 886–892.
[80] K. Petcharoen, A. Sirivat, Electrostrictive properties of thermoplastic polyurethane elastomer: Effects of urethane type and soft–hard segment composition, Curr. Appl. Phys. 13 (6) (2013) 1119–1127.
[81] D. Jaaoh, C. Putson, N. Muensit, Deformation on segment-structure of electrostrictive polyurethane/polyaniline blends, Polymer 61 (2015) 123–130.
[82] G. Inzelt, A.J. Bard, Electroanalytical Chemistry, vol. 18, Marcel Dekker, New York, 1994.
[83] H. Korri-Youssoufi, F. Garnier, P. Srivastava, P. Godillot, A. Yassar, Toward bioelectronics: specific DNA recognition based on an oligonucleotide-functionalized polypyrrole, J. Am. Chem. Soc. 119 (31) (1997) 7388–7389.
[84] Y. Wang, G.A. Sotzing, R.A. Weiss, Preparation of conductive polypyrrole/polyurethane composite foams by in situ polymerization of pyrrole, Chem. Mater. 20 (7) (2008) 2574–2582.
[85] J.Y. Lee, D.Y. Kim, C.Y. Kim, Synthesis of soluble polypyrrole of the doped state in organic solvents, Synth. Met. 74 (2) (1995) 103–106.

[86] C. Basavaraja, N.R. Kim, E. Jo, R. Pierson, D.S. Huh, A. Venkataraman, Transport properties of polypyrrole films doped with sulphonic acids, Bull. Korean Chem. Soc. 30 (11) (2009) 2701–2706.
[87] H.L. Wang, J.E. Fernandez, Blends of polypyrrole and poly (vinyl alcohol), Macromolecules 26 (13) (1993) 3336–3339.
[88] M. Ghorbani, H. Eisazadeh, Nanoparticles preparation of aniline and acrylonitrile copolymer using various surfactants, World Appl. Sci. J. 5 (2) (2008) 204–210.
[89] G. Wegner, M. Monkenbusch, G. Wieners, R. Weizenhöfer, G. Lieser, W. Wernet, New routes to conducting polymers and new insights into structure-properties relations, Mol. Cryst. Liq. Cryst. 118 (1) (1985) 85–94.
[90] W. Wernet, M. Monkenbusch, G. Wegner, A new series of conducting polymers with layered structure: polypyrrole n-alkylsulfates and n-alkylsulfonates, Die Makromolekulare Chemie, Rapid Commun. 5 (3) (1984) 157–164.
[91] Q. Lü, Unstirred preparation of soluble electroconductive polypyrrole nanoparticles, Microchim. Acta 168 (3–4) (2010) 205–213.
[92] P.K. Maji, P.K. Guchhait, A.K. Bhowmick, & ACS, A. M. I. 1: 289–300 (2009). CrossRef, PubMed, CAS, Web of Science® Times Cited, 11.
[93] S.H. Hosseini, A.A. Entezami, Conducting polymer blends of polypyrrole with polyvinyl acetate, polystyrene, and polyvinyl chloride based toxic gas sensors, J. Appl. Polym. Sci. 90 (1) (2003) 49–62.
[94] A. Shakoor, P.J.S. Foot, T.Z. Rizvi, Conductive poly (methyl methacrylate)-polypyrrole dodecylbenzenesulfonate (PMMA-PPy. DBSA) blends prepared in solution in the presence of hydroquinone, J. Mater. Sci. Mater. Electron. 21 (12) (2010) 1270–1276.
[95] M. Omastová, J. Pavlinec, J. Pionteck, F. Simon, Synthesis, electrical properties and stability of polypyrrole-containing conducting polymer composites, Polym. Int. 43 (2) (1997) 109–116.
[96] R. Faez, A.D. Reis, M.A. Soto-Oviedo, M.C. Rezende, M.A. De Paoli, Microwave absorbing coatings based on a blend of nitrile rubber, EPDM rubber and polyaniline, Polym. Bull. 55 (4) (2005) 299–307.
[97] E. Ruckenstein, J.H. Chen, Polypyrrole conductive composites prepared by coprecipitation, Polymer 32 (7) (1991) 1230–1235.
[98] S. Radhakrishnan, D.R. Saini, Structure and electrical properties of polypyrrole–thermoplastic elastomer blends, Polym. Int. 34 (1) (1994) 111–117.
[99] M. Kotal, S.K. Srivastava, A.K. Bhowmick, Thermoplastic polyurethane and nitrile butadiene rubber blends with layered double hydroxide nanocomposites by solution blending, Polym. Int. 59 (1) (2010) 2–10.
[100] M. Kotal, S.K. Srivastava, S.K. Manu, Layered double hydroxide as nanofiller in the development of polyurethane nanocomposites, J. Nanosci. Nanotechnol. 10 (9) (2010) 5730–5740.
[101] A. Pattanayak, S.C. Jana, Synthesis of thermoplastic polyurethane nanocomposites of reactive nanoclay by bulk polymerization methods, Polymer 46 (10) (2005) 3275–3288.
[102] M. Kotal, S.K. Srivastava, B. Paramanik, Enhancements in conductivity and thermal stabilities of polypyrrole/polyurethane nanoblends, J. Phys. Chem. C 115 (5) (2011) 1496–1505.
[103] M. Morita, I. Hashida, M. Nishimura, Conducting polypyrrole composite thin films chemically prepared by spreading on surface of aqueous solution containing oxidizing agent, J. Appl. Polym. Sci. 36 (7) (1988) 1639–1650.
[104] H.S. Chan, T.S.A. Hor, P.K.H. Ho, K.L. Tan, B.T.G. Tan, Chemical preparation and characterization of conductive polypyrrole composite thin films, J. Macromol. Sci. Chem. 27 (8) (1990) 1081–1094.

[105] S.E. Lindsey, G.B. Street, Conductive composites from polyvinyl alcohol and polypyrrole, Synth. Met. 10 (1) (1984) 67–69.
[106] R.M. Penner, C.R. Martin, Preparation and electrochemical characterization of ultramicroelectrode ensembles, Anal. Chem. 59 (21) (1987) 2625–2630.
[107] M. Watanabe, H. Shirai, T. Hirai, Wrinkled polypyrrole electrode for electroactive polymer actuators, J. Appl. Phys. 92 (8) (2002) 4631–4637.
[108] M. Watanabe, N. Wakimoto, H. Shirai, T. Hirai, Bending electrostriction and space-charge distribution in polyurethane films, J. Appl. Phys. 94 (4) (2003) 2494–2497.
[109] Y. Fu, R.A. Weiss, P.P. Gan, M.D. Bessette, Conductive elastomeric foams prepared by in situ vapor phase polymerization of pyrrole and copolymerization of pyrrole and N-methylpyrrole, Polym. Eng. Sci. 38 (5) (1998) 857–862.
[110] S.L. Shenoy, I. Kaya, C. Erkey, R.A. Weiss, Synthesis of conductive elastomeric foams by an in situ polymerization of pyrrole using supercritical carbon dioxide and ethanol cosolvents, Synth. Met. 123 (3) (2001) 509–514.
[111] S.L. Shenoy, D. Cohen, R.A. Weiss, C. Erkey, Supercritical carbon dioxide aided preparation of conductive polyurethane–polypyrrole composites, J. Supercrit. Fluid 28 (2) (2004) 233–239.
[112] B. Sari, M. Talu, F. Yildirim, E.K. Balci, Synthesis and characterization of polyurethane/polythiophene conducting copolymer by electrochemical method, Appl. Surf. Sci. 205 (1) (2003) 27–38.
[113] R. Sattar, A. Kausar, M. Siddiq, Influence of conducting polymer on mechanical, thermal and shape memory properties of polyurethane/polythiophene blends and nanocomposites, Adv. Mater. Lett. 7 (2016) 282–288.
[114] B. Oh, D. Vissers, Z. Zhang, R. West, H. Tsukamoto, K. Amine, New interpenetrating network type poly (siloxane-g-ethylene oxide) polymer electrolyte for lithium battery, J. Power Source 119 (2003) 442–447.
[115] K. Nagaoka, H. Naruse, I. Shinohara, M. Watanabe, J. Polym. Sci. Polym. Lett. Ed. 22 (1984) 659.
[116] J. Ni, Y. Lin, F. Wang, Synthesis of a novel polysiloxane-based polymer electrolyte and its ionic conductivity, Polym. Adv. Technol. 4 (2–3) (1993) 80–84.
[117] Y. Kang, W. Lee, D.H. Suh, C. Lee, Solid polymer electrolytes based on cross-linked polysiloxane-g-oligo (ethylene oxide): ionic conductivity and electrochemical properties, J. Power Source 119 (2003) 448–453.
[118] S. Wang, K. Min, Solid polymer electrolytes of blends of polyurethane and polyether modified polysiloxane and their ionic conductivity, Polymer 51 (12) (2010) 2621–2628.
[119] T.C. Wen, S.L. Hung, M. Digar, Effect of polypyrrole on the morphology and ionic conductivity of TPU electrolyte containing LiClO 4, . Synth. Met. 118 (1) (2001) 11–18.
[120] N. Ahmed, A. Kausar, B. Muhammad, Advances in shape memory polyurethanes and composites: a review, Polym. Plast. Technol. Eng. 54 (13) (2015) 1410–1423.
[121] S. Mondal, P. Memmott, D. Martin, Preparation and characterization of spinifex resin-based bio-polyurethane/thermoplastic polyurethane blends, Polym. Plast. Technol. Eng. 52 (15) (2013) 1535–1541.
[122] M.R. Rahnama, M. Barikani, M. Barmar, H. Honarkar, An investigation into the effects of different nanoclays on polyurethane nanocomposites properties, Polym. Plast. Technol. Eng. 53 (8) (2014) 801–810.
[123] M. Behl, M.Y. Razzaq, A. Lendlein, Multifunctional shape-memory polymers, Adv. Mater. 22 (31) (2010) 3388–3410.

[124] Z.G. Wei, R. Sandstroröm, S. Miyazaki, Shape-memory materials and hybrid composites for smart systems: Part I Shape-memory materials, J. Mater. Sci. 33 (15) (1998) 3743–3762.
[125] P. Ping, W. Wang, X. Chen, X. Jing, The influence of hard-segments on two-phase structure and shape memory properties of PCL-based segmented polyurethanes, J. Polym. Sci. Part B: Polym. Phys. 45 (5) (2007) 557–570.
[126] B.K. Kim, S.Y. Lee, M. Xu, Polyurethanes having shape memory effects, Polymer 37 (26) (1996) 5781–5793.
[127] J. Hu (Ed.), Advances in Shape Memory Polymers, Elsevier, Amsterdam, 2013.
[128] K.E. Schurch, K.H.G. Ashbee, A near perfect shape-memory ceramic material, Nature 266 (1977) 706–707.
[129] Z. Hu, X. Zhang, Y. Li, Synthesis and application of modulated polymer gels, Science 269 (5223) (1995) 525.
[130] B. Martel, R. Villwock, H. Stone, U.S. Patent Application No. 11/711,654; 2007.
[131] A.M. Issam, G. Sankar, New nonlinear polyurethane: Synthesis and optical properties, Polym. Plast. Technol. Eng 49 (2010) 1606–1610.
[132] A. Lendlein, S. Kelch, Shape-memory polymers, Angew. Chem. Int. Ed 41 (2002) 2034–2057.
[133] S.V. Ahir, A.R. Tajbakhsh, E.M. Terentjev, Self-assembled shape-memory fibers of triblock liquid-crystal polymers, Adv. Funct. Mater. 16 (4) (2006) 556–560.
[134] M.H. Li, P. Keller, B. Li, X. Wang, M. Brunet, Light-driven side-on nematic elastomer actuators, Adv. Mater. 15 (7-8) (2003) 569–572.
[135] J. Naciri, A. Srinivasan, H. Jeon, N. Nikolov, P. Keller, B.R. Ratna, Nematic elastomer fiber actuator, Macromolecules 36 (22) (2003) 8499–8505.
[136] T.T. Reddy, A. Kano, A. Maruyama, M. Hadano, A. Takahara, Thermosensitive transparent semi-interpenetrating polymer networks for wound dressing and cell adhesion control, Biomacromolecules 9 (4) (2008) 1313–1321.
[137] M.A. Rogulska, W. Kultys, W. Podkoscielny, Studies on thermoplastic polyurethanes based on new diphenylethane-derivative diols. II. Synthesis and characterization of segmented polyurethanes from HDI and MDI, Eur. Polym. J 43 (2007) 1402–1414.
[138] G. Gultekin, C. Atalay-Oral, S. Erkal, F. Sahin, D. Karastova, S.B. Tantekin-Ersolmaz, et al., Fatty acid-based polyurethane films for wound dressing applications, J. Mater. Sci. Mater. Med. 20 (1) (2009) 421–431.
[139] K. Hearon, K. Gall, T. Ware, D.J. Maitland, J.P. Bearinger, T.S. Wilson, Post-polymerization crosslinked polyurethane shape memory polymers, J. Appl. Polym. Sci. 121 (1) (2011) 144–153.
[140] K.A. Davis, J.A. Burdick, K.S. Anseth, Photoinitiated crosslinked degradable copolymer networks for tissue engineering applications, Biomaterials 24 (14) (2003) 2485–2495.
[141] H. Yang, V. Mittal, Functional Polymer Blends Synthesis, Properties, and Performance, 2012.
[142] C. Prisacariu, Polyurethane Elastomers: From Morphology to Mechanical Aspects, Springer Science & Business Media, New York, 2011.
[143] S. Farzaneh, J. Fitoussi, A. Lucas, M. Bocquet, A. Tcharkhtchi, Shape memory effect and properties memory effect of polyurethane, J. Appl. Polym. Sci. 128 (5) (2013) 3240–3249.
[144] N.G. Sahoo, Y.C. Jung, N.S. Goo, J.W. Cho, Conducting shape memory polyurethane-polypyrrole composites for an electroactive actuator, Macromol. Mater. Eng. 290 (11) (2005) 1049–1055.

[145] H.S. Nalwa, Handbook of Organic Conductive Molecules and Polymers, Volume 4, Conductive Polymers: Transport, Photophysics and Applications, John Wiley & Sons Ltd., Chichester, 1997.
[146] R.T. Fox, V. Wani, K.E. Howard, A. Bogle, L. Kempel, Conductive polymer composite materials and their utility in electromagnetic shielding applications, J. Appl. Polym. Sci. 107 (4) (2008) 2558–2566.
[147] S. Geetha, K.K. Satheesh Kumar, C.R. Rao, M. Vijayan, D.C. Trivedi, EMI shielding: methods and materials—a review, J. Appl. Polym. Sci. 112 (4) (2009) 2073–2086.
[148] K.J. Vinoy, R.M. Jha, Radar Absorbing Materials- From Theory to Design and Characterization (Book), Kluwer Academic Publishers, Boston, MA, 1996, p. 1996.
[149] J. Jagur-Grodzinski, Electronically conductive polymers, Polym. Adv. Technol. 13 (9) (2002) 615–625.
[150] Z. Liu, G. Bai, Y. Huang, F. Li, Y. Ma, T. Guo, et al., Microwave absorption of single-walled carbon nanotubes/soluble cross-linked polyurethane composites, J. Phys. Chem. C 111 (37) (2007) 13696–13700.
[151] A.N. Yusoff, M.H. Abdullah, S.H. Ahmad, S.F. Jusoh, A.A. Mansor, S.A.A. Hamid, Electromagnetic and absorption properties of some microwave absorbers, J. Appl. Phys. 92 (2) (2002) 876–882.
[152] G. Oertel, Polyurethane Handbook: Chemistry, Raw Materials, Processing, Application, Hanser Publishing, Munich–Vienna–New York, 1985.
[153] M. Szycher (Ed.), Szycher's Handbook of Polyurethanes, CRC Press, Boca Raton, 1999.
[154] S. Brady, D. Diamond, K.T. Lau, Inherently conducting polymer modified polyurethane smart foam for pressure sensing, Sens. Actuat. A Phys. 119 (2) (2005) 398–404.
[155] C.L. Choong, M.B. Shim, B.S. Lee, S. Jeon, D.S. Ko, T.H. Kang, et al., Highly stretchable resistive pressure sensors using a conductive elastomeric composite on a micropyramid array, Adv. Mater. 26 (21) (2014) 3451–3458.
[156] S.P. Lacour, J. Jones, S. Wagner, T. Li, Z. Suo, Stretchable interconnects for elastic electronic surfaces, Proc. IEEE 93 (8) (2005) 1459–1467.
[157] S.P. Lacour, C. Tsay, S. Wagner, An elastically stretchable TFT circuit, Electron Device Lett. IEEE 25 (12) (2004) 792–794.
[158] D.H. Kim, J.H. Ahn, W.M. Choi, H.S. Kim, T.H. Kim, J. Song, et al., Stretchable and foldable silicon integrated circuits, Science 320 (5875) (2008) 507–511.
[159] R. Faez, W.A. Gazotti, M.A. De Paoli, An elastomeric conductor based on polyaniline prepared by mechanical mixing, Polymer 40 (20) (1999) 5497–5503.
[160] H. Stoyanov, M. Kollosche, S. Risse, R. Waché, G. Kofod, Soft conductive elastomer materials for stretchable electronics and voltage controlled artificial muscles, Adv. Mater. 25 (4) (2013) 578–583.
[161] F. El-Tantawy, N.A. Aal, A.A. Al-Ghamdi, E.H. El-Mossalamy, New smart conducting elastomer blends of Bi-based superconductor ceramics nanoparticles reinforced natural rubber/low-density polyethylene for double thermistors, antistatic protectors, and electromagnetic interference shielding effectiveness applications, Polym. Eng. Sci. 49 (3) (2009) 592–601.
[162] U. Khan, P. May, A. O'Neill, J.N. Coleman, Development of stiff, strong, yet tough composites by the addition of solvent exfoliated graphene to polyurethane, Carbon 48 (14) (2010) 4035–4041.
[163] S. Shang, W. Zeng, X.M. Tao, High stretchable MWNTs/polyurethane conductive nanocomposites, J. Mater. Chem. 21 (20) (2011) 7274–7280.
[164] J.B. Lee, D.Y. Khang, Electrical and mechanical characterization of stretchable multi-walled carbon nanotubes/polydimethylsiloxane elastomeric composite conductors, Compos. Science Technol. 72 (11) (2012) 1257–1263.

[165] Y.R. Lee, A.V. Raghu, H.M. Jeong, B.K. Kim, Properties of waterborne polyurethane/functionalized graphene sheet nanocomposites prepared by an in situ method, Macromol. Chem. Phys. 210 (15) (2009) 1247–1254.
[166] M. Chen, T. Tao, L. Zhang, W. Gao, C. Li, Highly conductive and stretchable polymer composites based on graphene/MWCNT network, Chem. Commun. 49 (16) (2013) 1612–1614.
[167] T.S. Hansen, K. West, O. Hassager, N.B. Larsen, Highly stretchable and conductive polymer material made from poly (3, 4-ethylenedioxythiophene) and polyurethane elastomers, Adv. Funct. Mater. 17 (16) (2007) 3069–3073.
[168] J.S. Noh, Highly conductive and stretchable poly (dimethylsiloxane): poly (3, 4-ethylenedioxythiophene): poly (styrene sulfonic acid) blends for organic interconnects, RSC Adv. 4 (4) (2014) 1857–1863.

CHAPTER 9

Electrospun Polyurethane Nanofibrous Mats for Wound Dressing Applications

Afeesh R. Unnithan, Arathyram R.K. Sasikala, Chan H. Park and Cheol S. Kim
Chonbuk National University, Jeonju, Republic of Korea

Contents

9.1 INTRODUCTION

When the size of a material is reduced from micrometer to nanometer, the resultant properties can change dramatically. These properties include amazing characteristics such as a very large surface area to volume ratio, flexibility in surface functionalities, and superior mechanical performance (e.g. stiffness and tensile strength) compared with any other known form of the material. These exceptional properties make the polymer nanofibers ideal nominees for many significant applications [1]. These one-dimensional (1D) nanostructured organic materials have gained constant scientific, technological, and industrial attention, with applications in fields such as air and water filtration [2], drug delivery [3], tissue engineering and regenerative medicine [4], active materials for photonics or electronics etc [5]. Electrospinning has gained popularity in the last decade due in a large part to an increased interest in nanoscale properties and technologies. This method permits the production of polymer fibers with diameters changing from 3 nm to bigger than 5 μm [6]. Electrospinning is considered a unique technology, not only due to its unmatched operational easiness, but also because of its effective upscaling, opening actual perspectives for industrial

Polyurethane Polymers: Blends and Interpenetrating Polymer Networks
DOI: http://dx.doi.org/10.1016/B978-0-12-804039-3.00009-9

fabrication. The industrial applications of electrospinning have acquired special interest because the nanofibers have shown great potential in many fields due to its intriguing peculiarities including their in-principle extreme length (up to km), high surface area and tunable porosity, inherent three-dimensional (3D) topography, and functional properties [7,8].

The most interesting feature of electrospinning is the easiness and low-cost nature of the setup; a typical electrospinning setup involves a syringe pump, a high-voltage source, and a collector. During the electrospinning practice, a polymer solution is held at a needle tip by surface tension. When an electric field was applied using the high-voltage source, it causes charge to be induced within the polymer, resulting in charge repulsion within the solution [9,10]. This electrostatic force opposes the surface tension. Ultimately the charge repulsion overcomes the surface tension, causing the initiation of a jet. The jet undergoes substantial stretching and whipping processes induced by Coulombic repulsion among surface charges, and associated solvent evaporation [11]. During the jet flight, the stream cross-section can be reduced up to six orders of magnitude and later the polymer nanofibers are collected on target electrodes. Once adequate throughput and process stability and reliability are accomplished, fibers are then available to be used in technical applications [11–13].

Various polymers, both synthetic and natural, were electrospun to make the nanofibers. Among them, polyurethanes (PU) were considered to be the most biocompatible and blood compatible material for wound dressing due to their structure and property diversity [14,15]. Some of the notable properties like durability, fatigue resistance in tensile, shear resistance, and elasticity are worth mentioning [16,17]. Moreover the modification of PU by hydrophilic–hydrophobic balance or by attachments of biologically active materials such as growth factors, anticoagulants, antibacterial agents and cell targeting agents or function groups, are possible through the hydrogen contain functional groups [18,19]. Polyurethanes have a vast range of applications in diverse industrial areas such as automotive, construction, furniture, and footwear. In addition a number of studies have reported on the successful electrospinning of polyurethane solutions, for the purpose of applications in protective textiles, tissue engineering, wound dressing, sensor, and filter applications [20].

9.2 WOUND DRESSING APPLICATION OF PU NANOFIBROUS MATS

Electrospun nanofibrous membranes for wound dressing applications can accomplish the necessities such as higher gas permeation and protection

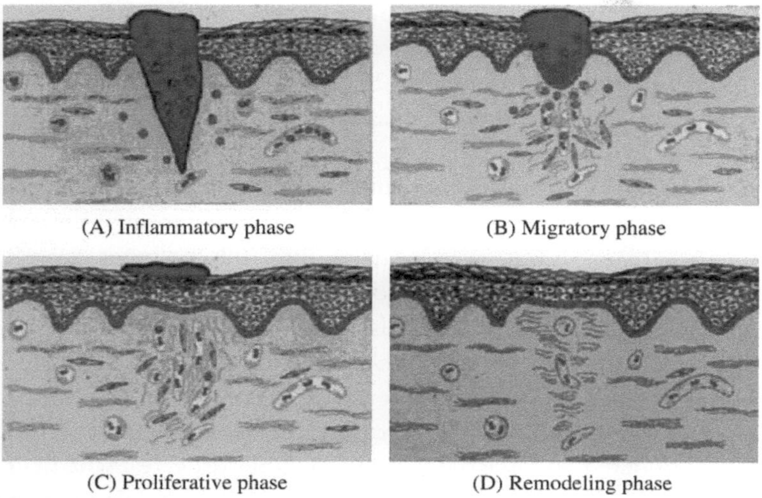

Figure 9.1 Schematic representation of the phases of wound healing: (A) infiltration of neutrophils into the wound area; (B) invasion of wound area by epithelial cells; (C) epithelium completely covers the wound; (D) many of the capillaries and fibroblasts, formed at early stages have all disappeared. *Reprinted from Ref. P. Zahedi, I. Rezaeian, S.O. Ranaei-Siadat, S.H. Jafari, P. Supaphol. A review on wound dressings with an emphasis on electrospun nanofibrous polymeric bandages. Polym. Adv. Technol. 21 (2010) 77–95., Copyright Wiley-Liss, Inc. and the American Pharmacists Association.*

of wound from infection and dehydration [21,22]. The goal of wound dressing is the production of an ideal structure that gives higher porosity and good barrier. To reach this goal, wound dressing materials must be selected carefully, and the structure must be controlled to confirm that it has good barrier properties and oxygen permeability [23]. PU is regularly used in wound dressings because of its good barrier properties and oxygen permeability (see Fig. 9.1).

Many researchers also reported that semipermeable wound dressings made of PU improved the wound healing. The permeability to water is also important so that fluid from the wound does not build up between the wound and the dressing and that wound desiccation does not occur. The wound dressings made from electrospun PU nanofibers can meet the requirements such as higher gas permeation rate and protection of wound from infection and dehydration [22,24]. Wound healing is a special biological process related to physiological parameters. Selection of a suitable wound dressing material for a specific type of wound requires comprehensive knowledge of the wound healing process. There are numerous reports describing the various biological and physiological stages of the healing process of a wound. These can be summarized into five consecutive cascades of events of hemostasis, inflammation, migration, proliferation, and maturation [23,25]. Some of the most interesting PU nanofibers based wound dressing works are summarized in this chapter.

PU nanofibers integrating inorganic nanoparticles for wound dressing applications have received a great deal of attention recently, because the introduction of inorganic nanoparticles into the polymer matrix has proved to be an effective and low-cost method of improving the performance of existing polymeric materials. It is already well-known that silver ions or silver nanoparticles are highly toxic to microorganisms [21,22]. When silver comes in contact with microorganisms, it leads to immediate distortion of cell wall, which later causes the death of these organisms. Silver is considered to be active against multiple drug resistant microbial strains. Remarkably, there are minimal chances for development of bacterial resistance due to immediate death of microorganisms upon contact with silver ions [26]. Beside its antimicrobial activity, it has been found that the use of silver based dressing materials enhances epithelialization in clean wounds of animal model which indicates synergistic effect of silver ions as antimicrobials and as well as wound healing agents [27]. Therefore, silver nanoparticles or ions, as an antibacterial component, have been applied in the preparation of various biomedical materials such as wound dressings by compounding them into PU nanofiber membranes or coating them on the surfaces.

In a study by Chen et al. [22], antimicrobial PU nanofibers were prepared by a simple method, where N,N′ -dimethylformamide (DMF) was used as a solvent for PU, as well as a reducing agent for Ag^+ ions. The Ag^+ ions in the PU solutions were directly reduced to produce silver nanoparticles and the resulting solution could be electrospun into nanofiber membranes. The silver nanoparticles containing PU nanofibers were grafted with Type I collagen by reacting with the carboxyl and peroxide groups generated on PU surface from DC pulsed oxygen plasma modification to produce a functional bioactive wound dressing. After modification, the nanofiber membranes antimicrobial activity improved to $\sim$100% inhibition of bacterial growth with associated increase of membrane water absorption ability, which enables its use as a functional wound dressing. This composite electrospun PU nanofiber membrane will have a potential application as a wound dressing for skin regeneration.

Sheikh et al. reported that the nano-biotechnologically engineered silver nitrate/PU sol – gel can be easily electrospun to form smooth nanofibers [28]. The presence of DMF can result in reduced $AgNO_3$to silver nanoparticles (NPs) in a simple and biologically safe way. Moreover, the obtained silver nanoparticles do have strong stagnancy as these NPs have not been released after many successive washing processes of the nanofiber mats by big amounts of fresh water for a long time. Two model organisms, *E. coli* and *S. typhimurium*, were used to analyze the antimicrobial influence of these nanofiber mats. Subsequently, antimicrobial tests indicated that the prepared nanofibers have a high bactericidal effect. Accordingly, these results highlight the potential use of these nanofiber mats as antimicrobial agents. According to their results, they claimed that the prepared silver NPs/PU nanofibrous matrix could be properly employed as a suggested candidate for many biological applications such as for sustained antimicrobial wound dressing agents (see Fig. 9.2).

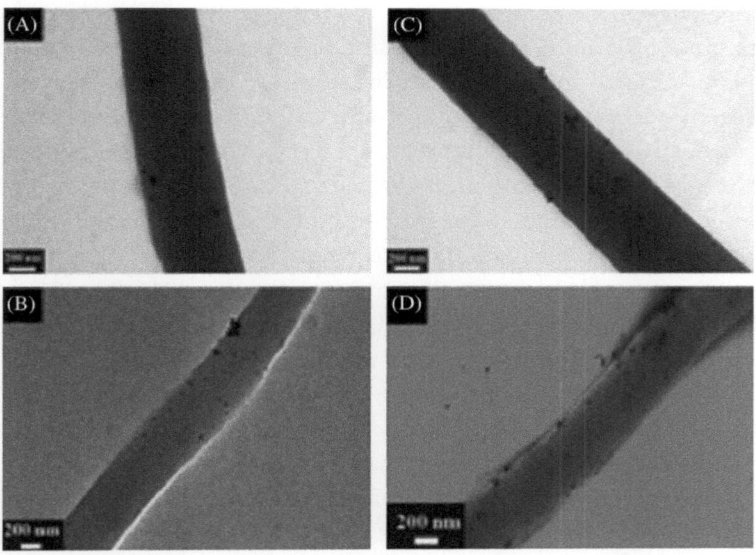

Figure 9.2 Transmission electron micrographs of the prepared nanofibers obtained from silver nitrate/PU sol – gels containing different amounts of silver nitrate, 2% (A), 5% (B), 7% (C), and 10% (D) silver nitrate with respect to PU. *Reprinted from Ref. F.A. Sheikh, N.A.M. Barakat, M.A. Kanjwal, A.A. Chaudhari, I.H. Jung, J.H. Lee, et al. Electrospun antimicrobial polyurethane nanofibers containing silver nanoparticles for biotechnological applications. Macromol. Res. 17 (2009) 688–696. Copyright 2009, with permission from Springer.*

Similarly, Kim et al. [29] demonstrated the fabrication of wound dressings that contain badger oil as a healing agent, AgNPs as an antibacterial agent, and polyurethane as the foundation polymer. In this study, silver nanoparticles and badger oil were embedded into nanofibrous PU mats via electrospinning. Badger oil is mainly composed of fatty acids and is used as a traditional medicine to heal wounds and the silver nanoparticles contribute to the wound healing process by reducing wound contamination. The composite mats exhibited good bactericidal activity against both of Gram-positive and Gram-negative bacteria, with better cell attachment and proliferation. The results of this study indicate that the proposed composite mats can be used for various biomedical purposes, including as dressings for burn wounds or to treat skin disease. In short all these studies propose that the PU nanofibers and silver nanoparticles make a good combination for the fabrication of vital wound dressing material.

Drug loaded nanofibers are also a good choice for making the wound dressing material. Ciprofloxacin HCl (CipHCl), a fluoroquinolone antibiotic, is one of the most widely used antibiotics in wound healing because of its low minimal inhibitory concentration for both Gram-positive and Gram-negative bacteria that cause wound infections and the frequency of spontaneous resistance to ciprofloxacin is very low. An electrospun nanofiber membrane containing antibiotic agents can be used as a barrier to prevent the post wound infections. In study done by Afeesh et al., an antibacterial electrospun scaffold was prepared by electrospinning of a solution composed of dextran, PU, and ciprofloxacin HCl (CipHCl) [30]. The addition of dextran into the PU increased the cell attachment and viability. These composite nanofiber mats showed very good antibacterial activity toward both Gram-positive and Gram-negative bacteria. This work provides a basic understanding of the design of efficient nanofiber-based antibacterial wound dressing material (see Figs. 9.3 and 9.4).

In another study, an antibacterial electrospun nanofibrous scaffolds were prepared by physically blending polyurethane with two biopolymers such as cellulose acetate (CA) and zein [31]. Here, PU was used as the foundation polymer, was blended with CA and zein to achieve desirable properties such as better hydrophilicity, excellent cell attachment, proliferation and blood clotting ability. To prevent common clinical infections, an antimicrobial agent, streptomycin sulfate, was incorporated into the electrospun fibers and its antimicrobial ability against the Gram-negative and Gram-positive bacteria were examined. The interaction between

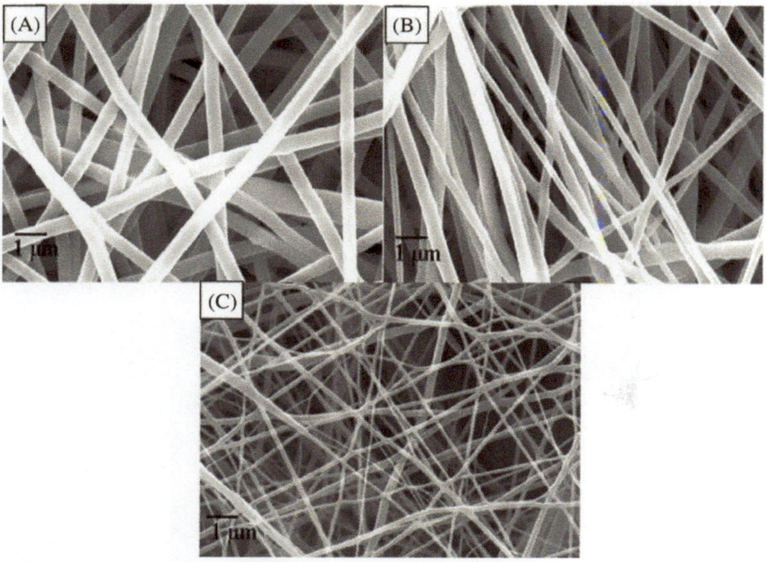

Figure 9.3 SEM images of electrospun (A) PU; (B) PU–dextran; (C) PU–dextran–drug nanofibrous mat. *Reprinted from Ref. A.R. Unnithan, N.A.M. Barakat, P.B.T. Pichiah, G. Gnanasekaran, R. Nirmala, Y.S. Cha, et al. Wound-dressing materials with antibacterial activity from electrospun polyurethane-dextran nanofiber mats containing ciprofloxacin HCl. Carbohydr. Polym. 90 (2012): 1786–1793. Copyright 2012, with permission from Elsevier.*

fibroblasts and the PU–CA and PU–CA–zein-drug scaffolds such as viability, proliferation, and attachment were characterized. PU–CA–zein-drug composite nanoscaffold showed enhanced blood clotting ability in comparison to pristine PU nanofibers. The presence of CA and zein in the nanofiber membrane improved its hydrophilicity, bioactivity and created a moist environment for the wound, which can accelerate wound recovery (see Fig. 9.5).

Natural wound healing material incorporating PU nanofibers were also studied for their ability to act as the wound dressing materials. In one such work, PU-Emu oil blended nanofibrous membranes scaffolds were fabricated successfully via electrospinning with different composition ratios with PU. PU was used as the electrospinning driven polymer to ease the spinning of emu oil [32]. Emu oil is derived from the emu (*Dromaius novaehollandiae*), which originated in Australia, and has been reported to have antiinflammatory properties. It has been observed that 10 wt% is the optimum oil content in the electrospun solution to get

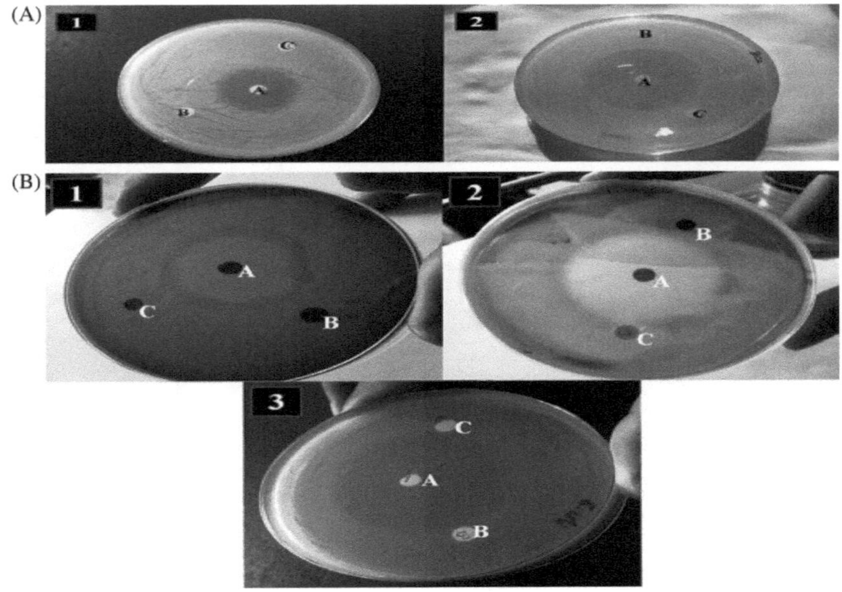

Figure 9.4 (A) Bactericidal activity of PU–dextran–drug nanofibrous mat with Gram-positive *Staphylococcus aureus* (1), *Bacillus subtilis* (2), respectively. PU–dextran–drug, PU–dextran and pristine PU disks were denoted as A, B, and C respectively in the Petri plates. (B) Bactericidal activity of PU–dextran–drug nanofibrous mat with Gram-negative *Escherichia coli* (1), *Salmonella typhimurium* (2), and *Vibrio vulnificus* (3), respectively. PU–dextran–drug, PU–dextran and pristine PU disks were denoted as A, B, and C respectively in the petri plates. *Reprinted from Ref. A.R. Unnithan, N.A.M. Barakat, P.B.T. Pichiah, G. Gnanasekaran, R. Nirmala, Y.S. Cha, et al. Wound-dressing materials with antibacterial activity from electrospun polyurethane-dextran nanofiber mats containing ciprofloxacin HCl. Carbohydr. Polym. 90 (2012): 1786–1793. Copyright 2012, with permission from Elsevier.*

good morphology ultrafine PU/emu oil blended nanofibers. The composite material could support long-term cell growth, form 3D networks of the nanofibrous structure, and provide good antibacterial activity. The antibacterial activity of this composite material was found to be active in both Gram-positive and Gram-negative species. Acceleration of fibroblast proliferation during wound closure is crucial for an effective skin healing. It is well-known that emu oil greatly stimulates proliferation and migration of skin tissues and thereby enhancing the early wound closure. As emu oil possesses all these properties it indicates that emu oil loaded with a PU nanofiber matrix is an ideal wound dressing material. Taking into account all these data, emu oil

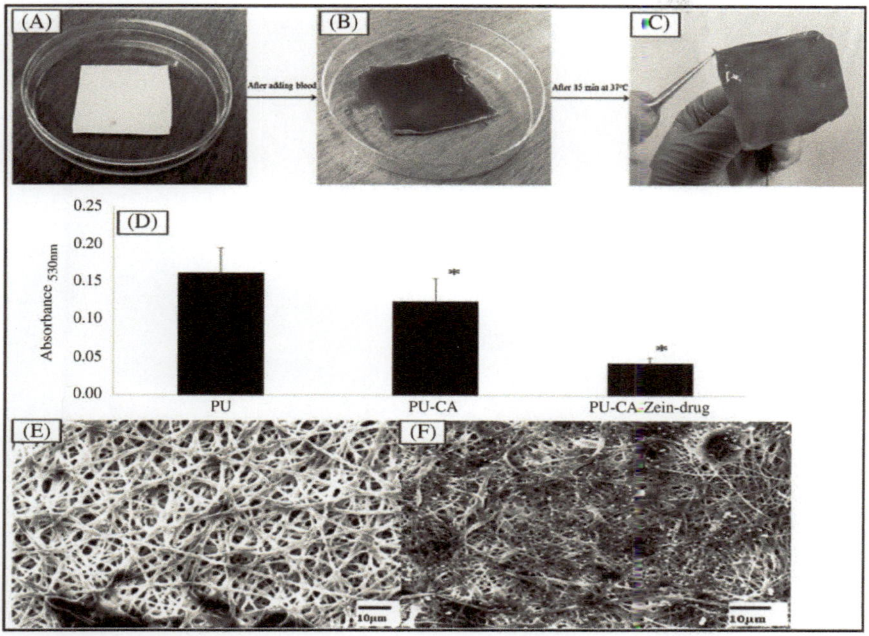

Figure 9.5 Photographs of the (A) PU—CA—zein nanofibrous mat, (B) blood on the PU—CA—zein composite mat, (C) clotted blood on the composite mat, (D) whole-blood clotting evaluation of nanofibrous mats (star symbols (*) represent the $P < 0.05$ level, indicating that the means are significantly different, compared to the control), SEM images of platelet activation of (E) PU nanofibrous mat, and (F) PU—CA—zein-drug composite nanofibrous mat (white dots indicate the platelet adhesion on composite mat) respectively. *Reprinted from Ref. A.R. Unnithan, G. Gnanasekaran, Y. Sathishkumar, Y.S. Lee, C.S. Kim. Electrospun antibacterial polyurethane-cellulose acetate-zein composite mats for wound dressing. Carbohydr. Polym. 102 (2014) 884—892. Copyright 2014, with permission from Elsevier.*

immobilization on PU nanofibrous matrices was vital in terms of a potential wound healing substrate (see Fig. 9.6).

Amna et al. introduced a unique natural and synthetic hybrid micro-nanofiber wound dressing [33]. The composite nanofiber comprised virgin olive oil-copper oxide nanocrystals and PU, developed via facile electrospinning. Considering the potential of nanofibrous PU membrane to enhance epithelial growth and the benefits of olive oil and copper, an organic/inorganic hybrid micro-nanofibrous dressing will aid in quick healing as well as decrease the risk of wound infection. Cells grown on the designed nanomatrix were very similar to the unexposed control cells in morphology and function. Moreover, the hybrid PU micro-nanofibers

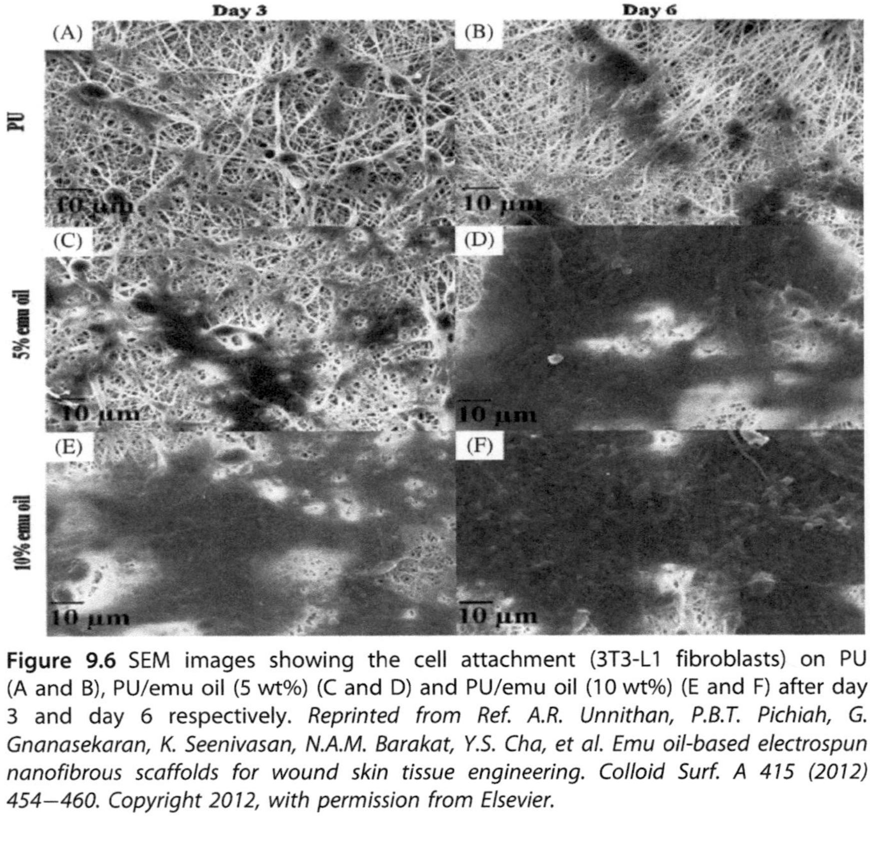

Figure 9.6 SEM images showing the cell attachment (3T3-L1 fibroblasts) on PU (A and B), PU/emu oil (5 wt%) (C and D) and PU/emu oil (10 wt%) (E and F) after day 3 and day 6 respectively. *Reprinted from Ref. A.R. Unnithan, P.B.T. Pichiah, G. Gnanasekaran, K. Seenivasan, N.A.M. Barakat, Y.S. Cha, et al. Emu oil-based electrospun nanofibrous scaffolds for wound skin tissue engineering. Colloid Surf. A 415 (2012) 454–460. Copyright 2012, with permission from Elsevier.*

are bactericidal, therefore, on the basis of the results; it is supposed that the hybrid micro-nanofibers will support accelerated wound healing. The electrospun composite dressing possesses multifunctionality, such as obstructing access of infectious microbes to wounds, maintaining a hydrated environment, deodorizing the wounds, preventing apoptosis of wound repair cells. Copper oxide–olive oil PU wound dressing may exert its positive beneficial effects at every stage during wound healing progression, and these micro-nanofibers may serve diverse biomedical applications, such as tissue regeneration, damaged skin treatment, and wound healing applications.

Wang et al. combined the merits of polyurethane, keratin, and silver nanoparticles (AgNPs) together and developed a novel nanofibrous mat for wound dressing [34]. The modified keratin was then introduced and coelectrospun with PU to improve hydrophilicity, cell viability, cell attachment, and proliferation. Sequentially, PU/keratin (PU/K) mats were immersed with silver nitrate solution and treated with sodium citrate to form PU/keratin/AgNP (PU/K/AgNP) mats in situ. These fabricated PU/K/AgNP mats showed good cell viability and antibacterial property in all tested mats. Moreover, PU/K/AgNP mats did not elicit TNF-α secretion and exerted very mild inflammatory responses. From the results of animal testing, these mats could promote wound healing. PU/keratin/silver nanoparticle mats have the potential to be a good wound dressing material for skin repair.

In a recent study, PU nanofibers were successfully applied as a postmenopausal wound dressing material [35]. Here the authors introduced a new material for wound tissue dressing, in which a polyurethane–dextran composite nanofibrous wound dressing material loaded with β-estradiol was obtained through electrospinning. In that preliminary work, a composite nanofibrous wound dressing material loaded with β-estradiol was fabricated through electrospinning of PU. The continuous systemic estrogen release from the estradiol loaded PU nanofibrous mat at the wound area can accelerate healing of acute cutaneous wounds, linked to its potent antiinflammatory activity. Using a validated in vivo model, they have proven that incisional wounds take far longer time to heal in the absence of estradiol loaded PU–dextran mat. All the results indicated that this advanced estradiol loaded PU-dextran composite mats can be successfully applied as the potential wound dressing material for the postmenopausal stage (see Fig. 9.7).

CONCLUSION

PU has been widely used for wound dressing applications due to its superior biocompatibility over other synthetic polymers. For wound healing, electrospun, interconnected porous nanofibers are advantageous for oozing fluids, controlling drainage and facilitating oxygen permeation. The main purpose of wound dressings is to remain close to native skin, inhibit the loss of blood, assist in removing exudates, block external microorganisms from creating an infection, and advancing the appearance of the wound site. Moreover the flexibility of the electrospinning process allows the addition of drugs in the nanofibers, for medical treatment and antibacterial purposes. PU is frequently used in wound dressings because of its good barrier properties and oxygen permeability. It has been reported that semipermeable dressings, many of which are PU, enhance wound healing. The permeability to water is also important so that fluid from the wound

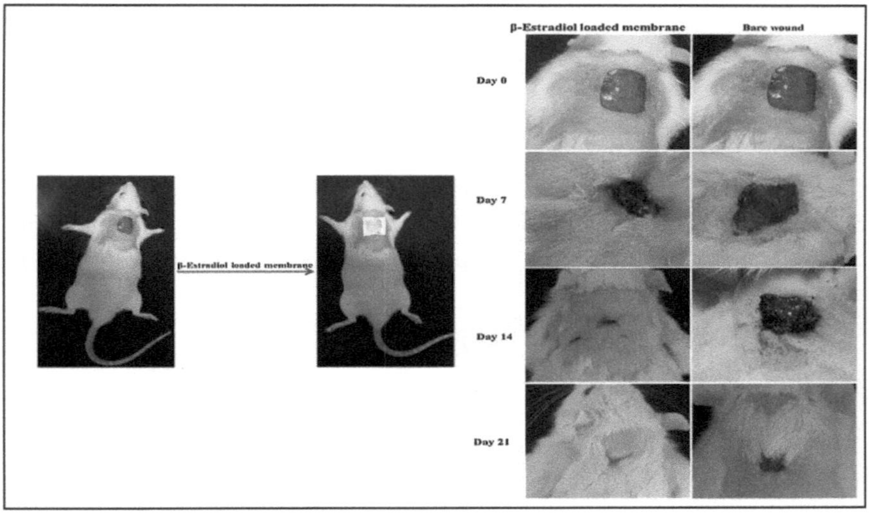

Figure 9.7 Photographs of the in vivo wound healing study. Note the extent of wound closure in the wounds treated with estradiol loaded PU-dextran composite mat compared to bare wound. *Reprinted from Ref. A.R. Unnithan, A.R.K. Sasikala, P. Murugesan, M. Gurusamy, D.M. Wu, C.H. Park, et al. Electrospun polyurethane-dextran nanofiber mats loaded with Estradiol for post-menopausal wound dressing. Int. J. Biol. Macromol. 77 (2015) 1–8. Copyright 2015, with permission from Elsevier.*

does not build up between the wound and the dressing and the wound desiccation does not occur. Electrospun polymeric PU nanomembranes are the most advanced and efficient wound dressing materials compared to other modern bandages such as hydrocolloids, hydrogels, and alginates. These membranes, due to their very high surface area to volume ratio, have extraordinary abilities such as controlled release of drugs, cultivation in the physiological environment of the cells, and a very high increase in the reproduction and growth of live cells. All these characteristics have highly extended their applications on different wounds as compared to traditional wound dressing materials. This research showed that the combination of functionalized nanofibers fabricated through electrospinning process can open up new doors for the development of the next generation of wound dressing materials with highly desirable properties.

ACKNOWLEDGMENT

This work was supported by Korea Research Fellowship Program through the National Research Foundation of Korea (NRF) funded by the Ministry of Science, ICT and Future Planning (KRF-2016H1D3A1938077 and NRF-2016R1D1A1B03934124).

REFERENCES

[1] Z.M. Huang, Y.Z. Zhang, M. Kotaki, S. Ramakrishna, A review on polymer nanofibers by electrospinning and their applications in nanocomposites, Compos. Sci. Technol. 63 (2003) 2223–2253.

[2] R. Gopal, S. Kaur, Z.W. Ma, C. Chan, S. Ramakrishna, T. Matsuura, Electrospun nanofibrous filtration membrane, J. Membr. Sci. 281 (2006) 581–586.

[3] T. Okuda, K. Tominaga, S. Kidoaki, Time-programmed dual release formulation by multilayered drug-loaded nanofiber meshes, J. Control Release 143 (2010) 258–264.

[4] D. Grafahrend, K.H. Heffels, M.V. Beer, P. Gasteier, M. Moller, G. Boehm, et al., Degradable polyester scaffolds with controlled surface chemistry combining minimal protein adsorption with specific bioactivation, Nat. Mater. 10 (2011) 67–73.

[5] L. Persano, A. Camposeo, C. Tekmen, D. Pisignano, Industrial upscaling of electrospinning and applications of polymer nanofibers: a review, Macromol. Mater. Eng. 298 (2013) 504–520.

[6] T. Subbiah, G.S. Bhat, R.W. Tock, S. Pararneswaran, S.S. Ramkumar, Electrospinning of nanofibers, J. Appl. Polym. Sci. 96 (2005) 557–569.

[7] W. Sambaer, M. Zatloukal, D. Kimmer, 3D modeling of filtration process via polyurethane nanofiber based nonwoven filters prepared by electrospinning process, Chem. Eng. Sci. 66 (2011) 613–623.

[8] H.F. Yang, W. Hong, L. Dong, A controlled biochemical release device with embedded nanofluidic channels, Appl. Phys. Lett. (2012) 100.

[9] D.H. Reneker, A.L. Yarin, H. Fong, S. Koombhongse, Bending instability of electrically charged liquid jets of polymer solutions in electrospinning, J. Appl. Phys. 87 (2000) 4531–4547.

[10] A.L. Yarin, S. Koombhongse, D.H. Reneker, Taylor cone and jetting from liquid droplets in electrospinning of nanofibers, J. Appl. Phys. 90 (2001) 4836–4846.

[11] M.M. Hohman, M. Shin, G. Rutledge, M.P. Brenner, Electrospinning and electrically forced jets. I. Stability theory, Phys. Fluids 13 (2001) 2201–2220.

[12] S.A. Theron, E. Zussman, A.L. Yarin, Experimental investigation of the governing parameters in the electrospinning of polymer solutions, Polymer 45 (2004) 2017–2030.

[13] S.V. Fridrikh, J.H. Yu, M.P. Brenner, G.C. Rutledge, Controlling the fiber diameter during electrospinning, Phys. Rev. Lett. (2003) 90.

[14] M.S. Khil, D.I. Cha, H.Y. Kim, I.S. Kim, N. Bhattarai, Electrospun nanofibrous polyurethane membrane as wound dressing, J. Biomed. Mater. Res. B 67b (2003) 675–679.

[15] C. Yao, X.S. Li, K.G. Neoh, Z.L. Shi, E.T. Kang, Surface modification and antibacterial activity of electrospun polyurethane fibrous membranes with quaternary ammonium moieties, J. Membr. Sci. 320 (2008) 259–267.

[16] S. Kidoaki, K. Kwon, T. Matsuda, Structural features and mechanical properties of in situ-bonded meshes of segmented polyurethane electrospun from mixed solvents, J. Biomed. Mater. Res. B 76b (2006) 219–229.

[17] H.T. Zhuo, J.L. Hu, S.J. Chen, L.P. Yeung, Preparation of polyurethane nanofibers by electrospinning, J. Appl. Polym. Sci. 109 (2008) 406–411.

[18] K.H. Hsieh, D.C. Liao, C.Y. Chen, W.Y. Chiu, Interpenetrating polymer networks of polyurethane and maleimide-terminated polyurethane for biomedical applications, Polym. Adv. Technol. 7 (1996) 265–272.

[19] D. Macocinschi, C. Tanase, D. Filip, S. Vlad, A. Oprea, Study of the relationship between new polyurethane composites for biomedical applications and fungal contamination, Mater. Plast. 47 (2010) 286–291.
[20] R. Nirmala, D. Kalpana, R. Navamathavan, Y.S. Lee, H.Y. Kim, Preparation and characterizations of silver incorporated polyurethane composite nanofibers via electrospinning for biomedical applications, J. Nanosci. Nanotechnol. 13 (2013) 4686–4693.
[21] H.J. Jeon, J.S. Kim, T.G. Kim, J.H. Kim, W.R. Yu, J.H. Youk, Preparation of poly (epsilon-caprolactone)-based polyurethane nanofibers containing silver nanoparticles, Appl. Surf. Sci. 254 (2008) 5886–5890.
[22] J.P. Chen, Y. Chiang, Bioactive electrospun silver nanoparticles-containing polyurethane nanofibers as wound dressings, J. Nanosci. Nanotechnol. 10 (2010) 7560–7564.
[23] P. Zahedi, I. Rezaeian, S.O. Ranaei-Siadat, S.H. Jafari, P. Supaphol, A review on wound dressings with an emphasis on electrospun nanofibrous polymeric bandages, Polym. Adv. Technol. 21 (2010) 77–95.
[24] R. Chen, C. Huang, Q.F. Ke, C.L. He, H.S. Wang, X.M. Mo, Preparation and characterization of coaxial electrospun thermoplastic polyurethane/collagen compound nanofibers for tissue engineering applications, Colloid Surf. B 79 (2010) 315–325.
[25] J.S. Boateng, K.H. Matthews, H.N.E. Stevens, G.M. Eccleston, Wound healing dressings and drug delivery systems: a review, J. Pharm. Sci. US 97 (2008) 2892–2923.
[26] S. Silver, Bacterial silver resistance: molecular biology and uses and misuses of silver compounds, Fems Microbiol. Rev. 27 (2003) 341–353.
[27] J. Tian, K.K.Y. Wong, C.M. Ho, C.N. Lok, W.Y. Yu, C.M. Che, et al., Topical delivery of silver nanoparticles promotes wound healing, Chemmedchem 2 (2007) 129–136.
[28] F.A. Sheikh, N.A.M. Barakat, M.A. Kanjwal, A.A. Chaudhari, I.H. Jung, J.H. Lee, et al., Electrospun antimicrobial polyurethane nanofibers containing silver nanoparticles for biotechnological applications, Macromol. Res. 17 (2009) 688–696.
[29] J.H. Kim, A.R. Unnithan, H.J. Kim, A.P. Tiwari, C.H. Park, C.S. Kim, Electrospun badger (Meles meles) oil/Ag nanoparticle based anti-bacterial mats for biomedical applications, J. Ind. Eng. Chem. 30 (2015) 254–260.
[30] A.R. Unnithan, N.A.M. Barakat, P.B.T. Pichiah, G. Gnanasekaran, R. Nirmala, Y.S. Cha, et al., Wound-dressing materials with antibacterial activity from electrospun polyurethane-dextran nanofiber mats containing ciprofloxacin HCl, Carbohydr. Polym. 90 (2012) 1786–1793.
[31] A.R. Unnithan, G. Gnanasekaran, Y. Sathishkumar, Y.S. Lee, C.S. Kim, Electrospun antibacterial polyurethane-cellulose acetate-zein composite mats for wound dressing, Carbohydr. Polym. 102 (2014) 884–892.
[32] A.R. Unnithan, P.B.T. Pichiah, G. Gnanasekaran, K. Seenivasan, N.A.M. Barakat, Y.S. Cha, et al., Emu oil-based electrospun nanofibrous scaffolds for wound skin tissue engineering, Colloid Surf. A 415 (2012) 454–460.
[33] T. Amna, M.S. Hassan, J. Yang, M.S. Khil, K.D. Song, J.D. Oh, et al., Virgin olive oil blended polyurethane micro/nanofibers ornamented with copper oxide nanocrystals for biomedical applications, Int. J. Nanomed. 9 (2014) 891–898.
[34] Y.F. Wang, P.F. Li, P. Xiang, J.T. Lu, J. Yuan, J. Shen, Electrospun polyurethane/keratin/AgNP biocomposite mats for biocompatible and antibacterial wound dressings, J. Mater. Chem. B 4 (2016) 635–648.
[35] A.R. Unnithan, A.R.K. Sasikala, P. Murugesan, M. Gurusamy, D.M. Wu, C.H. Park, et al., Electrospun polyurethane-dextran nanofiber mats loaded with Estradiol for post-menopausal wound dressing, Int. J. Biol. Macromol. 77 (2015) 1–8.

CHAPTER 10

Poly(urethane-methacrylate) Copolymers Prepared by the Atom Transfer Radical Polymerization Methods as a New Material for Hydrophobic Coatings

Paweł Chmielarz and Piotr Król
Rzeszów University of Technology, Rzeszów, Poland

Contents

10.1 INTRODUCTION

Atom transfer radical polymerization (ATRP) is currently one of the most often used synthetic polymerization methods to prepare well-defined (co)polymers with complex architecture [1–20]. This chapter summarizes recent research activities in ATRP for copolymers with particular emphasis on coatings applications. The highly robust and flexible ATRP techniques are particularly suited for the preparation of protective coatings with increased hydrophobicity. Research activities in the past decade indicate that ATRP has become an essential tool for the design and synthesis of advanced, noble, and novel copolymer coatings.

Recent needs in a range of protective coatings permanently connected with natural materials such as wood and paper or with ceramics and metal articles forced the search for new polymeric materials. However raw material limited possibilities and environmental issues are directing the

Polyurethane Polymers: Blends and Interpenetrating Polymer Networks
DOI: http://dx.doi.org/10.1016/B978-0-12-804039-3.00010-5

attention of researchers to improving polymerization methods of the well-known and widely applied such as vinyl monomers, as well as raw materials useful for producing condensation polymers and additive polymers e.g., polyurethanes (PU) or poly(meth)acrylates [19]. In recent years, these needs have been additionally enhanced by the need to develop new synthetic biomaterials that cooperate with the tissues of the human body. The latest methods in this regard, already well developed from the side of preparative and theoretical explaining kinetics and mechanism of their progress, but still poorly used in materials engineering, are methods known under the general name of atom transfer radical polymerization.

In our recent review about applying PU in the modern materials engineering we paid attention to the importance of the ability of these polymers for preparation of polymer coatings with increased hydrophobicity. It is possible to obtain it by building linear PU segments containing fluorine or different alkylammonium groups present in the ionomer coatings produced on the base of ecological waterborne PU lacquers [21]. ATRP methods are creating additional possibilities in this area, because they have been used extensively in the preparation of various tri-block copolymer surfaces for use as a functional biomaterials as well as special protective coatings to polar materials (old prints, wood, or ceramics) [22].

10.1.1 Atom Transfer Radical Polymerization

ATRP is one of the most rapidly developing areas of polymer science, allowing effective control over molecular weights (MWs), preparation of polymers with narrow molecular weight distributions (MWDs), incorporation of site selected functionalities, fabrication of various architectures as well as synthesis of well-defined hybrid compositions [1,3,5,7–14, 16–20,22–38]. It has been successfully used to prepare various (co)polymers with essentially any desired complex macromolecular architectures. This includes polymers with controlled topology, ranging from linear chains with precisely controlled dimensions and controlled dispersity (*DI*), to various branched structures formed by copolymerization of macromonomers, grafting onto functional backbones, grafting from multifunctional backbones, and copolymers formed using multifunctional initiators. ATRP was also used to precisely control chain composition and prepare segmented, block, and graft copolymers as well as periodic and gradient systems [33]. The list of monomers successfully homopolymerized by ATRP includes various styrenes, (meth)acrylates, (meth)acrylamides, and acrylonitrile as well as vinyl chloride [24,33,39–50].

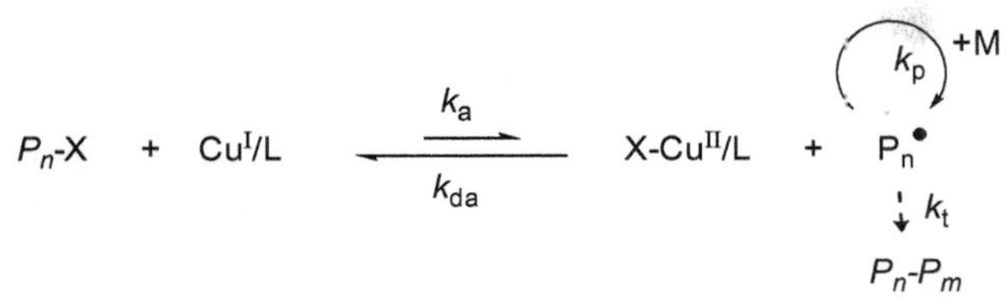

Scheme 10.1 ATRP processes (k_a = activation rate constant, k_{da} = deactivation rate constant, k_p = propagation rate constant, k_t = termination rate constant, M = monomer, L = ligand, X = halogen, and $P_n - X$ = alkyl halide) [18,53,54].

The essential feature of ATRP is controlled by an equilibrium between a low concentration of active propagating species and a larger number of dormant chains, predominately in the form of initiating alkyl halides/macromolecular species ($P_n - X$). In this case the dormant species periodically react with the rate constant of activation (k_a) with transition metal complexes in their lower oxidation state (Cu^I/L). After which, these species acting as activators to intermittently form growing radicals ($P_n^•$), and deactivators—transition metal complexes in their higher oxidation state, coordinated with halide ligands $X - Cu^{II}$/L (Scheme 10.1) [18,48,49,51,52].

The deactivator reacts with the propagating radical in a reverse reaction (k_{da}) to reform the dormant species and the activator. Upon addition of the intermediate radicals to monomers, polymer chains grow with the rate constant of propagation (k_p). Termination reactions (k_t) also occur in ATRP, mainly through radical coupling (P_n-P_m) or disproportionation ($P_n =$ and P_m-H). However, a very small fraction of polymer chains undergo termination in an ATRP procedure. This is due to the low concentration of active propagating radicals and higher concentration of dormant species, which minimizes termination [18,48,51,52].

10.1.2 Block Copolymers Prepared by the ATRP

Block copolymers are an interesting class of materials that possess different properties [55–60]. These copolymers are usually composed of two or more chemically distinct polymer chains linked together at one or more junction points through covalent or noncovalent bonds. As a consequence of the inherent immiscibility of different polymer segments, block copolymers undergo microphase separation in the bulk phase and in thin films. In solution, block copolymers will form micelles when the solvent is selective for one of the blocks. Most of the desirable properties of block copolymers originate from their ability to form well-defined nanostructures with different morphologies of tunable periodicity or size, and this provides the primary driving force for the intensive interest in this field [61]. As block length plays a major role on the properties of the block copolymers, effective control of the block lengths is important and this can easily be achieved using different controlled radical polymerization (CRP) methods [58]. Most of the desirable properties of narrow MWD block copolymers originate from their ability to form well-defined nanostructures with different morphologies of tunable periodicity or size, and this provides the primary driving force for the intensive interest in field of coating applications over the polydisperse copolymers synthesized by conventional radical polymerization. Narrow MWD block copolymers are more useful for coating applications rather than polydisperse random copolymers synthesized by conventional radical polymerization. It results from a regular arrangements of narrow MWD block copolymers, in which arranging structures results from intermolecular interactions, exists a possibility of the simpler crystallization as a result of stronger intermolecular interactions [22].

However, tri-block copolymers have sparked much interest and their potential has been realized in many areas involving the fields of chemistry, physics, and materials science, as well as the biological and medical sciences. Tri-block copolymers are an interesting class of materials that possess different properties compared to those of each individual homopolymer segments they are composed of.

Tri-block copolymers using different PU based macroinitiators were already obtained through ATRP classical mechanisms [55–60]. However, this method was arduous because of the necessity to degas the reaction mixture. Recently, we have demonstrated that ARGET ATRP can be successfully used for the synthesis of tri-block copolymers using PU as a macroinitiator in the presence of limited amounts of air, with omitting the deoxygenation process [57,58]. Realizing this concept, new types of poly(urethane-methacrylate) copolymers were synthesized, using the specially made urethane macroinitiators: bromine [58] and tetraphenylethane [57], capable of reacting with methacrylate monomers.

In the first case, 2-methyl-2-bromopropionate terminated (tertiary bromine-terminated) PU macroinitiator (MBP-PU-MBP) was synthesized from one mole of poly(oxytetramethylene) glycol (PTMO), two moles of 4,4″-methylene diphenyl diisocyanate (MDI), and two moles of 2-hydroxyethyl-2′ -methyl-2′-bromopropionate (HMB) [58]. When tetraphenylethane terminated PU macroinitiator (TPE-PU-TPE) was synthesized from one mole of PTMO, two moles of MDI, and two moles of 1,1,2,2-tetraphenylethane-1,2-diol (TPED) [57].

PU structures affect the improvement in mechanical properties of copolymers, while segments of PMMA are used to hydrophobize the PU surface [22]. Poly(urethane-acrylate) copolymers are finding application as biomaterials, including biocompatible implants in tissue engineering [62], hydrophobic antimicrobial coating [63], and drug delivery in the form of appropriately formed micellar structures [55].

The comparative study of these materials in terms of their resistance to hydrolytic degradation in the environment of physiological liquid similar to the blood plasma was conducted [64]. Performed examinations confirm that replacement of the urethane segments by more hydrophobic methacrylic segments contributes to increase the resistance to hydrolysis of such block copolymers, because of high resistance to biodegradation in conditions of their long contact with physiological liquids and tissues of living organisms. Meanwhile, by using the confocal microscopy (Fig. 10.1) method, it was determined that it is highly variable, which has a large influence to the

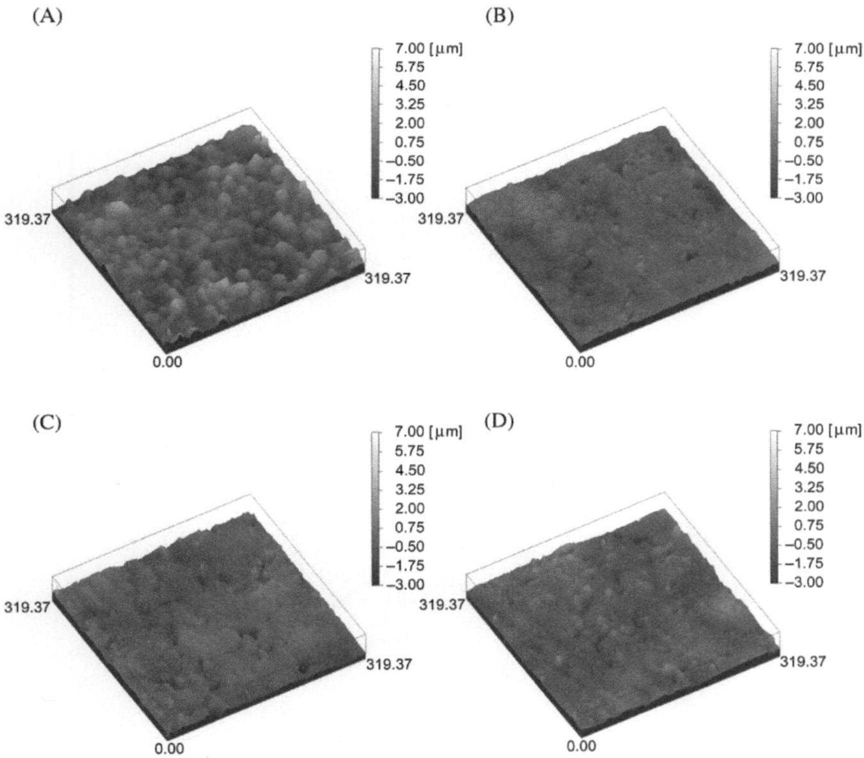

Figure 10.1 Pictures of the surface of poly(urethane-methacrylate) copolymers obtained by using MBP-PU-MBP macroinitiator before (A) and after (B) exposure in physiologic salt solution, as well as surface of copolymers obtained by using TPE-PU-TPE macroinitiator before (C) and after (D) exposure in physiologic salt solution, were taken for the objects 320 × 320 μm and with magnification 50 × [64].

quality of biomaterial, because in places of the inequality or structural dislocation, faults of phases built from segments of different polarity can be initiated degradation processes, causing pathological changes in the organ tissues being in contact with these areas [22].

Chatterjee and coworkers [65] describe the synthesis of different block copolymers of unsaturated and saturated polyesters with methyl methacrylate with well-defined molecular weight and narrow *DI* by combination of condensation polymerization and ATRP (Scheme 10.2).

In the case of unsaturated-polyester-based block copolymers the main chain double bond in the polyester backbone remains almost unaffected during ATRP. The unsaturated block copolymers are crosslinkable and can form networks. These copolymers might be interesting candidates for coatings with better overall properties than those based on neat polyesters.

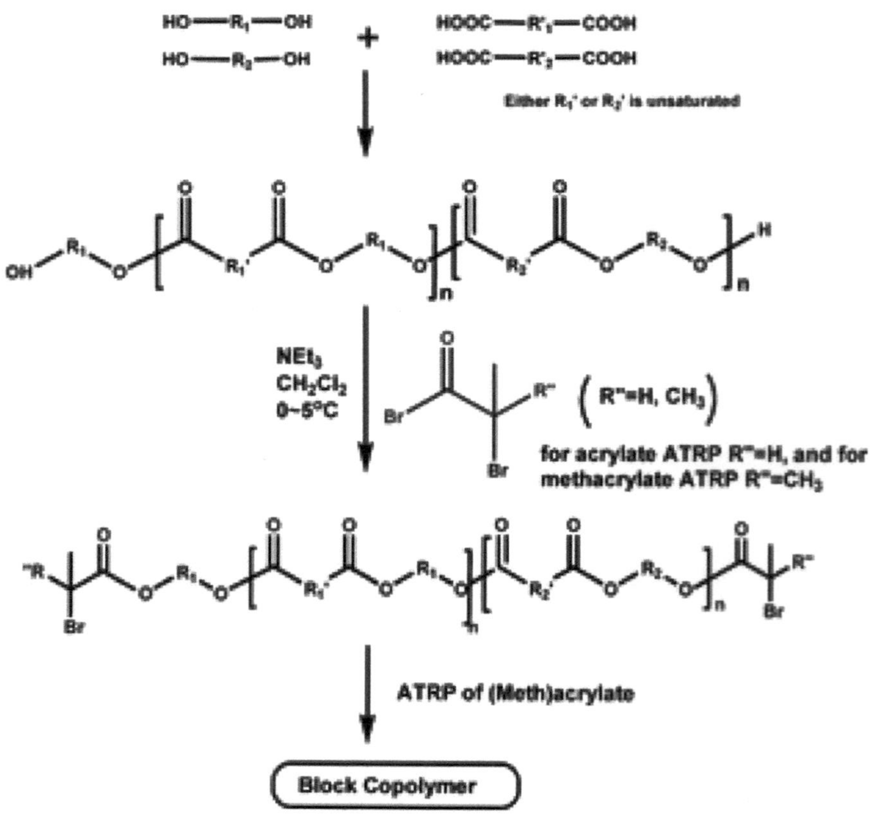

Scheme 10.2 Synthesis of polyester-poly(meth)acrylates block copolymers [65].

High-molecular-weight polyesters or polyester/polyurethanes have appreciable mechanical properties such as high tensile strength together with flexibility. But the presence of ester linkages in the backbone of polyesters makes them vulnerable to acids/bases. On the other hand, poly(meth) acrylates are better abrasion resistant, hydrolytically stable with superior mechanical properties [65].

More recently, Melinte and coworkers [66] describe the achievement of polymeric coatings enclosing noble metal nanoparticles (NPs) that can be easily produced by simply irradiation of the triazene polymeric films with embedded noble metal precursors ($AgNO_3$ or $AuBr_3$ metal salts). The Ag/Au nanoparticles were equally formed in the hydrophobic random copolymer based on poly(isodecyl methacrylate-*co*-triazene urethane-methacrylate) (COP-1) as well as in the amphiphilic block copolymer poly(isodecyl methacrylate-*co*-triazene urethane-methacrylate)-*block*-poly (acrylic acid) (COP-2), as both polymers comprised the radical-forming triazene pendent units (Scheme 10.3).

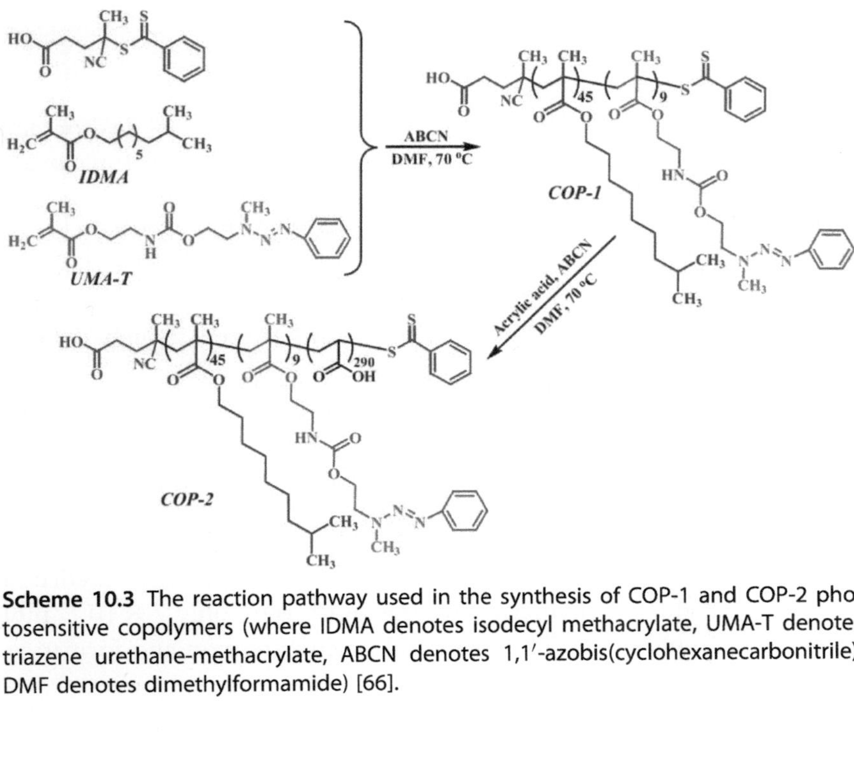

Scheme 10.3 The reaction pathway used in the synthesis of COP-1 and COP-2 photosensitive copolymers (where IDMA denotes isodecyl methacrylate, UMA-T denotes triazene urethane-methacrylate, ABCN denotes 1,1′-azobis(cyclohexanecarbonitrile), DMF denotes dimethylformamide) [66].

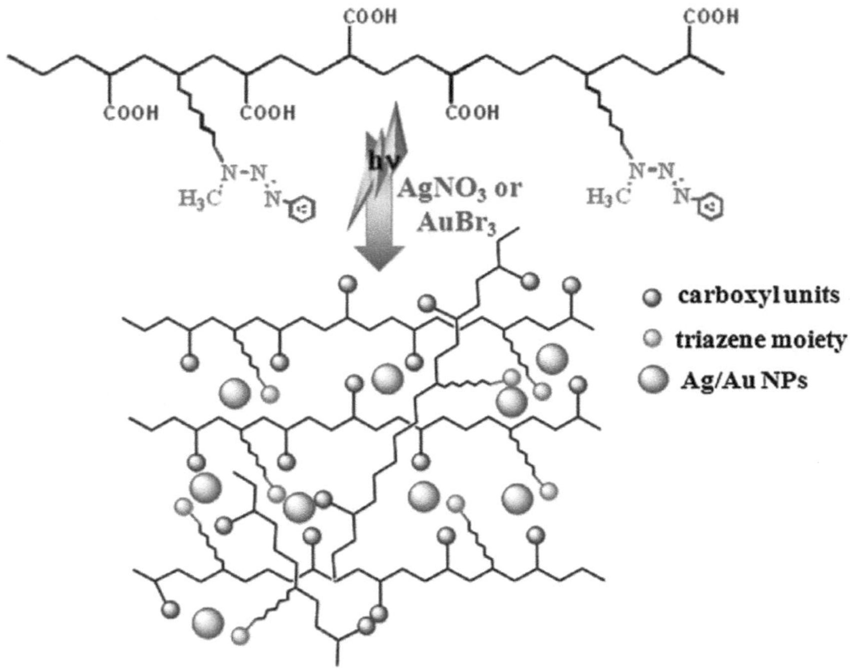

Scheme 10.4 The formation and stabilization of silver/gold NPs in COP-2 block copolymer matrix [66].

However, especially for the silver NPs, the presence of hydrophilic poly(acrylic acid) block favored the stabilization of the resulting nanoparticles, preventing their agglomeration (Scheme 10.4) [66].

The TEM images allowed the visualization of the silver NPs (sizes of 4–16 nm in COP-1 and of 2–6 nm in COP-2) as well as of the gold NPs (sizes between 10 nm and 20 nm in COP-1 and from 15 to 25 nm in COP-2), which are mainly spherical in shape, even though there is some triangular or hexagonal gold nanoparticles (Fig. 10.2).

Such a light-induced approach to in situ prepared hybrid organic/inorganic nanocomposites may be applied to a variety of polymer structures that contain photocleavable radical-forming entities, of interest for the field of catalysts, biosensors or nanocarriers [66].

Poly(*n*-butyl acrylate)-*block*-polyurethane-*block*-poly(*n*-butyl acrylate) copolymers were also prepared via electrochemically mediated atom transfer radical polymerization (*e*ATRP) using only 7×10^{-6} mol % of Cu^{II} complex [53]. The successful chain extension and formation

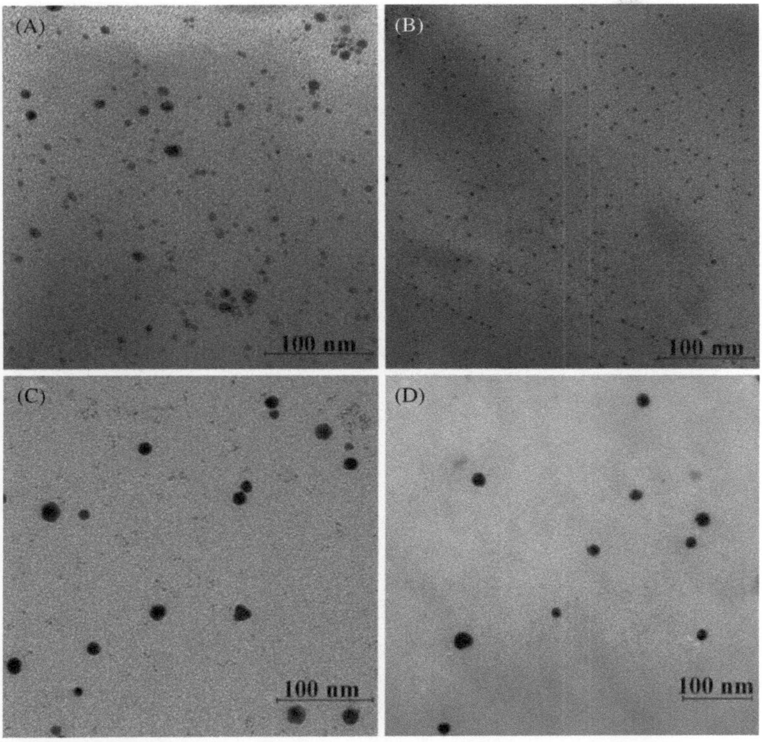

Figure 10.2 TEM images of Ag nanoparticles dispersed in COP-1 (A) and COP-2 (B) polymeric matrices along with Au nanoparticles photogenerated in COP-1 (C) and COP-2 (D) polymers [66].

of penta-block poly(*tert*-butyl acrylate-*block*-poly(*n*-butyl acrylate)-*block*-polyurethane-*block*-poly(*n*-butyl acrylate)-*block*-poly(*tert*-butyl acrylate) copolymers confirmed the living nature of the poly(alkyl acrylates) prepared by *e*ATRP (Scheme 10.5).

The tri-block and penta-block urethane-acrylate copolymers were synthesized for the first time by using tertiary bromine-terminated polyurethane macro initiators as transitional products reacting with *n*-butyl acrylate, and subsequently with *tert*-butyl acrylate in the presence of the $Cu^{II}Br_2$/TPMA catalyst complex [53]. It is also possible to obtain urethane-methacrylate copolymers using the same procedure. Those polymers are promising precursors for anionic polyelectrolyte coating materials, with potential application in the biomedical application field.

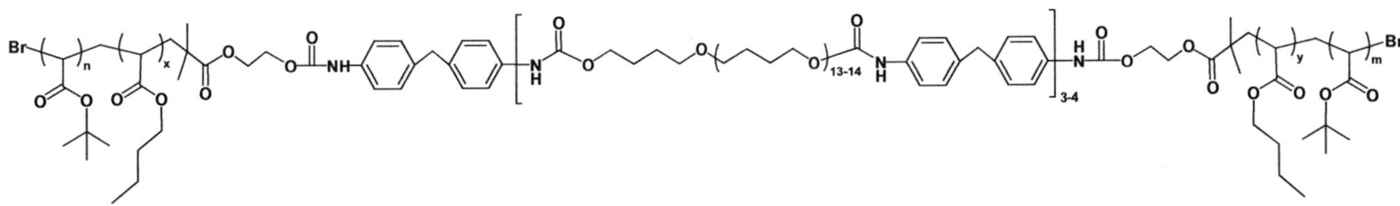

Scheme 10.5 Structure of P*t*BA-*b*-PBA-*b*-PU-*b*-PBA-*b*-P*t*BA penta-block copolymers prepared *via* *e*ATRP [53].

CONCLUSIONS

This chapter summarized recent research activities in ATRP for urethane-methacrylate copolymers with particular emphasis on coatings applications. The highly robust and flexible ATRP techniques are particularly suited for the preparation of protective coatings with increased hydrophobicity. The research activities in the last decade indicate that ATRP has become an essential tool for the design and synthesis of advanced, noble and novel copolymer coatings.

REFERENCES

[1] J.S. Wang, K. Matyjaszewski, Controlled living radical polymerization-Atom-transfer radical polymerization in the presence of transition-metal complexes, J. Am. Chem. Soc. 117 (1995) 5614–5615.

[2] H. Uegaki, Y. Kotani, M. Kamigaito, M. Sawamoto, Nickel-mediated living radical polymerization of methyl methacrylate, Macromolecules 30 (1997) 2249–2253.

[3] T.E. Patten, K. Matyjaszewski, Atom transfer radical polymerization and the synthesis of polymeric materials, Adv. Mater. 10 (1998) 901–915.

[4] D.A. Shipp, K. Matyjaszewski, Kinetic analysis of controlled/"living" radical polymerizations by simulations. 1. The importance of diffusion-controlled reactions, Macromolecules 32 (1999) 2948–2955.

[5] R.R. Shah, D. Merreceyes, M. Husemann, I. Rees, N.L. Abbott, C.J. Hawker, et al., Using atom transfer radical polymerization to amplify monolayers of initiators patterned by microcontact printing into polymer brushes for pattern transfer, Macromolecules 33 (2000) 597–605.

[6] K. Matyjaszewski, D.A. Shipp, G.P. McMurtry, S.G. Gaynor, T. Pakula, Simple and effective one-pot synthesis of (meth)acrylic block copolymers through atom transfer radical polymerization, J. Polym. Sci. Part A Polym. Chem. 38 (2000) 2023–2031.

[7] V. Coessens, T. Pintauer, K. Matyjaszewski, Functional polymers by atom transfer radical polymerization, Prog. Polym. Sci. 26 (2001) 337–377.

[8] G.L. Cheng, A. Boker, M.F. Zhang, G. Krausch, A.H.E. Muller, Amphiphilic cylindrical core-shell brushes *via* a "grafting from" process using ATRP, Macromolecules 34 (2001) 6883–6888.

[9] J. Pyun, T. Kowalewski, K. Matyjaszewski, Synthesis of polymer brushes using atom transfer radical polymerization, Macromol. Rapid Commun. 24 (2003) 1043–1059.

[10] S.H. Qin, D.Q. Oin, W.T. Ford, D.E. Resasco, J.E. Herrera, Polymer brushes on single-walled carbon nanotubes by atom transfer radical polymerization of *n*-butyl methacrylate, J. Am. Chem. Soc. 126 (2004) 170–176.

[11] K. Min, H.F. Gao, K. Matyjaszewski, Preparation of homopolymers and block copolymers in miniemulsion by ATRP using activators generated by electron transfer (AGET), J. Am. Chem. Soc. 127 (2005) 3825–3830.

[12] W.A. Braunecker, K. Matyjaszewski, Controlled/living radical polymerization: features, developments, and perspectives, Prog. Polym. Sci. 32 (2007) 93–146.

[13] J.G. Jiang, X.Y. Lu, Y. Lu, Preparation of carboxyl-end-group polyacrylamide with low polydispersity by ATRP initiated with chloroacetic acid, J. Polym. Sci. Pol. Chem. 45 (2007) 3956–3965.

[14] N.V. Tsarevsky, K. Matyjaszewski, "Green" atom transfer radical polymerization: From process design to preparation of well-defined environmentally friendly polymeric materials, Chem. Rev. 107 (2007) 2270–2299.
[15] K. Nakatani, M. Ouchi, M. Sawamoto, Terminal umpolung in metal-catalyzed living radical polymerization: quantitative end-capping of carbon – halogen bond *via* a modifier monomer, Macromolecules 41 (2008) 4579–4581.
[16] H.-i Lee, J. Pietrasik, S.S. Sheiko, K. Matyjaszewski, Stimuli-responsive molecular brushes, Prog. Polym. Sci. 35 (2010) 24–44.
[17] K. Matyjaszewski, Atom transfer radical polymerization: from mechanisms to applications, Isr. J. Chem. 52 (2012) 206–220.
[18] D.J. Siegwart, J.K. Oh, K. Matyjaszewski, ATRP in the design of functional materials for biomedical applications, Prog. Polym. Sci. 37 (2012) 18–37.
[19] P. Król, P. Chmielarz, Recent advances in ATRP methods in relation to the synthesis of copolymer coating materials, Prog. Org. Coat. 77 (2014) 913–948.
[20] K. Matyjaszewski, N.V. Tsarevsky, Macromolecular engineering by atom transfer radical polymerization, J. Am. Chem. Soc. 136 (2014) 6513–6533.
[21] P. Król, Synthesis methods, chemical structures and phase structures of linear polyurethanes. Properties and applications of linear polyurethanes in polyurethane elastomers, copolymers and ionomers, Progr. Mater. Sci. 52 (2007) 915–1015.
[22] F. di Lena, K. Matyjaszewski, Transition metal catalysts for controlled radical polymerization, Prog. Polym. Sci. 35 (2010) 959–1021.
[23] M. Kato, M. Kamigaito, M. Sawamoto, T. Higashimura, Polymerization of methylmethacrylate with the carbon-tetrachloride dichlorotris(Triphenylphosphine)ruthenium(Ii) methylaluminum bis(2,6-Di-Tert-Butylphenoxide) initiating system - possibility of living radical polymerization, Macromolecules 28 (1995) 1721–1723.
[24] K. Matyjaszewski, T.E. Patten, J.H. Xia, Controlled/"living" radical polymerization. Kinetics of the homogeneous atom transfer radical polymerization of styrene, J. Am. Chem. Soc. 119 (1997) 674–680.
[25] K. Matyjaszewski, P.J. Miller, N. Shukla, B. Immaraporn, A. Gelman, B.B. Luokala, et al., Polymers at interfaces: Using atom transfer radical polymerization in the controlled growth of homopolymers and block copolymers from silicon surfaces in the absence of untethered sacrificial initiator, Macromolecules 32 (1999) 8716–8724.
[26] J.H. Xia, K. Matyjaszewski, Controlled/"living" radical polymerization. Atom transfer radical polymerization catalyzed by copper(I) and picolylamine complexes, Macromolecules 32 (1999) 2434–2437.
[27] J.F. Hester, P. Banerjee, Y.Y. Won, A. Akthakul, M.H. Acar, A.M. Mayes, ATRP of amphiphilic graft copolymers based on pvdf and their use as membrane additives, Macromolecules 35 (2002) 7652–7661.
[28] R. Venkatesh, L. Yajjou, C.E. Koning, B. Klumperman, Novel brush copolymers *via* controlled radical polymerization, Macromol. Chem. Phys. 205 (2004) 2161–2168.
[29] H. Kong, C. Gao, D.Y. Yan, Controlled functionalization of multiwalled carbon nanotubes by in situ atom transfer radical polymerization, J. Am. Chem. Soc. 126 (2004) 412–413.
[30] T. Meng, X. Gao, J. Zhang, J. Yuan, Y. Zhang, J. He, Graft copolymers prepared by atom transfer radical polymerization (ATRP) from cellulose, Polymer 50 (2009) 447–454.
[31] K. Matyjaszewski, N.V. Tsarevsky, Nanostructured functional materials prepared by atom transfer radical polymerization, Nat. Chem. 1 (2009) 276–288.
[32] D. Wang, J. Tan, H. Kang, L. Ma, X. Jin, R. Liu, et al., Synthesis, self-assembly and drug release behaviors of pH-responsive copolymers ethyl cellulose-*graft*-PDEAEMA through ATRP, Carbohydr. Polym. 84 (2011) 195–202.

[33] K. Matyjaszewski, Atom transfer radical polymerization (ATRP): current status and future perspectives, Macromolecules 45 (2012) 4015–4039.
[34] Y. Sui, Z. Wang, X. Gao, C. Gao, Antifouling PVDF ultrafiltration membranes incorporating PVDF-*g*-PHEMA additive *via* atom transfer radical graft polymerizations, J. Membr. Sci. 413–414 (2012) 38–47.
[35] J. Ran, L. Wu, Z. Zhang, T. Xu, Atom transfer radical polymerization (ATRP): A versatile and forceful tool for functional membranes, Prog. Polym. Sci. 39 (2014) 124–144.
[36] K. Pangilinan, R. Advincula, Cyclic polymers and catenanes by atom transfer radical polymerization (ATRP), Polym. Int. 63 (2014) 803–813.
[37] F. Joubert, O.M. Musa, D.R.W. Hodgson, N.R. Cameron, The preparation of graft copolymers of cellulose and cellulose derivatives using ATRP under homogeneous reaction conditions, Chem. Soc. Rev. 43 (2014) 7217–7235.
[38] M. Ding, X. Jiang, L. Zhang, Z. Cheng, X. Zhu, Recent progress on transition metal catalyst separation and recycling in ATRP, Macromol. Rapid Commun. 36 (2015) 1702–1721.
[39] J.-S. Wang, K. Matyjaszewski, Controlled/"living" radical polymerization. halogen atom transfer radical polymerization promoted by a Cu(I)/Cu(II) redox process, Macromolecules 28 (1995) 7901–7910.
[40] J. Qiu, K. Matyjaszewski, Polymerization of substituted styrenes by atom transfer radical polymerization, Macromolecules 30 (1997) 5643–5648.
[41] D.A. Shipp, J.-L. Wang, K. Matyjaszewski, Synthesis of acrylate and methacrylate block copolymers using atom transfer radical polymerization, Macromolecules 31 (1998) 8005–8008.
[42] M. Teodorescu, K. Matyjaszewski, Atom transfer radical polymerization of (meth) acrylamides, Macromolecules 32 (1999) 4826–4831.
[43] T. Kowalewski, N.V. Tsarevsky, K. Matyjaszewski, Nanostructured carbon arrays from block copolymers of polyacrylonitrile, J. Am. Chem. Soc. 124 (2002) 10632–10633.
[44] V. Percec, A.V. Popov, E. Ramirez-Castillo, M. Monteiro, B. Barboiu, O. Weichold, et al., Aqueous room temperature metal-catalyzed living radical polymerization of vinyl chloride, J. Am. Chem. Soc. 124 (2002) 4940–4941.
[45] K. Min, W. Jakubowski, K. Matyjaszewski, AGET ATRP in the presence of air in miniemulsion and in bulk, Macromol. Rapid Commun. 27 (2006) 594–598.
[46] D.A.Z. Wever, P. Raffa, F. Picchioni, A.A. Broekhuis, Acrylamide homopolymers and acrylamide-*N*-isopropylacrylamide block copolymers by atomic transfer radical polymerization in water, Macromolecules 45 (2012) 4040–4045.
[47] P. Chmielarz, S. Park, A. Simakova, K. Matyjaszewski, Electrochemically mediated ATRP of acrylamides in water, Polymer 60 (2015) 302–307.
[48] V.A. Williams, T.G. Ribelli, P. Chmielarz, S. Park, K. Matyjaszewski, A silver bullet: elemental silver as an efficient reducing agent for atom transfer radical polymerization of acrylates, J. Am. Chem. Soc. 137 (2015) 1428–1431.
[49] P. Chmielarz, P. Krys, S. Park, K. Matyjaszewski, PEO-*b*-PNIPAM copolymers *via* SARA ATRP and *e*ATRP in aqueous media, Polymer 71 (2015) 143–147.
[50] L.T. Strover, J. Malmström, L.A. Stubbing, M.A. Brimble, J. Travas-Sejdic, Electrochemically-controlled grafting of hydrophilic brushes from conducting polymer substrates, Electrochim. Acta 188 (2016) 57–70.
[51] P. Chmielarz, S. Park, A. Sobkowiak, K. Matyjaszewski, Synthesis of β-cyclodextrin-based star polymers *via* a simplified electrochemically mediated ATRP, Polymer 88 (2016) 36–42.
[52] P. Chmielarz, Synthesis of α-D-glucose-based star polymers through simplified electrochemically mediated ATRP, Polymer 102 (2016) 192–198.

[53] P. Chmielarz, A. Sobkowaik, Synthesis of urethane-acrylic multiblock copolymers *via* electrochemically mediated ATRP, Chem. Pap. 70 (2016) 1228–1237.
[54] P. Chmielarz, A. Sobkowiak, Synthesis of poly(butyl acrylate) using a electrochemically mediated ATRP, Polimery 61 (2016) 585–590.
[55] H. Verma, T. Kannan, Novel telechelic 2-methyl-2-bromopropionate terminated polyurethane macroinitiator for the synthesis of ABA type tri-block copolymers through atom transfer radical polymerization of methyl methacrylate, Polym. J. 40 (2008) 867–874.
[56] K. Tharanikkarasu, H. Verma, W. Jang, S.K. Lee, J. Seo, S. Baek, et al., Novel poly (methyl methacrylate)-*block*-polyurethane-*block*-poly(methyl methacrylate) tri-block copolymers through atom transfer radical polymerization, J. Appl. Polym. Sci. 108 (2008) 1538–1544.
[57] H. Verma, T. Kannan, Atom transfer radical polymerization of methyl methacrylate using telechelic tribromo terminated polyurethane macroinitiator, J. Macromol. Sci. A. 47 (2010) 407–415.
[58] P. Król, P. Chmielarz, Synthesis of PMMA-*b*-PU-*b*-PMMA tri-block copolymers through ARGET ATRP in the presence of air, Express Polym. Lett. 7 (2013) 249–260.
[59] S. Nayak, H. Verma, T. Kannan, Synthesis and characterization of amphiphilic and hydrophobic ABA-type tri-block copolymers using telechelic polyurethane as atom transfer radical polymerization macroinitiator, Colloid Polym. Sci. 288 (2010) 181–188.
[60] P. Chmielarz, P. Król, PSt-*b*-PU-*b*-PSt copolymers using tetraphenylethane-urethane macroinitiator through SARA ATRP, Express Polym. Lett. 10 (2016) 302–310.
[61] F.H. Schacher, P.A. Rupar, I. Manners, Functional block copolymers: nanostructured materials with emerging applications, Angew. Chem. Int. Ed. 51 (2012) 7898–7921.
[62] S. Sharifpoor, R.S. Labow, J.P. Santerre, Synthesis and characterization of degradable polar hydrophobic ionic polyurethane scaffolds for vascular tissue engineering applications, Biomacromolecules 10 (2009) 2729–2739.
[63] P. Król, P. Chmielarz, Controlled radical polymerization (CRP) methods in the synthesis of polyurethane copolymers, Polimery 56 (2011) 530–540.
[64] P. Król, P. Chmielarz, B. Krol, K. Pielichowska, Comparison of hydrolytic resistance of polyurethanes and poly(urethane-methacrylate) copolymers in terms of their use as polymer coatings into contact with the physiological liquid, Pol. J. Chem. Technol. 16 (2014) 16–26.
[65] U. Chatterjee, X. Wang, S.K. Jewrajka, M.D. Soucek, Polyester/poly(meth)acrylate block copolymers by combined polycondensation/ATRP: characterization and properties, Macromol. Chem. Phys. 212 (2011) 1879–1890.
[66] V. Melinte, A. Chibac, T. Buruiana, G. Hitruc, E. Buruiana, Triazene UV-triggered photogeneration of silver/gold nanoparticles in block copolymer templates, J. Nanopart. Res. 17 (2015) 1–15.

CHAPTER 11

Ageing Behavior of Polyurethane Based Blends and Interpenetrating Polymer Networks

Alpesh U. Vala[1], Ajay V. Rane[2], Krishnan Kanny[2], Vayyaprontavida K. Abitha[3] and Sabu Thomas[3]

[1]Institute of Chemical Technology, Mumbai, India
[2]Durban University of Technology, Durban, South Africa
[3]Mahatma Gandhi University, Kottayam, Kerala, India

Contents

Polyurethane Polymers: Blends and Interpenetrating Polymer Networks
DOI: http://dx.doi.org/10.1016/B978-0-12-804039-3.00011-7

11.1 INTRODUCTION

Polyurethane (PUR or PU) is a polymer composed of organic units joined by carbonate (urethane) links. Polyurethanes are thermosetting polymers as well as thermoplastic. Polyurethanes are commonly formed by the polymerization of diisocyanate and diol. Isocyanates must have two or more isocyanate groups on each molecule. The most commonly used isocyanates are the aromatic diisocyanate, toluene diisocyanate (TDI), and methylene diphenyl diisocyanate (MDI). PU is a remarkable class of polymers that exhibit a wide range of chemical and mechanical properties.

Polymer ageing may involve physical ageing without chemical reaction occurring; chemical changes such as crosslinking during curing of a thermoset; thermal conditioning at elevated temperature; or photochemical ageing, as occurs in weathering [1]. Physical ageing occurs when a polymer is in a nonequilibrium state and is caused by molecular relaxations that are biased in the direction required to drive the material closer to equilibrium. Crosslinking is a method where molecular attachment increases and glass transition temperature also increases. If an elevated temperature is applied to a polymer in the presence of an aggressive chemical agent (often oxygen), then this may give rise to chemical reactions that may occur only very slowly, or not at all, at ambient temperature. The many changes that occur are collectively known as 'thermal degradation'. Weathering, or more specifically, photo-oxidation of polymers, refers to the chemical and physical changes that occur when radiation is absorbed by a polymer. Chemical degradation involves specific chemical reaction of the polymer with the fluid with the most common mode of failure being hydrolysis by water, acids, and alkalis. Ionizing radiation covers a wide range of different forms of radiation including x-rays, gamma rays, neutrons, alpha particles, and beta particles. When a polymer is irradiated, the ionizing radiation induces degradation by the formation of free radicals or ions in the polymer. These reactive intermediates are capable of initiating chemical reactions that occur by free radical or ionic mechanisms and result in scission as well as in crosslinking reactions. Biological degradation is not a common form of degradation as most commonly used thermoplastics are resistant to microbiological attack. The only case where biological attack has influenced life expectancy has been with certain polyurethanes.

11.2 AGEING MECHANISMS

11.2.1 Thermal Ageing

Thermal stability of polyurethanes has been thoroughly investigated by many researchers, as a function of the nature (aliphatic/aromatic) of the diisocyanate, of the polyol (polyester/polyether) and the chain extender [2]. It has been demonstrated that primary alcohol functions reacted with aliphatic diisocyanate should give the most stable materials. Hydroxyl terminated polybutadiene (HTPB) is certainly the most famous in transformation of polyurethane. Thermal stability of polybutadiene, irrespective of the molecular weight, has been thoroughly investigated [3].

Thermal degradation refers to the chemical and physical processes in polymers that occur at elevated temperatures. Increased temperature accelerates most of the degradation processes that occur in polymers such as oxidation.

$$P - P \rightarrow P^*$$
$$P^* + O_2 \rightarrow POO^*$$

The influence of temperature on oxidative degradation is dependent on the chemical structure of the polymer. Thermal degradation initiates with generation of the free radical at high temperature. This free radical is generated because of weaker bond in the polymer chain. Then in presence of environmental oxygen, the free radical is propagated and forms an unstable compound.

Ageing experiments were performed at 150°C both in an air-circulating oven and in a vacuum oven, for variable periods up to 1 month, on polyurethane sheets of ca. 1 mm thickness. Some testing methods were used to analyze the samples after heating to a nominal specification. Fourier transform infrared radiations are used to analyze the change in peak of specific group due to thermal ageing, before and after ageing is undertaken. Two more techniques to determine thermal ageing are the use of differential scanning calorimetry (DSC) and dynamic mechanical thermal analysis (DMTA), to calculate the glass transition temperature (T_g). DMTA is used to give change in T_g as well as mechanical properties. The shear and loss modulus versus temperature is to be recorded. Sometimes shore hardness is also used for determine the degradation.

11.2.2 Chemical Ageing

Chemical degradation of polymers involves specific chemical reactions like hydrolysis with the water, acid, or alkalis. Esters, imides, carbonates, and amide groups are susceptible to break the bond due to chemical reactions [4]. It causes the reduction of the molecular weight, chain scission; deterioration of mechanical and other properties (Fig. 11.1).

11.2.3 Biological Ageing

Biological degradation occurs due to microbiological attack. Biodegradation influences life expectancy. After years of production of PUs, manufacturers found that they were susceptible to degradation. Variations in the degradation patterns of different samples of PUs were attributed to the many properties of PUs such as topology and chemical composition [5]. The regularity in synthetic polymers allows the polymer chain to pack easily, resulting in the formation of crystalline regions. This limits accessibility of the polymer chains to degradative agents. Microbial degradation of polyurethanes is dependent on the many properties of the polymer such as molecular orientation, crystallinity, crosslinking, and chemical groups present in the molecular chains which determine the

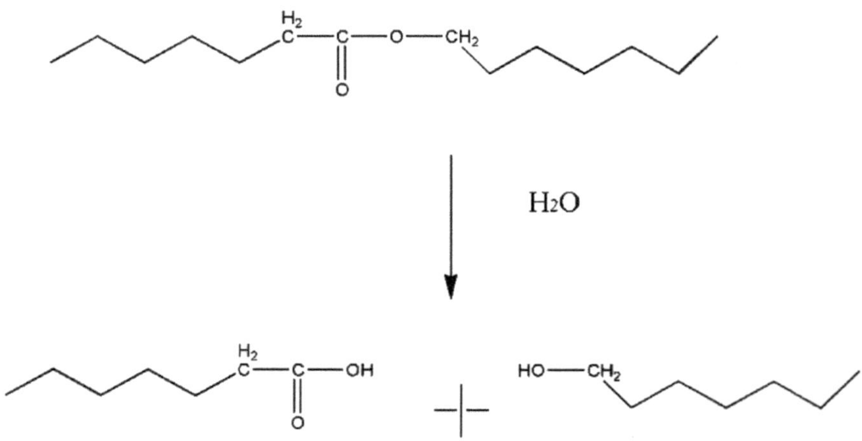

Figure 11.1 Hydrolysis of polyester.

accessibility to degrading-enzyme systems. Esterase activity (both membrane-bound and extracellular) has been noted in microbes which allow them to utilize polyurethane. Microbial degradation of polyester polyurethane is hypothesized to be mainly due to the hydrolysis of ester bonds by these esterase enzymes [6].

11.2.4 Weathering

Degradation due to weathering contains much factors for consideration. Weathering or photo degradation can be initiated due to UV absorption. The UV sensitive group get excited and form free radicals. The chain then breaks. When a polymer is exposed to solar radiation, the energy absorbed by the polymer results in the formation of free radicals within the polymer by the dissociation of the C-H bonds in the polymer chains.

$$PH \rightarrow P^* + H^*$$

The propagation of these chain occure due to an attack of the atmospheric oxygen, and hence is seen as both chemical and physical degradation.

$$P^* + O_2 \rightarrow POO^*$$
$$POO^* + PH \rightarrow POOH + P^*$$

11.2.5 Other Ageing

Environmental stress cracking is the most prominent in the failure of a polymer. It is caused by liquid or chemical being diffused into the polymer. It becomes locally dissolved, promoting cracking and crazing into the polymer. The degradation due to ionizing radiation comes from a wide range of forms including X-ray, gamma rays, neutrons, and beta particles. Some reactive intermediates can be ionized by this high radiation and it causes chain scission or crosslinking of the polymer.

11.3 AGEING OF POLYURETHANE BLEND

11.3.1 PU/PVC, PVDF Blend

At sea, polyurethane ageing occurs due to chemical attack and environmental constraints. To increase the chemical resistance, PU is blended with some halogenated polymers such as PVC, PVDF. This blend gives the resistance in marine environments against NaCl, UV light and other chemicals. The different weight ratio between 0 to 100% of PVC blended with PU and the consequent properties were studied. An increase in the energetic characteristics of hydrogen bonds networks dramatically reduces the size of dispersed PVC thermoplastic phase in an elastomer matrix. This micro- and nano-structurized poly (urethane)/poly(vinyl chloride) thermoplastic elastomer has excellent physical and mechanical properties [7]. A 50/50 PU/PVDF blend showed lower mechanical properties. Mechanical strength increased proportionally to PVDF.

11.3.2 PU/Epoxy Blend

Extended exposure time at high temperatures to epoxy caused a decrease in molecular weight. As a result, mechanical, thermodynamic, and physical properties are affected in ways that can compromise the reliability of epoxy-based engineering components and structures [8].

Adhesion in coating systems with polyurethane topcoats applied on aged epoxy coatings depends strongly on the degradation degree of epoxy intermediate coatings and the value of generated internal stresses. Coatings with good adhesion retention in corrosive environments have good protective properties even when temporary blistering has occurred. Intermediate adhesion depends on the degree of epoxy coating degradation (active groups on the coating surface and chalking of the coating). There is no significant influence on washing of coatings after ageing on improvement of adhesion, except for strongly chalked coatings. Adhesion of polyurethane coatings to aged epoxy coatings depends on the type of epoxy paint used. However intermediate adhesion decreases in all systems with aged intermediate coating [9].

11.3.3 PU/Olefins Blend

Thermoplastic polyurethane (TPU) is a highly versatile material with superior physical properties but a high cost. Melt blending of TPU with polyolefins (PO) can lower the cost and improve mechanical and chemical properties. As TPU and PO are completely immiscible polymers, property enhancement cannot be attained. Polypropylene copolymer (PPCP) is used as a compatibilizer for improving miscibility of the blends. Calcium carbonate is used as a filler to enhance the mechanical properties. The results show that the with 20% loading of polyurethane (PU) with suitable compatibilizers and calcium carbonate gives excellent performance in all aspects [10].

The total degree of crystallinity of the blends and the crystallization temperature of PP decreased with increasing TPU content. On the other hand, the addition of elastomeric TPU to PP significantly increased the spherulite size of PP.

11.3.4 PU/Acrylate Blend

Acrylic polyurethane were exposed in two artificial weathering environments produced by a fluorescent UV/condensation weathering device and a xenon lamp exposure and weathering equipment for different time periods. The results showed that the Xenotest protocol had a significant effect on the thickness loss and gloss. However, the yellow index measurements and low frequency impedance showed that greater degradation was caused by the UV exposure protocol. These findings were supported by the results obtained via FTIR analysis. In the two artificial weathering environments, the low frequency impedance modulus decreased exponentially with increase of ageing time, and correlated with the changes of other properties of ageing, which can be used to predict a product's lifetime [11].

Polyurethane and polyurethane-poly (methylmethacrylate) (PMMA) blend membranes were used in gas separation studies. The effects of blend composition, temperature, and pressure on the permeability, diffusivity, and solubility of CO_2, H_2, O_2, CH_4, and N_2 were investigated [12].

11.3.5 PU/Polyether Blend

The amorphous polyurethane (APU) was synthesized from butanediol and a mixture of the 2, 4- and 2,6-isomers of tolyldiisocyanate. A poly (ethylene oxide-co-propylene oxide) copolymer (EPO) containing 70 wt % ethylene oxide was simply blended with an amorphous polyurethane polymer. This interaction and formed bond was diagnosed by infrared spectroscopy. This system was chosen because of its relevance to the more complicated segmented polyurethanes and in its essentials is focused upon the spectroscopic characteristics of the interaction of urethane N-H with ether functional groups. The concordance between the theoretical predictions and gross experimental observations of the miscibility of these polymer blend systems is very encouraging [13].

11.3.6 PU/Polyester Blend

The preparation of polyurethanes with a particular hard segment from toluene diisocyanate and hexane diol, and soft-segment from poly (methylene adipate glycol) was blended with a series of poly (vinyl chloride-vinyl acetate) (P(VC-VA)). Systematic studies have been done on the compatibility of series of blends. A solubility parameter calculation shows that the polymer blends are partially miscible with each other, and this was confirmed by optical micrograph. The changes in mechanical properties with change in blend composition also show that up to 40% of P(VC-VA) there is an improvement in the elasticity of the material [14].

11.3.7 PU/SAN Blend

Blends of crosslinked polyurethane (PUR) and styrene—acrylonitrile (SAN) copolymer, were prepared by reactive blending with polymer polyols. The results by different techniques indicate that the degree of micro phase separation of PUR into hard segment (HS) micro domains and soft-segment (SS) micro phase increases on addition of SAN. The results show that for the dynamic glass transition of the PUR SS micro phase, the characteristic timescales at the same temperature are similar for DMTA, DSC, thermally stimulated depolarization currents (TSDC) and shorter for dielectric relaxation spectroscopy (DRS). In terms of fragility, the PUR/SAN blends are classified as fragile systems [15].

11.3.8 PU/EVA Blend

Heat shrinkability of the polymer, which depends on the elastic memory, is being utilized in various applications, mainly in the field of encapsulation. The elastic memory is introduced into the system in the form of an elastomeric phase. The blends of ethylene vinyl acetate and polyurethane were studied with reference to their shrinkability, introducing crosslinking in both the phases. It is found that with an increase in elastomer content the shrinkage increased to a certain level and then decreased. With increase in cure time, shrinkage was decreased [16].

11.4 ACCELERATED TEST METHODS

11.4.1 Liquid absorption test method

ISO 175 [17] and ASTM D543 [18] cover the absorption of chemical liquids into polymers and ISO 62 [19] and ASTM D570 [20] discusses the absorption of water. As per ISO and ASTM principles for determining [ ] the procedure is as follows:

1. Changes in the mass, dimensions, and appearance immediately after removal of liquid and drying.

   Weigh each conditioned specimen separately and measure thickness at the center, length, and width, or two diameters at right angles to each other, to the nearest 0.025 mm (0.001in.). Place the specimens in appropriate containers for the reagents being used and allow the specimens to be totally immersed in fresh reagent for seven days in the standard laboratory atmosphere. Stir the reagents every 24 h by moderate manual rotation of the containers. After 168 h, or other agreed upon period of time, individually remove each specimen from the reagent, immediately weigh, dry and re-measure its dimensions.
2. Changes in the physical properties.

   Immerse and handle the mechanical test specimens in accordance with instructions given in Step (1). Determine the mechanical properties of identical nonimmersed and immersed specimens in accordance with the standard methods for tensile tests.
3. The amount of liquid absorbs.

   Samples can be returned to the medium for continuing exposure providing that the time out of the medium is minimized. The balance used should have a resolution of 0.1 mg or better, which can be relaxed to 1 mg if water absorption is greater than 1%. The increase in mass divided by the initial mass, measured at regular time intervals, is plotted against time in order to define the absorption curve. Results can be reported either as the mass gain after a particular period of immersion or as the mass at saturation defined as the weight gain from 3 successive measurements that differ in value by less than 1% of the overall weight gain.

$$\text{Increase in weight\%} = \frac{\text{wet weight} - \text{conditioned weight}}{\text{conditioned weight}} * 100$$

$$\text{Soluble matter lost\%} = \frac{\text{conditioned weight} - \text{reconditioned weight}}{\text{conditioned weight}} * 100$$

11.4.2 Thermal stability test method

ISO 2578 [21] gives a standard procedure for the thermal stability of polymers using the Arrhenius expression. The standard enables the user to define the highest temperature to which a plastic can be subjected for a chosen length of time before a particular property will have degraded unacceptably. The property of the polymer and the level at which it is considered unacceptable will depend on the particular application. The significance of this standard is that it is the only one to specify the Arrhenius method or any other procedure for making long-term prediction from multipoint polymer ageing data.

ISO 176 [22] and ASTM D1203 [23] cover the stability of plasticizers in polymers, measuring the amount of plasticizer that is lost due to absorption on activated charcoal. These test methods are both essentially simple quality control tests. After conditioning, weigh each test specimen to the nearest 0.001 g and determine its mean thickness to the nearest 0.01 mm. Place one specimen on the bottom of a metal container and spread about 120 cm^3 of activated carbon over this specimen. Place two additional specimens in the container, each covered by 120 cm^3 of carbon. Finally, put the lid on the container. Only test specimens of the same composition shall be placed in one container, in order to avoid the possibility of plasticizers or other volatile components migrating from one specimen to another. Place the container in the oven or thermostatic bath controlled at a temperature of 70°C ± 1°C. After 24 h, remove the container from the oven or bath and allow it to cool at room temperature. Remove the specimens from the container, carefully brush them free from any trace of carbon particles and recondition under the same conditions as those to which they were subjected before the original weighing. Reweigh each specimen to the nearest 0.001 g.

$$\Delta m = \frac{m_0 - m_1}{m_0} * 100$$

Where

The change in mass, Δm, is expressed as a percentage

m_0 is the mass, in grams, of the test specimen after conditioning;

m_1 is the mass, in grams, of the test specimen after treatment in the oven or thermostatic bath and reconditioning.

11.4.3 ESC Test Method

The definition of stress cracking according to ASTM D 883 is 'an external or internal crack in a plastic caused by tensile stresses less than its short-term mechanical strength.' This type of cracking typically involves brittle cracking, with little or no ductile drawing of the polymeric material from its adjacent failure surfaces. Slow crack growth is another term commonly used to describe stress cracking. The best-known type of slow crack growth is 'environmental stress cracking' (ESC). These are instances involving cracking of stressed samples, generally in the presence of surface-active wetting agents such as alcohols, soaps, surfactants, or others. The surface-active agents do not chemically attack the polymer nor produce any effect other than microscopically brittle-appearing fractures. In the absence of the surface-active environment, these fractures would not occur in any reasonable period of time under the same stress conditions. These cracks are generally thought to initiate at microscopic imperfections and propagate through the crystalline regions of the polymer structure. The ability of a polymer to resist slow crack growth or environmental stress cracking is known as ESCR. Different polymers exhibit varying degrees of ESCR [24].

11.4.3.1 ASTM D1693—Bent Strip ESCR Test

Developed by Bell Labs in the late 1940s, this is one of the original and best-known ESCR tests. Ten rectangular-shaped specimens are cut from a molded plaque prepared with standard methods. A controlled notch is cut horizontally across each specimen, which serves as a crack initiation point. The specimens are bent and inserted into a "C" shaped bracket, creating a stress in the specimen. A diagram of this test method is shown in Fig. 11.2. The specimens and bracket are inserted into a tube filled with IGEPAL solution. The tube is then placed into a heated environment and inspected periodically for cracking (failures). Solution concentration, environment temperature, and sample dimensions vary with the test

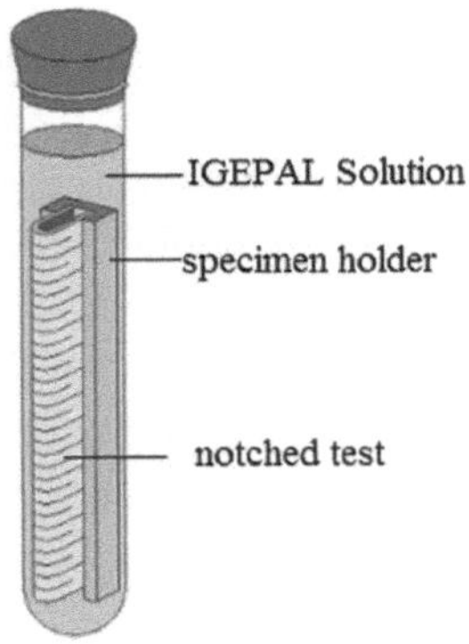

Figure 11.2 Bent strip techniques for polymers.

condition specified. These various test conditions introduce different stresses and strains and allow testing of different polymers and still obtain results in a timely manner. Today, this test is specified less frequently because it is not as aggressive towards modern resins. The Bent Strip test is a constant strain test, but polyethylene relaxes when strained. This stress relaxation allows testing to run without failure for very long periods (>1500 h) [25].

11.4.3.2 Ball and Pin Impression

The ball and pin impression test is used primarily for complex finished components. The method involves drilling a series of holes of a specific diameter into the polymer. A series of oversized balls or pins are inserted into the holes to induce a range of different stresses. One hour after the pins have been inserted the specimens are immersed in the environment for 20 h. The specimens are then dried and visually examined for crazing. The smallest ball to cause visible crazing is used to determine the ESC resistance of the polymer [26,27].

11.4.3.3 Constant Tensile Deformation

Measure, to the nearest 0.01 mm, the thickness and width of the specimen at the center and at each end of the parallel-sided portion of the specimen and determine the cross-sectional area using the minimum thickness and width from these measurements.

Calculate the force *F*, in Newton's, to be applied, using the following equation:

$$F = \sigma A$$

Where σ is the stress, in megapascals;

A is the cross-sectional area, in square millimeters.

Heat the temperature-controlled bath or oven to the selected test temperature. Insert a specimen in the clamps of the test device and immerse it in the test medium. After 15 min, apply to the specimen, within 2 min, the force F corresponding to the initial stress chosen, using a device such as a worm-gear by which the deformation produced in the specimen can be kept constant. Record the force until 200 h have elapsed. Plot a first stress relaxation curve, curve 1. Repeat steps with the other test specimens in the set. Conduct the test with a second initial stress which is slightly higher than the first, and plot a second stress relaxation curve, curve 2. If the second stress relaxation curve is located below the first

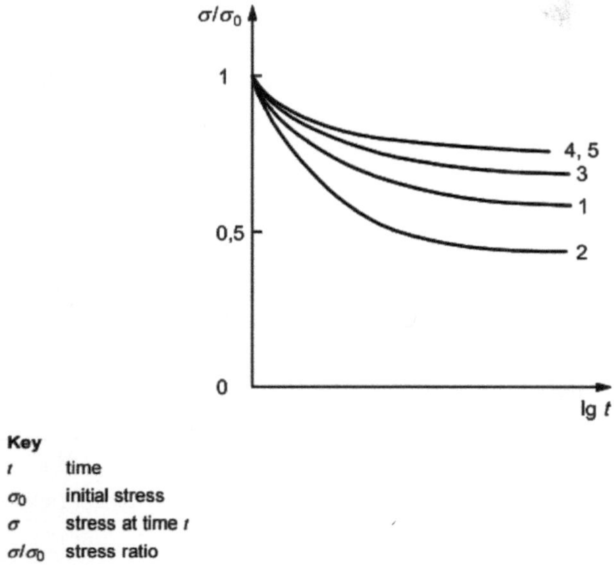

Figure 11.3 Stress relaxation curves obtained using progressively smaller levels of deformation [28].

one, perform the next test with an initial stress lower than the first one. Otherwise, carry out successive tests using initial stresses lower than previous stress until successive stress relaxation curves are superimposed. Record this stress as the critical stress for the test medium used [28] (see also Fig. 11.3).

11.4.3.4 Slow Strain Rate Testing

Set the crosshead speed on the tensile-testing machine to obtain the required strain rate in the specimen. The initial strain rate that has been used most frequently in the slow strain rate testing of plastics is $9 \times 10^{-6}\ s^{-1}$. The required crosshead speed (CHS) can be calculated as follows:

$$\mathrm{CHS} = \varepsilon lo$$

Where ε is the strain rate;

lo is the gage length.

Insert the specimen in the grips of the tensile-testing machine, ensuring that the specimen is axially aligned and completely enclosed within the environmental chamber. Load the specimen rapidly (<30 s) to a small load (<100 N) just sufficient to take up any slack in the machine. Fill the

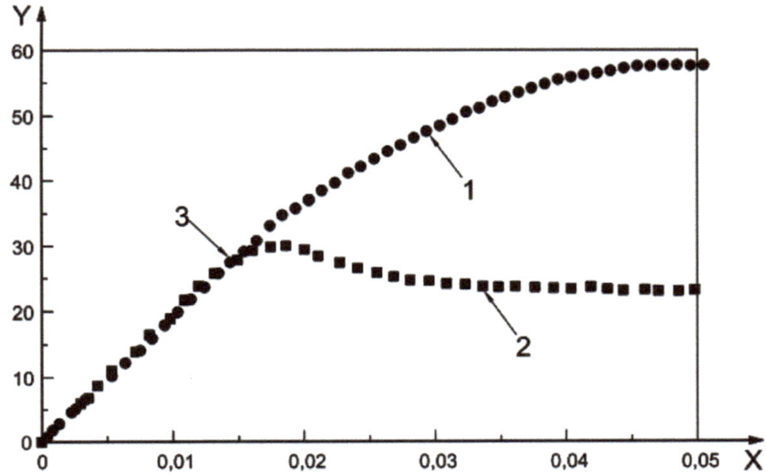

Key
X strain
Y stress (MPa)
1 specimen tested in air
2 specimen tested in test medium
3 departure point

Figure 11.4 Typical stress-strain plot showing difference in stress/strain profile for material exposed in air and in the test medium [29].

environmental chamber with the chemical under investigation to immerse the specimen. Record the initial load and displacement, and then switch on the data-logger. Start the drive motor on the tensile-testing machine. Allow the test to run until the specimen fractures or necks. Inspect the specimen to ensure that it has not slipped or broken in the grips. If slippage or fracture has occurred within the grips, record this and discard the results. Unless otherwise agreed by the interested parties, conduct at least one repeat test. To provide reference data, carry out the test in air in the same way as in the test medium. Unless otherwise agreed by the interested parties, the relative humidity of the air shall be (50 ± 10) % [29] (see also Fig. 11.4).

11.4.3.5 Constant Tensile Stress Test

The distinctive feature of this test is that a constant load is applied to the specimens, thereby avoiding the problem of stress relaxation that is found in the constant strain test methods. The technique involves subjecting the specimen under investigation to a constant tensile stress at a stress below the tensile yield stress of the polymer. This is usually achieved using a dead weight that is suspended from one end of the specimen. The specimen is then immersed in the stress cracking agent and inspected at regular intervals to establish the onset of crazing. The time required for crazes/cracks to appear after the specimen has been exposed, or the threshold stress below which no crazes appear in a specific time period (typically 100 h) can be used as a measure of the ESC resistance [30].

11.4.4 Weathering Test Method

11.4.4.1 Natural Exposure

As the ultimate aim of accelerated ageing is to improve lifetime prediction under service conditions, the most appropriate conditions are those that match the service environment exactly. These are most easily obtained by naturally exposing specimens at outdoor exposure sites. The problem with this is that the exposure period required would have to be at least as long as the life expectancy of the product. Natural weathering is therefore normally accelerated by exposure to climatic conditions that are more severe than those that are expected in service. For this purpose there are a number of established test sites in Saudi Arabia, Australia, and the southern states of the USA. Arizona is popular for exposure in hot dry climates and Florida for hot humid climates. Acceleration can also be achieved by the use of Fresnel mirrors that concentrate the sunlight directed on to the specimen. Standard test procedures for the weathering of polymers in natural daylight, glass filtered daylight, and concentrated sunlight using Fresnel mirrors are given in ISO 877, ASTM D1435, ASTM D4364-02 and ASTM G24-05 [31,32,33,34].

11.4.4.2 Artificial Exposure

Artificial weathering involves exposing test specimens to an artificial light source in a cabinet in which the temperature, humidity, and water spray are controlled. The advantage of artificial weathering is that it is easier to accelerate the testing as all of the exposure conditions are controlled. The problem with this approach is to know how these different parameters interact in the weathering process. The key parameter in all accelerated weathering apparatus is the light source, which should ideally simulate solar radiation. The main types of artificial light sources that are used are carbon-arc, xenon-arc and fluorescent tube [35].

11.5 INTERPENETRATING POLYMER NETWORKS

An interpenetrating polymer network (IPN) is a polymer comprising two or more networks which are at least partially interlaced on a polymer scale but not covalently bonded to each other. The network cannot be separated unless chemical bonds are broken. The two or more networks can be envisioned to be entangled in such a way that they are concatenated and cannot be pulled apart, but not bonded to each other by any chemical bond.

11.5.1 Classification of IPNs

11.5.1.1 Classification Based on Chemical Bonding

Covalent crosslinking leads to formation of hydrogels with a permanent network structure, as irreversible chemical links are formed. This type of linkage allows absorption of water and/or bioactive compounds without dissolution and permits drug release by diffusion. Examples are:

- *Covalent Semi IPN*: contains two separate polymer systems that are crosslinked to form a single polymer network.
- *Noncovalent Semi IPN*: one in which only one of the polymer system is crosslinked.
- *Noncovalent Full IPN*: one in which the two separate polymers are independently crosslinked.

11.5.1.2 Classification Based on Arrangement Pattern

- *Sequential IPN*: the second polymeric component network is polymerized following the completion of polymerization of the first component network.
- *Novel IPN*: Polymer comprising two or more polymer networks which are at least partially interlocked on a molecular scale but not covalently bonded to each other and cannot be separated unless chemical bonds are broken.
- *Semi IPN*: If only one component of the assembly is crosslinked leaving the other in a linear form, the system is transferred as semi IPN.
- *Simultaneously IPN*: prepared by a process in which both component networks are polymerized concurrently, the IPN may be referred to as a simultaneously IPN [36].

11.5.2 Properties of IPN

1. IPN systems are known to increase the phase stability of the final product.
2. Thermodynamic incompatibility can be overcome due to the permanent interlocking of the network segments
3. High thermal stability
4. Good dielectric properties
5. Radar transparency
6. Nutrient permeability
7. Optical clarity: The percentage of light transmittance was found to be 90% and the refractive index was found to be 1.35.

11.5.3 IPN of Polyurethane and Poly(Methyl Methacrylate)

The polyurethane network was crosslinked with trimethylol propane and the poly (methyl methacrylate) network with trimethylol propane trimethacrylate. The theoretical crosslink densities of the above networks were in the range from 20,000 to 27,000 of molecular weight. Polyurethane-poly (methyl methacrylate) pseudo-IPNs, in which only one of the component polymers was crosslinked, were prepared for selected compositions as well as linear blends of the two polymers [37].

Linear Blends and Pseudo-IPNs: Linear PU was prepared in the same manner as for the polyurethane preparation, except that 1 equiv of PU prepolymer was reacted with 1 equiv of 1,4-BD instead of 1,4-BD-TMP mixture. Linear PMMA was prepared in the same manner as for the PMMA preparation except that the crosslinking agent (TMPTMA) was omitted. Linear polyblends were prepared for 75:25% and 50:50% PU PMMA compositions in the same manner as the IPN preparation except that the 1,4-BD-TMP mixture was replaced by pure 1,4-BD in the PU network and the TMPTMA was omitted in the PMMA network. Two pseudo-IPNs were prepared for the 75% PU:25% PMMA composition. They were prepared in the same manner as the IPN preparation except that one of the component networks was made linear by the procedure described above [37].

In general, the polyurethane-poly (methyl methacrylate) IPNs were heterogeneous with phase domain sizes ranging from 500°A to 5000°A. The polyurethane phase showed staining by osmium tetroxide although the exact staining mechanism is not known. Pure crosslinked polyurethane is partially crystalline (crystalline melting peak was observed in differential scanning calorimetry studies) because of its low crosslink density. The electron micrograph of pure crosslinked polyurethane shows the nonstained (white) partially crystalline area. Linear polyurethane exhibits extensive crystalline structure [38].

11.5.4 IPN of Polyurethane with Polyester

Reaction mechanisms involved in synthesizing a polyurethane-polyester IPN were studied using Fourier transform infrared spectroscopy. Results showed that the reaction kinetics of urethane polymerization is not the same in bulk as in the solution state. Polyurethane reaction in the IPN resembles a solution polymerization before gelation and a bulk polymerization after gelation. Copolymerization in typical unsaturated polyester resins tends to follow the azeotropic condition. On the other hand, the polyester reaction in IPN deviates substantially from that in a pure polyester resin in the presence of a polyurethane network.

A much higher mixing energy, indicated by a high Reynolds number in their study, was required for an IPN to reach the same extent of reaction reached in polyurethanes by reaction injection molding RIM process. Fourier transform infrared spectroscopy is a very useful tool for analyzing the kinetics of complicated polymerizations. Using a reduced conversion and reaction time, it was able to show that reaction kinetics of urethane polymerization is not the same in bulk as in the solution state. In PU-PES IPN, PES has a 'solvent effect' on the PU reaction, while PU has a 'solid effect' on the PES reaction. Such interactions may change the reaction mechanisms of both PU and PES polymerizations. Adding the PU phase changes the PES reaction mechanism, but varying the E″ composition and the type of initiator does not seem to have a strong effect [39].

11.5.5 IPN of Polyurethane with Unsaturated Polyester

The mechanical properties of the IPNs were determined to relate to their structure. In order to improve the mechanical properties of the resultant IPNs, it was found that the —OH group present in the UP should be employed to react with some isocyanate-terminated PU prepolymer. Thus, the PU and UP were, in fact, interpenetrated with each other and chemically bonded through a grafting reaction to form the graft-IPN.

PU prepolymers were prepared by the reaction of polyols and MDI. The equivalent ratio of MDI/polyol was set at 2/1. Reactions were carried out in a reaction kettle at 60°C for 5 h under dry nitrogen. For preparation of PU/UP copolymers and graft-IPNs, suitable amounts of PU prepolymer and UP were mixed with 0.5 wt% cobalt catalyst (based on UP). The mixture was vigorously stirred for 0.5 min and then degassed. One wt% MEKPO was added as initiator. The mixture was agitated and degassed again and casted onto an aluminum mold at 70°C and pressed less than 140 kg/cm^{-2} pressure for 2 h. Finally, the samples were post-cured at 90°C for 8 h. Samples were removed and kept in a desiccators at a relative humidity of 50% for at least 3 days before their properties were determined. For preparation of PU/UP graft-IPNs, the process was similar to that described above except that an excess of PU pre polymer was added at first and then suitable amounts of chain extender and cross-linking agent for PU must be added for complete reaction [40].

The tensile strength of the IPNs was increased to a maximum value and then decreased as the PU content increased. The izod impact strength and fracture energy of the IPNs increased more or less with an increase of the PU content, depending on the types, chain length of polyols used in the PU and the degree of the grafting reaction between PU and UP [40].

REFERENCES

[1] J.R. White, Polymer ageing: physics, chemistry or engineering? Time to reflect, C. R. Chim. 9 (2006) 1396–1408.

[2] F.R. Flandrin, J.M. Widmaier, J.J. Flat, Thermal ageing of polyurethane with hydrogenated polyisoprene soft segments, Polym. Degrad. Stability 57 (1997) 59–67.

[3] O. Chiantore, Luda di Cortemiglia, M. Guaita, G. Rendina, Makromoleculare Chemie 190 (1989) 3143.

[4] R.P. Brown, D. Kockott, P. Trubiroha, W. Ketola, J. Shorthouse, A review of accelerated durability tests, Versailles Project Adv. Mater. Stand. (1995).

[5] R.A. Pathirana, K.J. Seal, A potential biodeteriogen of polyester polyurethane elastomers, Biodeterioration 5 (1983) 679–689.

[6] G.T. Howard, Biodegradation of polyurethane: a review, Int. Biodeteriorat. Biodegrad. 49 (2002) 245–252.

[7] T.L. Malysheva, et al., The effect of interfacial interactions on a structure and properties of polyurethane elastomer/poly(vinyl chloride) blends, Open J. Org. Polym. Mater. 3 (2011) 1–7.

[8] G.M. Odegard, A. Bandyopadhyay, Physical aging of epoxy polymers and their composites, J. Polym. Sci. Part B Polym. Phys. 49 (24) (2011) 1695–1716.

[9] M. Zubielewicz, A. Królikowska, The influence of ageing of epoxy coatings on adhesion of polyurethane topcoats and protective properties of coating systems, Progr. Org. Coat. 66 (2009) 129–136.

[10] A. Kamble, V. Singh, M. Thomas, N. John, Effectiveness of compatibilizers and filler on the performance of the blends of thermoplastic polyurethane/polyolefins, Chem. Sci. J. (2011) Volume CSJ-34.

[11] J. Hu, et al., Ageing behaviour of acrylic polyurethane varnish coating in artificial weathering environments, Progr. Org. Coat. 65 (2009) 504–509.
[12] J.A. de Sales, et al., Systematic investigation of the effects of temperature and pressure on gas transport through polyurethane/poly(methylmethacrylate) phase-separated blends, J. Membr. Sci. 310 (2008) 129–140.
[13] M.M. Coleman, D.J. Skrovanek, H. Jiangbin, P.C. Painter, Hydrogen bonding in polymer blends. 1. FTIR studies of urethane-ether blends, Macromolecules 21 (1) (1988) 59–65.
[14] S.N. Jaisankar, Y. Lakshminarayana, G. Radhakrishnan, Compatibility of poly(vinyl chloride.vinyl acetate)/polyester-based polyurethane blends, Polym. Int. 43 (1997) 368–372.
[15] Vatalis, et al., Phase behavior and molecular mobility in polyurethane/styrene–acrylonitrile blends, J. Appl. Polym. Sci. 80 (2001) 1071–1084.
[16] S.R. Chowdhury, C.K. Das, Shrinkability of ethylene vinyl acetate/polyurethane blends and their characterization, Polym. Adv. Technol. Polym. Adv. Technol 11 (2000) 359–367.
[17] ISO 175: 1999; Plastics – Methods of test for the determination of the effects of immersion in liquid chemicals.
[18] ASTM D543-95; 2001; Standard Practices for Evaluating the Resistance of Plastics to Chemical Reagents.
[19] ISO 62: 1999; Plastics – Determination of water absorption.
[20] ASTM D570-98; Standard Test Method for Water Absorption of Plastics.
[21] ISO 2578; 1993; Plastics – Determination of time-temperature limits after prolonged exposure to heat.
[22] ISO 176; 2005; Determination of Loss of Plasticisers – Activated Carbon Method.
[23] ASTM D1203-94; 2003; Standard Test Methods for Volatile Loss from Plastics Using Activated Carbon Methods.
[24] ASTM D 883; 2012; Standard terminology relating to plastics.
[25] ASTM D1693 – 15; Standard Test Method for Environmental Stress-Cracking of Ethylene Plastics.
[26] ISO 4600; 1992; Determination of environmental stress cracking (ESC) by the ball or pin impression method.
[27] DIN 53449; 1992; Evaluation of Resistance of Thermoplastics to Environmental Stress Cracking, Steel Ball Impression Method.
[28] ISO DIS 22088-5; 2004; Plastics – Determination of resistance to environmental stress cracking (ESC) – Part 5: Constant tensile deformation method.
[29] ISO DIS 22088-6; 2004; Plastics – Determination of resistance to environmental stress cracking (ESC) – Part 6: Slow strain rate method.
[30] ISO 6252; 1992; Determination of Environmental Stress Cracking (ESC) by the constant tensile stress method.
[31] ISO 877; 1994; Plastics - Methods of exposure to direct weathering, to weathering using glass-filtered daylight, and to intensified weathering by daylight using Fresnel mirrors.
[32] ASTM D1435-99; Standard Practice for Outdoor Weathering of Plastics.
[33] ASTM D4364-02; Standard Practice for Performing Outdoor Accelerated Weathering Tests of Plastics Using Concentrated Sunlight.
[34] ASTM G24-05; Standard Practice for Conducting Exposures to Daylight Filtered Through Glass.
[35] G. Wypych, Weathering of Plastics, Testing to Mirror Real Life Performance, Plastic Design Library, New York, 1999, pp. 1–13.
[36] Shivashankar et al; A review on interpenetrating polymer network; Int. J. Pharm. Pharm. Sci.; 4 (5); 1–7.

[37] Klempner, et al., Polyurethane interpenetrating polymer networks. I. Synthesis and morphology of polyuret hane-poly (methyl methacrylate) interpenetrating polymer networks, Macromolecules 9 (2) (1976) 258–263.
[38] S.C. Kim, D. Klempner, K.C. Frisch, H.L. Frisch; SIM; IUPAC; Brazil; 1974; Paper B-4.2.
[39] Y.S. Yang, L. James Lee, Polymerization of polyurethane-polyester interpenetrating polymer network (IPN), Macromolecules 20 (1987) 1490–1495.
[40] K.H. Hsieh, J.S. Tsai, K.W. Chang, Interpenetrating polymer network of polyurethane and unsaturated polyester: mechanical properties, J. Mater. Sci. 26 (1991) 5877–5882.

CHAPTER 12

Poly(urethane-siloxane) Copolymers as New Coating Materials

Łukasz Byczyński
Rzeszow University of Technology, Rzeszow, Poland

Contents

12.1 INTRODUCTION

The transport, electronics, defense, and space industries have specialized requirements for materials. This has fed the dynamic progress recently occurring in material engineering, where there have been outstanding accomplishments related to synthesizing and application methods combining organic and inorganic polymer materials. Examples of such materials include copolymers synthesized with the use of polyurethanes and polysiloxanes [1,2]. In order to understand the reasons for the growing interest in poly(urethane-siloxane) copolymers it is necessary to investigate the properties of each homopolymer separately.

Polyurethanes (PU) are polymers built of alternating flexible polyol segments and rigid urethane-urea segments originating from diisocyanates and chain extenders [3–5]. Adequately selected diisocyanates, as well as polyols with known molecular weights, and small molecule chain extenders, make it possible to obtain polyurethanes with a wide range of physical properties which are important for a variety of applications. Because of their good mechanical properties, particularly high abrasion resistance and chemical resistance to organic solvents,

Polyurethane Polymers: Blends and Interpenetrating Polymer Networks
DOI: http://dx.doi.org/10.1016/B978-0-12-804039-3.00012-9

polyurethanes have been used in manufacturing various products, such as foam materials, elastomeric coatings, and anticorrosive varnishes, widely applied in the construction, furniture, and automotive industries [4]. These materials have been produced for over 70 years, a period that has witnessed rapid development in polyurethane technologies. Nevertheless, ongoing advancements in technology, medicine, electronics, and the economy constitute a growing challenge for these materials, connected, for instance, with environmental requirements or the necessity to increase thermal resistance.

Polymer chains of polysiloxane consist of alternating atoms of silicon and oxygen, and a typical representative of this group of polymers is polydimethylsiloxane (PDMS) [2,6]. Weak interactions mainly resulting from the activity of dispersion forces between chains of these polymers are the reason why siloxanes are characterized with low glass transition temperature $T_g < -120°C$. These polymers also demonstrate stable dielectric properties, which do not change in a wide spectrum of temperatures and humidity, as well as relatively low free surface energy values. Moreover, polysiloxanes are known for high thermal resistance even up to 300°C [7]. This is the effect of the aforementioned groups of Si-O, which are characterized with higher bonding energy (ca. 460 kJ mol^{-1}) in comparison with e.g., C−O bonds (ca. 350 kJ mol^{-1}) or C−C bonds (ca. 304 kJ mol^{-1}) occurring in typical organic compounds [8]. Introduction of aryl substituents instead of alkyl moieties enhances the thermal stability of siloxanes [9,10]. Because of all these properties, polysiloxanes are used for various unique purposes in such areas as electronics, automotive, and the paint and varnish industries. Furthermore, polysiloxane coatings, which are resistant to water and weather conditions, are useful for purposes connected with antiques restoration, hydrophobization of construction materials, and in protecting external walls against graffiti, which in our times are a frequently encountered problem [11]. Additionally, due to their neutral physiological effect, siloxanes are also used in cosmetology and pharmacy [6].

Reviewing research reports, one notices that polyurethanes modified with siloxanes are characterized by lower glass transition temperature and enhanced thermal stability with simultaneously decreased hydrophilicity [1,2,12,13]. When speaking about free surface energy (FSE) of poly(urethane-siloxane) copolymers, they are materials with medium values of FSE ($\gamma_S = 30-50$ mJ m^{-2}), which results from specific orientation of the urethane and siloxane structural units in copolymer chain. Chemical and physical interactions in the copolymers result in acquisition of new properties, which combine the positive features of polar polyurethane elastomers, e.g., tensile strength greater than 10 MPa with polysiloxanes properties. Siloxanes are generally nonpolar and therefore less mechanically stable (0.1–0.5 MPa), but they are resistant to temperatures of 200–250°C and are hydrophobic with the surface free energy of $\gamma_S = 30–45\ mJ \cdot m^{-2}$ [3,6]. Due to their properties, poly(urethane-siloxane) copolymers are used as protective coatings [11,14] but also as selective membranes [15–17] or in medical implants [18,19].

There are many types of poly(urethane-siloxane) copolymers (PU-Si) and many ways of obtaining them have been reported [1]. Therefore, the chapter is divided into three parts focused on the latest research on PU-Si coatings obtained from linear copolymers, waterborne, and crosslinked systems.

12.1.1 Poly(urethane-siloxane) Copolymers with a Linear Structure

A significant amount of research relates to the synthesis of poly(urethane-siloxane) copolymers (PU-Si) that were mainly obtained by replacing some flexible polyurethane segments with polydimethylsiloxane [20,21]. Polysiloxanes, which were used to synthesize this type of material, had a linear structure, on average, both sides were terminated with reactive hydroxyl or amino groups, and were capable of polyaddition reaction with isocyanate groups (Structure 12.1) [22,23].

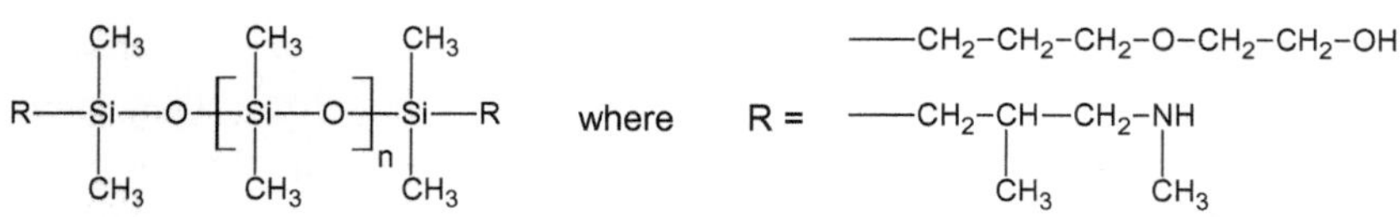

(12.1)

Significantly less frequently, modification of polyurethanes was conducted by employing polysiloxanes in which hydroxyl groups were present on one side of the chain (Structure 12.2) [24–28]. Because of the easier migration of PDMS chain toward the surface, copolymers of this type are found to have a lower free surface energy in comparison with similar PU-Si copolymers modified with conventional α,ω-dihydroxypolydimethylsiloxane. There were also reviews concerning PU-Si materials terminated with siloxane at one side of the polyurethane chain [29].

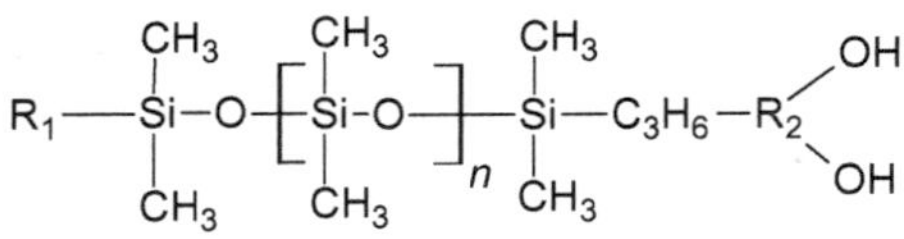
(12.2)

The synthesis of those kind of materials can be performed both by the one-stage or two-stage methods [30]. In the two-stage method, urethane-isocyanate or urethane-siloxane prepolymer is formed which at the next stage is extended with the use of 1,4-butenediol [30], 1,2-etanediol [31], etylenediamine [32,33], or N-ethyldietanolamine [34]. There are also examples of application of some specific chain extenders as 1,4-cyclohexane dimethanol for preparation of biomaterials for shape-adjustable cardiovascular devices [21]. Moreover, the mixture of chain extenders e.g., 1,4-butenediol and 4,4′-methylene-bis(3-chloro-2,6-diethylaniline) (M-CDEA) was employed, as well [20]. The introduction of M-CDEA in the presence of an aromatic chemical structure and chlorine moiety had a positive effect on thermal stability and self-extinguishing characteristics of the prepared PU-Si copolymers [20].

So far the siloxanes employed for poly(urethane-siloxane) synthesis have replaced a part of the soft polyurethane segments. However, PU-Si copolymers based only on polysiloxane soft segments are only mentioned in a very few reports [12,20,35]. For example, one prepared this type of copolymer in the microemulsion [36]. The resulting latex had a very low surface free energy ($\gamma_S = 26.3-28.9$ mJ m^{-2}). Replacing part of the siloxane with more hydrophobic dodecane did not result in the further lowering of free surface energy, but adversely affected the stability of the latex. These kind of copolymers showed phase separation between polar urethanes and nonpolar siloxanes [35]. However, application of silicone–carbonate oligomeric macrodiol as the soft segment resulted in enhancement of the intersegment mixing. Moreover, these PU-Si copolymers show relatively low mechanical modulus and have the potential for use in cardiac and orthopedic biomedical applications [35].

When considering PU, one should also mention polyhydroxyurethane, a polyurethane derivative, bearing hydroxyl groups in the side chain that was modified with diamine siloxanes (Structure 12.3) [37]. The introduction of siloxane moieties resulted in increase of water contact angle from 53° for unmodified PU to 96° for a PU-Si sample. There was also a drop in water absorption from 17% to 2% and glass point lowering from 40°C to 1°C. These values depended on siloxane chain length and was the lowest for siloxane with the long chain.

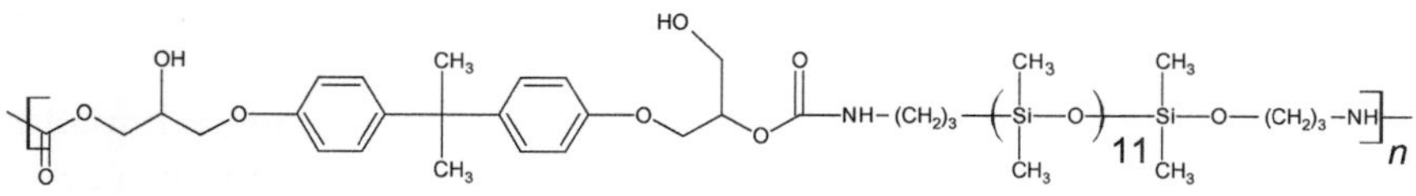

(12.3)

Hydroxy-terminated polyetherpolydimethylsiloxane-polyether (PE-PDMS-PE) triblock oligomers with different compositions were used for preparation of PU-Si materials in a two-step bulk polymerization method [30]. These copolymers showed lower glass point ($T_g < -120°C$) and higher temperature of 50% weight loss from 280 to 340°C as well as hydrophobicity comparing to pure polyurethane. Moreover, the mechanical properties were not strongly affected by the presence of siloxane. Tensile strength was maintained at 13 MPa and elongation was maintained at 300%.

Zhang et al. [38] proposed a synthesis of antibacterial PU-Si coatings, prepared by two-step condensation method with the use of polypropylene glycol (Mn = 400), isophorone diisocyanate, polyaryl polymethylene isocyanate (PAPI) and quaternary ammonium salt (QAS) hydroxyl-terminated polysiloxanes (PQMS) (Structure 12.4) with various siloxane lengths and QAS content.

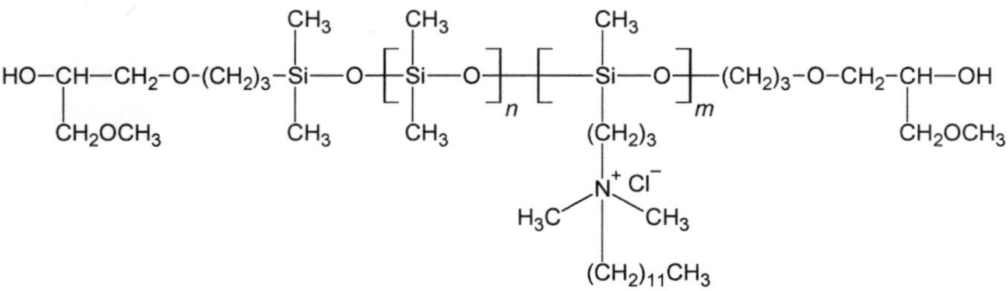

(12.4)

It was found that the covalently-bonded pendant QAS chains with high polarity tended to be compatible with the hydrophilic PU hard segments, which limits the migration of the hydrophobic polysiloxane segments. Additionally, a positive correlation was between the antibacterial activity and the content of N^+ on the surface, which indicates that the bactericidal effect of those coatings results from the quaternary ammonium salt side chain which can migrate toward surface.

Segmented poly(urethane-siloxane) derived from mixed hydroxyl-terminated polytetramethylene oxide and secondary aminoalkyl functional polydimethylsiloxane soft segment was investigated by Wang et al. [22] It was shown that 15% of the polydimethylsiloxane may reduce the cone calorimetry heat release rate by a factor of about 2/3 and hence improve fire resistance, while maintaining mechanical behavior. The low surface energy characteristics of PDMS promotes migration to the air—polymer interface to form a predominately PDMS-enriched surface. Thus the surface is oxidized at elevated temperatures in air to a silicate-like material and this serves as a protective layer, which further reduces the burning of the polyurethane.

The effect of a different soft segment, poly(oxytetramethylene)diol, poly (ethylene glycol), poly(propylene glycol), and α,ω-dihydroxyterminated polydimethylsiloxane on properties of PU-Si coatings based on side chain siloxane was also studied [24]. It was concluded that molecular mass, free surface energy and hardness of poly(urethane-siloxane) copolymers depended on the polyol that had been used. Thermal degradation of PU-Si in nitrogen was basically a two-stage process, but, in air, it was more complex. The coatings were generally soft with the relative hardness in the range of 0.027—0.120.

Polydimethylsiloxanes can be introduced into polyurethanes by direct transurethanetion of commercial PU [29]. To synthesize these PU-Si, polydimethylsiloxane was combined in different molecular weights of 1000, 5000, and 10,000 containing a hydroxyl group at one side of polymer chain. For these coatings, micromicellization, due to the self-assembling of the PU-Si copolymer chains in DMF-solution, was observed.

12.1.2 Waterborne Poly(urethane-siloxane)s

In the 21st century, technologies designed for manufacturing any kind of products must also be analyzed in terms of ecological aspects, at the stage of synthesis, safe application, and final waste management of the specific product. The previously reported modification of polyurethanes in order to obtain a hybrid polymer coating occurred mainly via a polyaddition reaction of reactive isocyanate groups originating from diisocyanates with hydroxyl or amino groups of respective polysiloxanes. Due to the high viscosity of the obtained copolymers, synthesis reactions usually require an organic solvent to be used in significant amounts, exceeding even 70% of the total weight of the basic raw materials. Significantly less frequently water, which is definitely less toxic, was the reaction medium [39–48].

The common polyurethanes, despite the relatively high surface free energy, show strongly hydrophobic characteristics [5]. To achieve at least their limited miscibility with water, suitable structural modification should be performed. Usually the urethane-isocyanate prepolymer based on mostly aliphatic diisocyanate as isophorone diisocyanate (IPDI) or 4,4′-diisocyanatomethylenedicyclohexane (HMDI), and hydrophilic nonionic polyols derivatives e.g., polyethylene glycol (PEG) with appropriate chain length is used [49]. The prepolymer is then extended in aqueous medium by means of suitable reactive compounds of low molecular weight, usually diamines. However, these dispersion systems are unstable in time [50].

An alternative method of modification is the introduction of a polar ionic group into usually rigid urethane segments [39,51–53]. After forming an ionomer dispersion in water in which a suitable counter-ion is present, one can get much more stable dispersion. In the case of polyurethane anionomers containing carboxyl groups as a neutralizating agent, triethylamine is usually used. For cationomers with the ionic groups of an alkylammonium cation a weak organic acid as counter-ion is employed [54,55].

It is relatively easy to introduce into the PU chain carboxyl groups that are dissociable in water during an extension of the urethane-isocyanate prepolymer using, e.g., 2,2-bis (hydroxymethyl) propionic acid (DMPA). The advantage of this monomer is the presence of a sterically blocked carboxyl group, which prevents the secondary reaction with isocyanate groups that can lead to unfavorable crosslinking by creating amide groups, resulting in a significant increase in viscosity of the reaction system [5]. In aqueous medium is preferred the reaction of the prepolymer with a much more reactive amino groups introduced as chain extenders. The ratio of $-COO^{-}$ an ions to alkylammonium cations, e.g., $NH^{+}(C_2H_5)_3$ has a significant effect on the physical properties of polyurethane anionomers as well [56,57].

For preparation of waterborne PU-Si, copolymers containing polysiloxanes with linear structure and both sides terminated with reactive groups as well as siloxanes in which reactive groups were present on one side of the chain. Due to their structure they form block or graft waterborne PU-Si, respectively [40]. The water absorption of the graft waterborne polyurethane–polysiloxane copolymers was lower than that of block ones by the same amount of siloxane. Additionally, the mechanical properties were different. The tensile strength of block waterborne polyurethane—polysiloxane copolymers decreased while the tensile strength of graft ones increased with an increase of polysiloxane.

There is some research concerning the application of waterborne poly (uretahane-siloxane)s as potential marine coatings [41,42]. Marine biofouling can be defined as the undesirable accumulation of microorganisms, plants, and animals on artificial surfaces that are exposed to the marine environment. This phenomenon can be observed on ship hulls, marine cages, pipelines, heat exchangers, and other structures including offshore platforms and bridges which are immersed in seawater [42].

Nagash et al. [41] obtained marine coatings from PU-Si anionomers with the use of novel side chain α-butyl ω-N,N-dihydroxyethyl aminopropylpolymethylhydrosiloxane (Structure 12.5). The decrease in growth of marine fouling after 90 days of immersion in seawater with increasing silicone content in waterborne coatings was noted. Similar results were obtained for coatings composed of linear polydimethylsiloxane [42]. However, in the case of side chain siloxane materials, the amount of siloxane in composition PU-Si was lower to maintain the same effect.

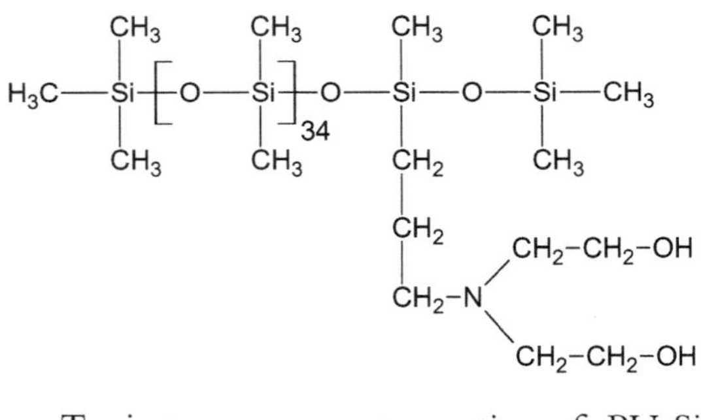

(12.5)

To improve some properties of PU-Si waterborne coatings e.g., hydrophobicity, water absorption, resistance to ultraviolet radiation or wear-ability, in addition to siloxane chains, fluorine moieties can be introduced [43,44,58]. The poly(urethane-siloxane) anionomer coatings modified with dihydroxy terminated perfluoropolyether revealed lower water swelling ratio (12.9%) and larger water contact angles (99°) comparing to unmodified PU-Si with similar siloxane content amounting to 17.5% and 87.7°, respectively. This phenomena results from the material surface enrichment with silicon and fluorine chain segments, which was confirmed by X-ray photoelectron spectroscopy [43].

Yu et al. obtained waterborne coatings modified with dodecafluoroheptyl methacrylate by using radical polymerization between vinyl groups which where present on 2-hydroxyehthyl acrylate terminated PU-Si chains and employed modifier [44]. It was noted that the particle size, elongation at break, thermostability, and contact angles of the coatings increased with increasing siloxane content. However, the viscosity of emulsion and tensile strength were significantly decreased.

An increase in mechanical properties of PU-Si anionomers was achieved by incorporation a hexamethylene isocyanate trimer into the structure [45]. Moreover, these copolymers showed excellent water- and xylene resistance by maintaining lower viscosity, which can have great potential for applications in coatings.

Some stable, water-dispersible poly(urethane-siloxane) anionomers were obtained with the use of amino-terminated polydimethylsiloxane [46]. It was noted that the particle size of the emulsion increased with an increase of polysiloxane content. Moreover, these dispersions had excellent freeze-thaw and high temperature stability. Compared with pure polyurethane, water absorption of poly(urethane-siloxane) decreased dramatically with increase of amino-terminated siloxane length and content, indicating that they have excellent water resistance.

Waterborne polyurethanes were also modified with the use of siloxane copolymers containing dimethyl and aminoethylaminopropyl-methyl units in a side chain [59]. The reactive pendant amino groups of the siloxane were employed for curing with isocyanate groups of urethane prepolymer at the stage of dispersion formation. Modification with polysiloxane had enlarged the molecular weight, narrowed the molecular weight distribution, and resulted in the microphase separation between the hard and soft segments of PU-Si, and this amplified the damping temperature of the materials.

To develop waterborne PU-Si copolymer with better water resistance than waterborne PU, and higher mechanical strength and abrasion characteristics than polydimethylsiloxane, a waterborne PU-Si system was often designed to form the crosslinking structure [47,48,60]. Introduction of siloxane containing ketone groups in side chain (Structure 12.6) allows for post-curing reaction with hydrazide at ambient temperature during coating formation [47].

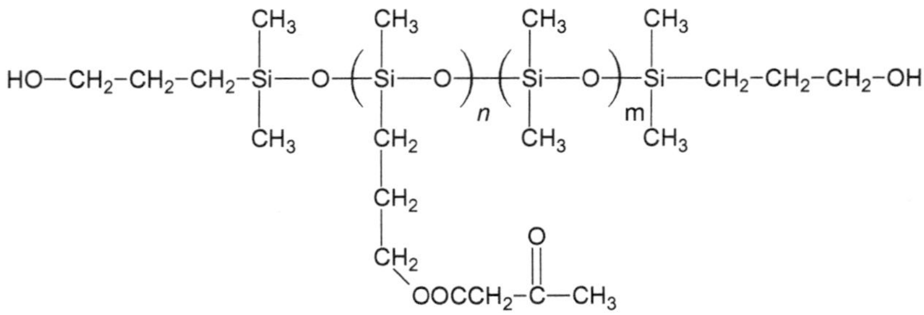

(12.6)

The waterborne poly(urethane-siloxane) coatings were also prepared via 'click' coupling [61]. This reaction was undertaken between alkyne-functionalized waterborne polyurethane and 3-azidopropyltriethoxysilane. The structure of hybrid crosslinking material was formed after the condensation reaction between silanol groups of polyurethane anionomer chains and silanol groups on the nanosilica, which was employed as the modifier. This way of modification resulted in improvements in the cross-link density and dispersibility of nanosilica.

12.1.3 Poly(urethane-siloxane) Polymer Networks

In recent literature, crosslinked poly(urethane-siloxane) structures cured with different the type of hardener and using various methods have been presented [62–97]. These coatings have very good mechanical properties e.g., tensile strength of up to 17 MPa, elongation at break greater than 700% and Shore hardness of up to 70 °Sh A. They also exhibit good water resistance and a thermal stability similar to siloxanes [1]. There has been growing focus on obtaining crosslinked PU modified by means of difunctional siloxane polymers with more complex structures that, in addition to dimethylsiloxane units $(-Si(CH_3)_2O-)_n$, contain various organic, e.g., acetoxyl or alkyl-fluorine substituents, which contribute to other beneficial properties in the obtained copolymers [68,69]. On the other hand, there have been very few reports concerning poly(urethane-siloxane) synthesis based on multifunctional polysiloxane containing reactive alkoxyhydroxyl groups in the side chain to enable production of cured poly(urethane-siloxane) coatings [64,65].

Moisture cured poly(urethane-siloxane) coatings are a topic of some research [66–69]. For preparation of this kind of material, the known high reactivity of isocyanate groups with water can be used. However, more often, reports concern moisture coatings obtained from compounds comprising hydrolizable alkoxy groups bonded to silicon by using the sol–gel process [66,70,71].

Xu et al. obtained moisture cured poly(urethane-siloxane) by employing reactivity with water of both isocyanate as well as methoxy groups [67]. These materials were synthesized with the use of polydimethylsiloxane comprising methoxy groups in the backbone (Structure 12.7) that could help form networks with the residual isocyanate under moisture, which was different from the conventional moisture-crosslinking polyurethane system. After the addition of siloxane, the thermal properties and water contact angle of the PU-Si materials were remarkably improved; and their microphase separation and water absorption was decreased from 2.3% to 0.5%. Some voids on the surface, a special surface morphology, have been observed. With the higher silicone content, more voids were generated and the diameter as well as the depth of them became larger.

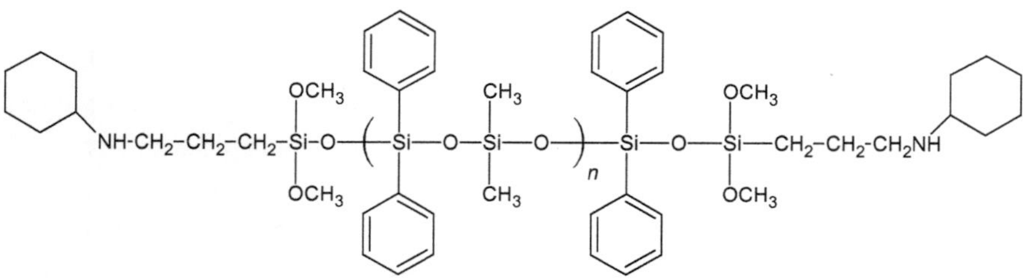

(12.7)

Moisture curing coatings were also performed with the use of fluorinated siloxane, i.e., α,ω-bis(3-aminopropyldiethoxylsilane) poly(trifluoropropylmethyl)siloxane (Structure 12.8) [68,69]. It was found that the equivalent ratio of reactants gives rise to a high molecular weight of copolymers and that low molecular weight siloxane in the copolymers can form a certain number of silicon–oxygen crosslinks resulting from silicon alkoxy to produce higher tensile strength elastomers [68]. Higher crosslink densities or the extent of hydrogen bonding that resulted from the higher hard segment concentration enhanced the mechanical properties of these materials. Moreover, the surface properties were associated with phase separation, hydrogen bonding, and crosslinkage [69].

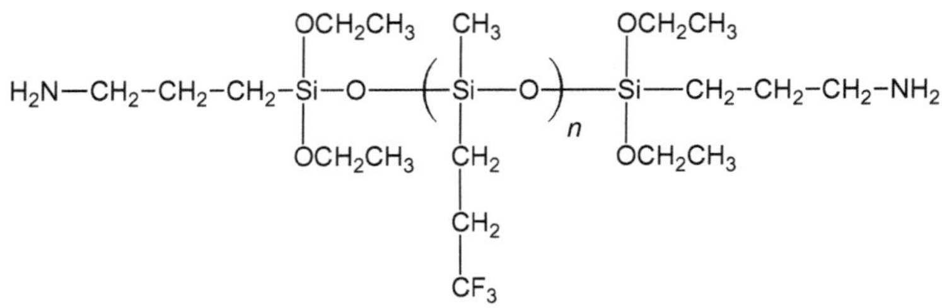

(12.8)

The major advantage of using a sol–gel method is the mild reaction conditions such as low temperature used to prepare silica nanoparticles [72]. The siloxane unit is formed during curing through a hydrolysis and condensation process of the methoxy group present in the silane moiety. These coatings are prepared with the use of diisocyanate, polyol and usually 3-aminopropyltriethoxysilane [73–75] or aminoethylaminopropyltrimethoxysilane [76]. They show good mechanical, adhesion, protective, and anticorrosion properties, as well as thermal stability [73,77]. The polyurethanes for these materials can be also formed from renewable resources e.g., castor oil replacing petrochemical polioles [72,78–80]. It was noted that increasing silane ratio, i.e., crosslinking, increased the hydrophobicity of the material. Moreover, the higher level of siloxane crosslinking decreased the surface roughness value [72]. The employed castor oil can be also modified with 3-mercaptopropionic acid and 3-mercaptopropyl trimethoxysilane through the thiol-ene coupling reaction [78]. This compound has a complex structure and can be characterized as an organo-functional alkoxysilane monomer that can undergo both sol–gel polymerization of the alkoxy groups and reaction with the –NCO functionality in the prepolymer to form a hybrid network with covalent bonds between the organic and inorganic phases. There are also reports on castor oil based coatings that exhibited excellent antibacterial activity and are composed with novel crosslinker, which was synthesized with 3-aminopropyl trimethoxy silane and 3-glycidoxypropyl trimethoxy silane [80].

Thevetia peruviana seed oil was also employed for preparation of the hybrid coatings, containing ZnO particles [81,82]. These materials show antibacterial resistance towards *Escherichia coli* and *Staphylococcus aureus* and exhibit good corrosion resistance properties [82].

For preparation of poly(urethane-siloxane) materials by the sol–gel process, siloxane with a more complex (Structure 12.9) was employed, as well [70]. The crosslink was obtained by using hydrochloric acid as catalyst in addition to water. For these materials shape memory capability was noted.

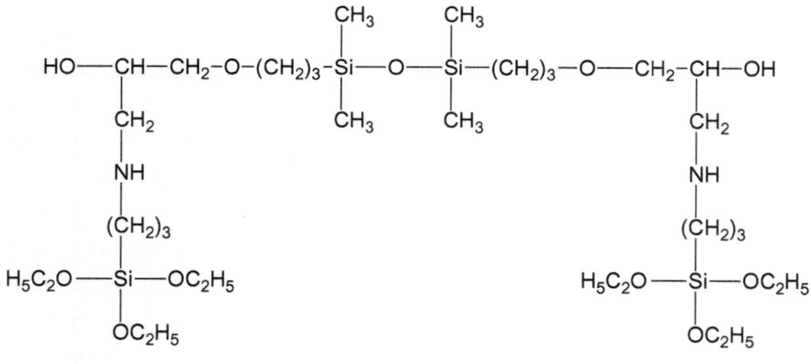

(12.9)

The sol–gel process was also employed by obtaining polyurethane waterborne coatings modified with TiO_2–SiO_2 particles [83]. New Ti–O–Si linkage into the organic polyurethane backbone was introduced. The chemical bonds between polyurethane anionomer and TiO_2–SiO_2 facilitated the homogenous dispersion and better compatibility during the sol–gel process.

UV radiation is another curing factor that allows for preparing of crosslinked coatings [84,85,98]. These kinds of materials, i.e., UV-curable coatings, is prepared by termination of isocyanate oligomer with acrylate monomer, e.g., hydroxyethylmethacrylate [84]. Moreover, some of the coatings combine benefits of both waterborne as well as UV-curing technologies [84]. Silicone moieties can be introduced by using 3-aminopropyltriethoxysilane or aqueous colloidal silica [84]. It was found that the incorporation of silica nanoparticles improved the modulus, tensile strength, and water resistance of poly(urethane-siloxane) coatings without damaging its surface hydrophobicity and transparency due to the homogeneous dispersion of silica clusters in the material matrix [84].

However, many reports on crosslinked PU-Si materials concern their curing with a multifunctional compound instead of moisture or UV radiation [86–92].

Pergal et al. obtained a set of crosslinked poly(urethane-siloxane) compositions with the use of hyperbranched polyesters [86–88]. For the samples prepared with polyesters of the fourth pseudo generation based on 2,2-bis(hydroxymethyl)propionic acid and ethoxylated pentaerythritol or di-trimethylolpropane as the core they found that the decrease of polydimethylsiloxane content increased crosslinking density, improved mechanical properties and the hardness of polyurethane [86]. The obtained materials showed phase separation. The first transition (at −128 to −115°C), corresponds to the glass transition of soft segments and the second one, located at 35–85°C, is related to the glass transition of hard segments. Moreover, the second relaxation peak during DMTA analysis located in the temperature region −80 to −65°C is ascribed to the movement of urethane groups close to the hyperbranched polyester. The surface of the polyurethane networks became more hydrophobic and the thermal stability was improved with increasing PDMS content. They also compared properties of crosslinked PU-Si based on hyperbranched polyester of the third pseudo generation and two siloxanes, hydroxyethoxy propyl terminated poly(dimethylsiloxane) and hydroxy propyl terminated poly(dimethylsiloxane) [87]. Higher crosslinking density and better waterproof properties were found for PU samples based on hydroxy propyl terminated poly(dimethylsiloxane). However, the surface energy values of the these samples were lower than values obtained for the samples based on hydroxyethoxy propyl terminated poly (dimethylsiloxane). The materials prepared from the same siloxane but crosslinked with hyperbranched polyesters of the second and third pseudogeneration show the difference only in the global properties of the structure, which is much less pronounced than the dependence of properties on the siloxane content [88].

For obtaining marine coatings, the zwitterionic polymers based on poly(sulfobetaine methacrylate) poly(SBMA) were employed [93]. These materials were synthesized using ABA-type triblock copolymers, poly(SBMA)-block-PDMS-block-poly(SBMA), having amine-termianted polydimethylsiloxane as the central block and poly(SBMA) as the flanking blocks with secondary amines at the junction points (Structure 12.10) which were crosslinked with polyisocyanate. It was noted that the coatings reveled excellent fouling-release performance toward the bacterium *Halomonas pacifica* and the diatom (microalga) *Navicula incerta* while the bacterium *Cellulophaga lytica* and sporelings of the green macroalga *Ulva linza* showed a relatively higher affinity for amphiphilic coating surfaces when compared to a series of standard control coatings.

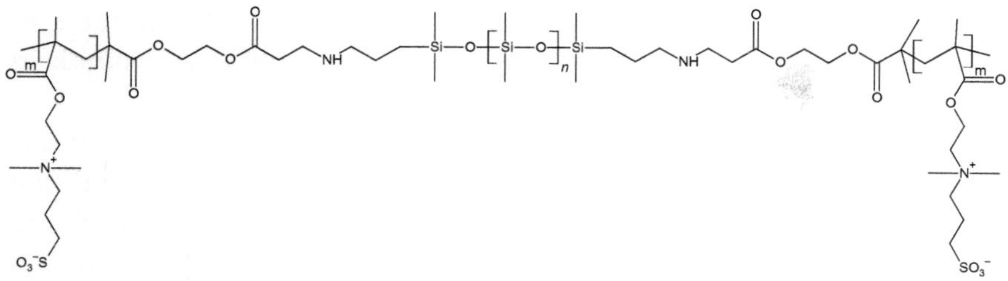

(12.10)

Some PU-Si fouling-release marine coatings were also obtained with the use of various polyisocyanates as crosslinking agent [94,99,100]. The coatings formulated with diisocyanate trimer comprised titanium dioxide pigment. However, they revealed similar properties to the coatings prepared without pigment, but in some cases more siloxane should be used.

The crosslinked poly(urethane-siloxane) coatings can be obtained with the use of polydimethylsiloxane of structures other than linear [65,95–97]. A comb-like structure of functionalized siloxanes with dimethyl units and reactive groups in a side chain were also employed (Structure 12.11) for the preparation of high-solid coatings with low amount of solvent [95,97]. Similar to the waterborne system, these coatings were environmentally friendly due to a reduced volatile organic content (VOC).

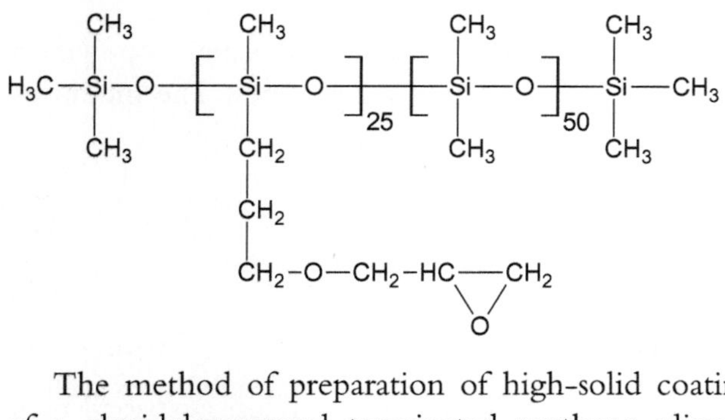

(12.11)

The method of preparation of high-solid coatings involved synthesis of a glycidyloxypropyl terminated urethane oligomer which was then cured with amine hardener in the presence of siloxane in a comb-like structure [95]. The materials showed phase separation confirmed by differential scanning calorimetry. An increased amount of siloxane resulted in an increase of thermal stability and water contact angle as well as a decrease of water absorption and of free surface energy. Due to the specific structure of the employed siloxane, the growth of hardness with its addition to polyurethane was observed.

Besides the typical crosslinked thermoset systems, also interpenetrating polymer networks (IPNs) composed of crosslinked polyurethane and polydimethylsiloxane can be distinguished [101]. The morphology and properties of the IPN depend on the degree of phase separation, which is connected to the mutual miscibility of the two polymer networks. There are in fact two-phase systems with two distinctive glass transition temperature T_g of the hard phase—in this case it will be phase constructed with rigid PU segments—approx. 200°C and a T_g of PDMS phase of approx. 60°C. The thermal stability of such IPN systems, as in the case of conventional thermoset, improved with increasing amount of PDMS. In addition to IPN there are also studies on semi-interpenetrating polymer networks (semi-IPNs) composed of linear polyurethane and crosslinked with diethylenetriamine 1,3-bis(glycidyloxypropyl)-1,1,3,3-tetramethyldisiloxane [102]. The semi-IPNs exhibited two-stage thermal decompsition mechanism in nitrogen. The thermal stability was also improved with an increase of siloxane in composition.

CONCLUSIONS

Analysis of the reviews shows that poly(urethane-siloxane) copolymers can be employed as new promising materials for coatings. The modification of PU-Si hybrid materials described in the literature involves the use of siloxanes with various chain structures, containing amino or hydroxyl groups, to form urethane-isocyanate prepolymers. Depending on the type of chain extender, its functionality and the method of synthesis, it is possible to synthesize thermoplastic PU-Si copolymers, ionomers, crosslinked thermosets or IPN systems. The poly(urethane-siloxane) coatings are characterized by low hydrophobicity and high thermal stability as well as antibacterial activity.

REFERENCES

[1] P. Król, Ł. Byczyński, Struktura, metody wytwarzania, właściwości i zastosowanie nowoczesnych kopolimerów poliuretanowo-polisiloksanowych [Structure, methods for production, properties and application of novel polyurethane-polysiloxane copolymers], Przem. Chem. 86 (2007) 643–651.

[2] E. Yilgör, I. Yilgör, Silicone containing copolymers: synthesis, properties and applications. Prog. Polym. Sci. 39 (2014) 1165–1195.

[3] P. Król, Synthesis methods, chemical structures and phase structures of linear polyurethanes, Properties and applications of linear polyurethanes in polyurethane elastomers, copolymers and ionomers. Prog. Mater. Sci. 52 (2007) 915–1015.

[4] D. Randall, S. Lee, The Polyurethanes Book, John Wiley & Sons, Hoboken, NJ, 2002.

[5] Z. Wirpsza, Poliuretany. Chemia Technologia Zastosowanie, WNT, Warszawa, 1991.

[6] P. Rościszewski, M. Zielecka, Silikony Właściwości i Zastosowanie, WNT, Warszawa 2002.

[7] S. Hamdani, C. Longuet, D. Perrin, J.M. Lopez-cuesta, F. Ganachaud, Flame retardancy of silicone-based materials. Polym. Degrad. Stab. 94 (2009) 465–795.

[8] G. Camino, S.M. Lomakin, M. Lazzari, Polydimethylsiloxane thermal degradation Part 1, Kinetic Aspects Polym. 42 (2001) 2395–2402.

[9] G. Deshpande, M.E. Rezac, Kinetic aspects of the thermal degradation of poly (dimethyl siloxane) and poly(dimethyl diphenyl siloxane). Polym. Degrad. Stab. 76 (2002) 17–24.

[10] G. Deshpande, M.E. Rezac, The effect of phenyl content on the degradation of poly(dimethyl diphenyl) siloxane copolymers. Polym. Degrad. Stab. 74 (2001) 363–370.

[11] D.K. Chattopadhyay, K.V.S.N. Raju, Structural engineering of polyurethane coatings for high performance applications. Prog. Polym. Sci 32 (2007) 352–418.

[12] J.T. Yeh, Y.C. Shu, Characteristics of the degradation and improvement of the thermal stability of poly(siloxane urethane) copolymers. J. Appl. Polym. Sci. 115 (2010) 2616–2628.

[13] F. Belva, S. Bourbigot, S. Duquesne, C. Jama, M. Le Bras, C. Pelegris, et al., Heat and fire resistance of polyurethane-polydimethylsiloxane hybrid material. Polym. Adv. Technol. 17 (2006) 304–311.

[14] K. Mequanint, R. Sanderson, Self-assembling metal coatings from phosphated and siloxane-modified polyurethane dispersions: an analysis of the coating-air interface. J. Appl. Polym. Sci. 88 (2003) 893–899.

[15] H.B. Park, Y.M. Lee, Separation of toluene/nitrogen through segmented polyurethane and polyurethane urea membranes with different soft segments. J. Membr. Sci. 197 (2002) 283–296.

[16] H.B. Park, C.K. Kim, Y.M. Lee, Gas separation properties of polysiloxane/polyether mixed soft segment urethane urea membranes. J. Membr. Sci. 204 (2002) 257–269.

[17] M.S. Yen, P.Y. Tsai, Effect on the structure and properties of membranes formed by blending polydimethylsiloxane polyurethane into different soft-segment waterborne polyurethanes. J. Appl. Polym. Sci. 102 (2006) 210–221.

[18] A.F. Osman, G.A. Edwards, T.L. Schiller, Y. Andriani, K.S. Jack, I.C. Morrow, et al., Structure – property relationships in biomedical thermoplastic polyurethane nanocomposites, Macromolecules 42 (2012) 198–210.

[19] D. Bezuidenhout, D.F. Williams, P. Zilla, Polymeric heart valves for surgical implantation, catheter-based technologies and heart assist devices. Biomaterials 36 (2015) 6–25.

[20] F. Askari, M. Barikani, M. Barmar, Siloxane-based segmented poly(urethane-urea) elastomer: synthesis and characterization. J. Appl. Polym. Sci. 130 (2013) 1743–1751.

[21] A. Silvestri, P.M. Serafini, S. Sartori, P. Ferrando, F. Boccafoschi, S. Milione, et al., Polyurethane-based biomaterials for shape-adjustable cardiovascular devices. J. Appl. Polym. Sci. 122 (2011) 3661–3671.

[22] L.F. Wang, Q. Ji, T.E. Glass, T.C. Ward, J.E. McGrath, M. Muggli, et al., Synthesis and characterization of organosiloxane modified segmented polyether polyurethanes. Polymer 41 (2000) 5083–5093.
[23] R. Adhikari, P.A. Gunatillake, S.J. McCarthy, G.F. Meijs, Mixed macrodiol-based siloxane polyurethanes: effect of the comacrodiol structure on properties and morphology. J. Appl. Polym. Sci. 78 (2000) 1071–1082.
[24] Ł. Byczyński, Effect of different polyethers on surface and thermal properties of poly (urethane-siloxane) copolymers modified with side-chain siloxane. J. Therm. Anal. Calorim. 114 (2013) 397–408.
[25] R.S. Chen, C.J. Chang, Y.H. Chang, Study on siloxane-modified polyurethane dispersions from various polydimethylsiloxanes. J. Polym. Sci. Part A: Polym Chem 43 (2005) 3482–3490.
[26] C.Y. Li, J.H. Chen, P.C. Chien, W. Chiu, R.S. Chen, T.M. Don, Preparation of poly(IPDI–PTMO–siloxanes) and influence of siloxane structure on reactivity and mechanical properties. Polym. Eng. Sci 47 (2007) 625–632.
[27] J. Zong, Q. Zhang, H. Sun, Y. Yu, S. Wang, Y. Liu, Characterization of polydimethylsiloxane–polyurethanes synthesized by graft or block copolymerizations. Polym. Bull. 65 (2010) 477–493.
[28] W.H. Chen, P.C. Chen, S.C. Wang, J.T. Yeh, C.H. Yuan, K.N. Chen, UV-curable PDMS-containing PU system for hydrophobic textile surface treatment. J. Polym. Res. 16 (2009) 601–610.
[29] F. Joki-Korpela, T.T. Pakkanen, Incorporation of polydimethylsiloxane into polyurethanes and characterization of copolymers. Eur. Polym. J 47 (2011) 1694–1708.
[30] Z. Dai, K. Yang, Q. Dong, Synthesis and characterization of hydroxy-terminated polyetherpolydimethylsiloxane-polyether (PE-PDMS-PE) triblock oligomers and their use in the preparation of thermoplastic polyurethanes. J. Appl. Polym. Sci. 132 (2015) 42521.
[31] P.L. Kuo, W.J. Liang, C.L. Lin, Solid polymer electrolytes, 2. Preparation and ionic conductivity of solid polymer electrolytes based on segmented polysiloxane-modified polyurethane. Macromol. Chem. Phys 203 (2002) 230–237.
[32] F.S. Chuang, W.C. Tsen, Y.C. Shu, The effect of different siloxane chain-extenders on the thermal degradation and stability of segmented polyurethanes. Polym. Degrad. Stab 84 (2004) 69–77.
[33] J.P. Sheth, A. Aneja, G.L. Wilkes, E. Yilgor, G.E. Atilla, I. Yilgor, et al., Influence of system variables on the morphological and dynamic mechanical behavior of polydimethylsiloxane based segmented polyurethane and polyurea copolymers: a comparative perspective. Polymer 45 (2004) 6919–6932.
[34] H.H. Wang, Y.T. Lin, Silicon-containing anionic water-borne polyurethane with covalently bonded reactive dye. J. Appl. Polym. Sci. 90 (2003) 2045–2053.
[35] T. Choi, K.A. Masser, E. Moore, J. Weksler, A. Padsalgikar, J. Runt, Segmented polyurethanes derived from novel siloxane–carbonate soft segments for biomedical applications. J. Polym. Sci. Part B Polym. Phys. 49 (2011) 865–872.
[36] K. Landfester, U. Pawelzik, M. Antonietti, Polydimethylsiloxane latexes and copolymers by polymerization and polyaddition in miniemulsion. Polymer 46 (2005) 9892–9898.
[37] B. Ochiai, H. Kojima, T. Endo, Synthesis and properties of polyhydroxyurethane bearing silicone backbone. J. Polym. Sci. Part A Polym. Chem. 52 (2014) 1113–1118.
[38] Q. Zhang, H. Liu, X. Zhan, F. Chen, J. Yan, H. Tang, Microstructure and antibacterial performance of functionalized polyurethane based on polysiloxane tethered cationic biocides. RSC Adv. 5 (2015) 77508–77517.
[39] P. Król, K. Pielichowska, Ł. Byczyński, Thermal degradation kinetics of polyurethane–siloxane anionomers., Thermochim. Acta 507–508 (2010) 91–98.

[40] Y. Yu, J. Wang, Synthesis and properties of block and graft waterborne polyurethane modified with α,ω-bis(3-(1-methoxy-2-hydroxypropoxy)propyl)polydimethylsiloxane and α-N,N-dihydroxyethylaminopropyl-ω-butylpolydimethylsiloxane. Polym. Eng. Sci 54 (2014) 805–811.
[41] H.J. Naghash, I. Mohammadidehcheshmeh, M. Mehrnia, Synthesis and characterization of a novel hydroxy terminated polydimethylsiloxane and its application in the waterborne polysiloxane–urethane dispersion for potential marine coatings. Polym. Adv. Technol 24 (2013) 307–317.
[42] M.M. Rahman, H.H. Chun, H. Park, Waterborne polysiloxane–urethane–urea for potential marine coatings. J. Coat. Technol. Res. 8 (3) (2011) 389–399.
[43] Y. Du, Z. Yang, C. Zhou, Study on waterborne polyurethanes based on poly (dimethyl siloxane) and perfluorinated polyether. Macromol. Res 23 (9) (2015) 867–875.
[44] F. Yu, X. Xu, N. Lin, X.Y. Liu, Structural engineering of waterborne polyurethane for high performance waterproof coatings. RSC Adv. 5 (2015) 72544.
[45] Q. Zhang, J. Hu, S. Gong, Preparation and characterization of aqueous polyurethane dispersions with well-defined soft segments. J. Appl. Polym. Sci. 122 (2011) (2011) 3064–3070.
[46] C. Gao, M. Chen, J. Ni, J. Hua, X. Xu, W. Lin, et al., Influence of siloxane cosegment length and content of waterborne polysiloxane-urethane copolymers on their water resistance, thermal stability and mechanical properties. Chin. J. Polym. Sci. 28 (2010) 219–229.
[47] M. Zhang, Y. Wu, H. Wu, Q. Zhang, Study on reactive polydimethylsiloxane-modified waterborne polyurethanes. J. Polym. Res. 19 (2012) 9807.
[48] M. Zhang, Y. Wu, H. Wu, Q. Zhang, Y. Xia, Synthesis and characterization of α,ω-bis(3-hydroxypropyl)-functionalized poly{dimethylsiloxane-co-methyl[3-(2-acetylacetoxy)]propylsiloxane}. J. Appl. Polym. Sci. 125 (2012) 595–607.
[49] K.L. Noble, Waterborne polyurethanes. Prog. Org. Coat. 32 (1997) 131–136.
[50] S. Subramani, J.M. Lee, I.W. Cheong, J.H. Kim, Synthesis and characterization of water-borne crosslinked silylated polyurethane dispersions. J. Appl. Polym. Sci. 98 (2005) 620–631.
[51] B.K. Kim, J.C. Lee, Polyurethane ionomer dispersions from poly(neopentylene phthalate) glycol and isophorone diisocyanate. Polymer 37 (1996) 469–475.
[52] J.S. Lee, B.K. Kim, Poly(urethane) cationomers from poly(propylene) glycol and isophorone diisocyanate: emulsion characteristics and tensile properties of cast films. Prog. Org. Coat 25 (1995) 311–318.
[53] C.Z. Yang, K.K.S. Hwang, S.L. Cooper, Morphology and properties of polybutadiene- and polyether-polyurethane zwitterionomers. Macromol. Chem. Phys. 184 (1983) 651–668.
[54] P. Król, B. Król, J. Pielichowski, Zastosowanie anionowych dyspersji poliuretanowych jako spoiw do proszków ceramicznych [Application of anionic polyurethane dispersions binders for ceramic pulvers]. Mater. Ceram. 56 (2004) 53–60.
[55] P. Król, B. Król, R. Stagraczyński, K. Skrzypiec, Waterborne cationomer polyurethane coatings with improved hydrophobic properties. J. Appl. Polym. Sci. 127 (2013) 2508–2519.
[56] S. Zhang, H. Lv, H. Zhang, B. Wang, Y. Xu, Waterborne polyurethanes: spectroscopy and stability of emulsions. J. Appl. Polym. Sci. 101 (2006) 597–602.
[57] A. Barni, M. Levi, Aqueous polyurethane dispersions: a comparative study of polymerization processes. J. Appl. Polym. Sci. 88 (2003) 716–723.
[58] T. Su, G.Y. Wang, X.D. Xu, C.P. Hu, Preparation and properties of waterborne polyurethaneurea consisting of fluorinated siloxane units. J. Polym. Sci Part A Polym. Chem. 44 (2006) 3365–3373.

[59] C. Xu, Z. Cai, J. Xing, Y. Ren, W. Xu, W. Shi, Synthesis of polypropylene carbonate polyol-based waterborne polyurethane modified with polysiloxane and its film properties. Fibers Polym. 15 (2014) 665–671.
[60] C. Bai, X. Zhang, J. Dai, Synthesis and characterization of a new UV cross-linkable waterborne siloxane-polyurethane dispersion. J. Macromol. Sci. Part A Pure Appl. Chem. 44 (2007) 1203–1208.
[61] X. Li, J. Hu, D. Sun, Y. Zhang, Nanosilica reinforced waterborne siloxane-polyurethane nanocomposites prepared via "click" coupling. J. Coat. Technol. Res. 11 (4) (2014) 517–531.
[62] Y. Ma, Q.Y. Tang, J. Zhu, L.H. Wang, C. Yao, Fluorescent and thermal properties of siloxane-polyurethanes based on 1,8-naphthalimide. Chin. Chem. Lett. 25 (2014) 680–686.
[63] B. Baradie, P.H.M. Lai, M.S. Shoichet, Synthesis and characterization of novel polysiloxane-grafted fluoropolymers can. J. Chem. 83 (2005) 553–558.
[64] K. Madhavan, B.S.R. Reddy, Poly(dimethylsiloxane-urethane) membranes: Effect of hard segment in urethane on gas transport properties. J. Membr. Sci. 283 (2006) 357–365.
[65] Ł. Byczyński, M. Dutkiewicz, H. Maciejewski, Thermal degradation studies of poly (urethane–siloxane) thermosets based on co-poly(dimethyl)(methyl, hydroxypolyoxyethylenepropyl)siloxane. Thermochim. Acta. 589 (2014) 252–261.
[66] H. Jiang, Z. Zheng, W. Song, X. Wang, Moisture-cured polyurethane/polysiloxane copolymers: effects of the structure of polyester diol and NCO/OH ratio. J. Appl. Polym. Sci. 108 (2008) 3644–3651.
[67] S. Xu, L. Xie, X. Yu, Y. Xiong, H. Tang, Synthesis and characterization of phenyl polysiloxane modified polyurea/polyurethanes. J. Polym. Sci., Part A Polym. Chem 53 (2015) 1794–1805.
[68] Z. Shi, Preparation and characterization of crosslinked polyurethane-block-poly(trifluoropropylmethyl)siloxane elastomers with potential biomedical applications. Polym. Int. 62 (2013) 1351–1357.
[69] Z. Shi, Preparation and characterization of high-strength elastomers with high poly (trifluoropropylmethyl)siloxane content into polyurethane urea. J. Polym. Res. 20 (2013) 57.
[70] M. Alexandru, M. Cazacu, M. Cristea, A. Nistor, C. Grigoras, B.C. Simionescu, poly(siloxane-urethane) crosslinked structures obtained by sol-gel technique. J. Polym. Sci. Part A Polym. Chem. 49 (2011) 1708–1718.
[71] F. Zhang, X. Wei, Z. Xiao, Study on high-solid content Si/PU polyurethane dispersion with PES/PPG composite soft segment. J. Appl. Polym. Sci. 127 (2013) 1730–1736.
[72] K.M.S. Meera, R.M. Sankar, S.N. Jaisankar, A.B. Mandal, Physicochemical studies on polyurethane/siloxane cross-linked films for hydrophobic surfaces by the sol – gel process. J. Phys. Chem. B 117 (2013) 2682–2694.
[73] A.M. Mikhailova, M. Tamboura, M.Q. Jia, Heat-resistant and anti-corrosion urethane-silicone-based coatings. Silicon 4 (2012) 197–208.
[74] A.M. Mikhailova, M. Tamboura, M.Q. Jia, Synthesis, characterization, and analyses of mechanical, adhesion, and thermal properties of polysiloxane resin modified with segmented polyurethane. J. Coat. Technol. Res. 10 (2013) 97–108.
[75] R. Gharibi, H. Yeganeh, H. Gholami, Z.M. Hassan, Aniline tetramer embedded polyurethane/siloxane membranes and their corresponding nanosilver composites as intelligent wound dressing materials. RSC Adv. 4 (2014) 62046.
[76] T. Zhang, K. Xi, H. Chen, X. Yu, Synthesis and properties of self-crosslinkable polyurethane–urea with silsesquioxane formation. J. Appl. Polym. Sci. 91 (2004) 190–195.

[77] E. Maya-Visuet, T. Gao, M. Soucek, H. Castaneda, The effect of TiO_2 as a pigment in a polyurethane/polysiloxane hybrid coating/aluminum interface based on damage evolution. Prog. Org. Coat. 83 (2015) 36–46.
[78] C. Fu, X. Hu, Z. Yang, L. Shen, Z. Zheng, Preparation and properties of waterborne bio-based polyurethane/siloxane cross-linked films by an in situ sol–gel process. Prog. Org. Coat. 84 (2015) 18–27.
[79] C. Fu, Z. Yang, Z. Zheng, L. Shen, Properties of alkoxysilane castor oil synthesized via thiol-ene and its polyurethane/siloxane hybrid coating films. Prog. Org. Coat. 77 (2014) 1241–1248.
[80] A. Shaik, R. Narayan, K.V.S.N. Raju, Synthesis and properties of siloxane-crosslinked polyurethaneurea/silica hybrid films from castor oil. J. Coat. Technol. Res. 11 (2014) 397–407.
[81] T.O. Siyanbola, K. Sasidhar, B.V.S.K. Rao, R. Narayan, O. Olaofe, E.T. Akintayo, et al., Development of functional polyurethane–ZnO hybrid nanocomposite coatings from thevetia peruviana seed oil. J. Am. Oil Chem. Soc. 92 (2015) 267–275.
[82] T.O. Siyanbola, K. Sasidhar, B. Anjaneyulu, K.P. Kumar, B.V.S.K. Rao, R. Narayan, et al., Anti-microbial and anti-corrosive poly (ester amide urethane) siloxane modified ZnO hybrid coatings from *Thevetia peruviana* seed oil. J. Mater. Sci. 48 (2013) 8215–8227.
[83] L. Zhai, Y. Wang, F. Peng, Z. Xiong, R. Liu, J. Yuan, et al., Synthesis of TiO_2–SiO_2/waterborne polyurethane hybrid with amino-siloxane terminated via a sol–gel process. Mater. Lett. 89 (2012) 81–85.
[84] S. Zhang, Z. Chen, M. Guo, H. Bai, X. Liu, Synthesis and characterization of waterborne UV-curable polyurethane modified with side-chain triethoxysilaneand colloidal silica. Colloids Surf. A. Physicochem. Eng. Aspects 468 (2015) 1–9.
[85] S. Rengasamy, V. Mannari, UV-curable PUDs based on sustainable acrylated polyol: Study of their hydrophobic and oleophobic properties. Prog. Org. Coat. 77 (2014) 557–567.
[86] M.V. Pergal, J.V. Džunuzović, R. Poręba, S. Ostojić, A. Radulović, M. Špírková, Microstructure and properties of poly(urethane-siloxane)s based on hyperbranched polyester of the fourth pseudo generation. Prog. Org. Coat. 76 (2013) 743–756.
[87] M.V. Pergal, J.V. Džunuzović, R. Poręba, D. Micić, P. Stefanov, L. Pezo, et al., Surface and thermomechanical characterization of polyurethane networks based on poly(dimethylsiloxane) and hyperbranched polyester. Exp. Polym. Lett. 7 (2013) 806–820.
[88] M.V. Pergal, J.V. Dzunuzović, R. Poręba, M. Steinhart, M.M. Pergal, V.V. Vodnik, et al., Structure – property correlation study of novel poly(urethane – ester – siloxane), Netw. Ind. Eng. Chem. Res. 52 (2013) 6164–6176.
[89] L.I. Makarova, N.V. Sergienko, O.G. Nikolskii, V.A. Martirosov, O.T. Gritsenko, Rus. Chem. Bull. Intern. Ed. 53 (2004) 2029–2034.
[90] Z. Ge, Y. Luo, Synthesis and characterization of siloxane-modified two-component waterborne polyurethane. Prog. Org. Coat. 76 (2013) 1522–1526.
[91] S. Oprea, Effect of the hard-segment structure on the dielectric relaxation of cross-linked polyurethanes. J. Appl. Polym. Sci. 119 (2011) 2196–2204.
[92] A. Santiago, L. Martin, J.J. Iruin, M.J. Fernández-Berridi, A. González, L. Irust, Microphase separation and hydrophobicity of urethane/siloxane copolymers with low siloxane content. Prog. Org. Coat. 77 (2014) 798–802.
[93] R.B. Bodkhe, S.J. Stafslien, J. Daniels, N. Cilz, A.J. Muelhberg, S.E.M. Thompson, et al., Zwitterionic siloxane-polyurethane fouling-release coatings. Prog. Org. Coat. 78 (2015) 369–380.

[94] S.A. Sommer, J.R. Byrom, H.D. Fischer, R.B. Bodkhe, S.J. Stafslien, J. Daniels, et al., Effects of pigmentation on siloxane–polyurethane coatings and their performance as fouling-release marine coatings. J. Coat. Technol. Res. 8 (6) (2011) 661–670.
[95] Ł. Byczyński, M. Dutkiewicz, H. Maciejewski, Synthesis and properties of high-solids hybrid materials obtainedfrom epoxy functional urethanes and siloxanes. Prog. Org. Coat. 84 (2015) 59–69.
[96] Ł. Byczyński, Thermal degradation studies of poly(urethane–siloxane) thermosets based on co-poly(dimethyl)(methyl, 3-glycidoxypropyl)siloxane and epoxy-terminated urethane oligomer. Thermochim. Acta. 592 (2014) 58–66.
[97] Ł. Byczyński, M. Dutkiewicz, H. Maciejewski, Thermal and surface properties of hybrid materials obtained from epoxy functional urethane and siloxane. Polym. Bull. 73 (2016) 1247–1265.
[98] J. Cheng, Y. Cao, S. Jiang, Y. Gao, J. Nie, F. Sun, Synthesis and performances of UV-curable polysiloxane – polyether block polyurethane acrylates for PVC leather finishing agents. Ind. Eng. Chem. Res. 54 (2015) 5635–5642.
[99] C.S. Lim, G.H. Dickinson, S. Sommer, S. Lay-Ming Teo, R.B. Bodkhe, D.C. Webster, et al., A small-scale waterjet test method for screening novel foul-release coatings. J. Coat. Technol. Res. 12 (2015) 533–542.
[100] R.B. Bodkhe, S.E.M. Thompson, C. Yehle, N. Cilz, J. Daniels, S.J. Stafslien, et al., The effect of formulation variables on fouling-release performance of stratified siloxane–polyurethane coatings. J. Coat. Technol. Res. 9 (2013) 235–249.
[101] S. Vlad, A. Vlad, S. Oprea, Interpenetrating polymer networks based on polyurethane and polysiloxane. Eur. Polym. J. 38 (2002) 829–835.
[102] Ł. Byczyński, M. Dutkiewicz, H. Maciejewski, Thermal degradation kinetics of semi-interpenetrating polymer network based on polyurethane and siloxane. Thermochim. Acta 560 (2013) 55–62.

CHAPTER 13

Polyurethane Blends for Powder Clear Coatings

Barbara Pilch-Pitera
Rzeszow University of Technology, Rzeszów, Poland

Contents

13.1 INTRODUCTION

Depending on the form in which they are manufactured, polyurethane coating systems can be divided into the following groups: powder, solvent-based, such as high-solids, as well as water-borne compositions. Due to the directives of the European Parliament defining the requirements to reduce emissions of volatile organic compounds (VOC), and consequently to decrease VOC contents in the manufactured and applied coatings, at present the solvent-based systems are being replaced with products based on high-solids, as well as water-borne and powder systems [1]. In recent years there has been growing interest in coatings obtained from powder systems, which constitute a dynamically developing assortment of anticorrosive products, primers, and top coats. In comparison to solvent-based and water-borne compositions, powder systems are more environmentally friendly because they do not contain toxic solvents, dispersing agents or biocides, the so-called "in-can preservatives" applied in the case of water-borne dispersions. Economic advantages of using powder coatings include the facts that they are easier to apply to the substrate without drips; nearly

Polyurethane Polymers: Blends and Interpenetrating Polymer Networks
DOI: http://dx.doi.org/10.1016/B978-0-12-804039-3.00013-0

100% of the powder can be applied as it is possible to reuse the material which sinks to the bottom of the chamber during application; the equipment is easy to clean—after it has been used, it can simply be wiped, or dusted, e.g., with compressed air; and it is possible to use automated methods to apply them to the substrate. In addition to their economic and environment-related advantages, powder products fully meet other new criteria specified for coating products by modern paints and lacquers industry: they have better functional properties, e.g., higher hydrophobicity, and better mechanical, thermal, chemical, weathering, corrosion, bacteria, and antifouling resistance [2].

Powder coatings also have some disadvantages. Manufacturing and powder application process requires specialized equipment which is expensive. Spraying powder in air may be explosive [3]. Melting and curing the powder takes place in an oven, and for this reason the coating surface must be thermally resistive [2].

Powder coatings, in accordance with ASTM, are classified as one-component polyurethane systems, together with three other types: moisture-cured systems, oil-containing compositions drying in contact with oxygen from atmospheric air, and thermoplastic PUR solutions [4]. The main advantages of one-component versus two-component systems result from the facts that: appropriate molar ratios of the specific components are retained while the system is being prepared at the production plant (rather than directly before the application); it is possible to avoid additional expenditures connected with special applicators; there is no waste resulting from mixing a greater quantity of components than can actually be used during the life of the mixture; finally, reduced toxicological risk linked with the use of resins with low contents of functional groups, crosslinking agents reactive at the temperature exceeding 100°C, and previously synthesized polymers.

Powder coatings are mainly used to provide anticorrosive protection and decorative paint for products, at the place where they are manufactured, i.e., directly in production facilities, which is of great importance in automotive industry. This type of protection can be applied to a wide range of products, such as agricultural machines exposed to very difficult working conditions, various automotive components, including car bodies and hubcaps, steel, and aluminum constructions designed for playgrounds, fencing, bridges, or furnishings, as well as metal grating, pipes, rods, iron ornaments, lighting fixtures, air-conditioners, and home appliances. They are also used to paint various elements used by the chemical, food, and pharmaceutical industries where it is necessary to ensure coatings with high resistance to influence of the natural environment, corrosion-inducing media, chemical substances, and biological factors. Polyurethane powder coatings produced currently are characterized by high abrasion resistance, hardness, adhesion to substrate, and good yellowing and chalking resistance.

In recent years significant progress has been achieved in the technologies of powder coating production and application. This is a result of the excellent chemical and mechanical properties of powder coatings and the long-term protection provided by them in comparison to the analogous solvent-based and water-borne systems. Given the considerable variety of raw materials, powder coatings may have a wide range of advantageous properties: they may be very hard or highly elastic, they may have very good abrasion, scratch, chemicals (including oils), salt spray, and oxidation resistance, they may have excellent adhesion to substrate, and be available in a complete range from high gloss to deep matt, in the full spectrum of colors and they can be used under water. They may be applied in extremely thin layers of 25 μm.

Growing demand of the market and challenging requirements defined by customers with regard to surface protection result in the necessity to look for solutions designed to enhance functional properties of coatings.

13.2 CHARACTERISTICS OF MANUFACTURED POLYURETHANE POWDER COATINGS

In terms of the curing methods, the currently manufactured polyurethane powder coatings can be divided into thermoplastic, thermosetting, as well as ultraviolet radiation (UV) or electron beam (EB) curable products [5,6].

Because of the availability of suitable polymers, thermoplastic powders were the first to emerge in the early 1950s. Progress in resins production towards the end of the 1950s contributed to the development of thermosetting powder related technology.

A thermoplastic product melts and flows at a higher temperature, yet it has the same chemical composition and structure when it solidifies after being cooled down. No chemical reactions occur in thermoplastic products which are subjected to heating, for instance in order to adjust them to the substrate. Thermoplastic powders are based on thermoplastic polymers of high molecular weight. Powder systems are manufactured with the use of thermoplastic polyester urethanes, for instance those marked

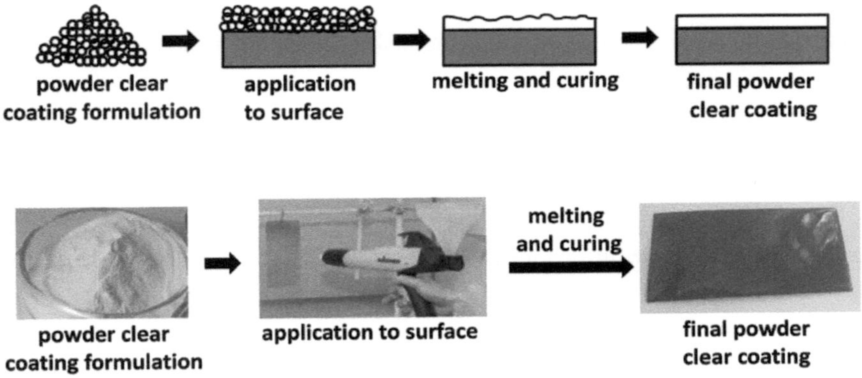

Figure 13.1 Formation process of the thermosetting powder clear coatings.

with Estane trade name, as well as obtained from a mixture of cycloaliphatic diisocyanate ($H_{12}MDI$) and carbodiimide bonds containing polyisocyanate $H_{12}MDI$ [7]. Thermoplastic polyurethanes are also used in combination with other thermoplastic powders, e.g., polyethylene [8] and polyethylene-vinyl [9] polymers.

Most powder systems (approx. 90%) are classified as thermosetting materials. They consist of solid resins containing reactive functional groups with molecular weights which are higher than in liquid products and lower than in thermoplastic systems. During the curing process, solid resin melts and reacts with a crosslinker to form a polymer network (Fig. 13.1). Because the curing reaction is irreversible, the obtained product does not melt and is insoluble. Thermosetting powders are cured at a lower temperature than thermoplastic systems.

The basic components of thermosetting polyurethane powder coating systems include polyester or polyacrylic resin with terminal hydroxyl groups, and blocked polyisocyanate used as a crosslinker [5]. Polyester resins used in polyurethane systems are produced through polycondensation of a mixture of tere- and isophthalic acid with neopentyl glycol, and with small quantities of dimethyl terephthalate, hexane-1,6-diol, propane-1,2-diol, 1,4-dimethylolcyclohexane, and trimethylolpropane [10]. The molar ratio of the raw materials is chosen to ensure OH number in the range of 25–60 mgKOH g^{-1}, glass transition temperature at 50–60°C, melting point at 70–90°C, and melt viscosity e.g., at 160°C in the range of 10–40 Pa•s. Polyester resins may also contain carboxyl groups (acid number=2–5 mgKOH g^{-1}), resulting from incomplete conversion of

Table 13.1 Example of polyester resin composition used in thermosetting polyurethane powder coating systems

Ingredient	wt%
Neopentyl glycol	38
Terephtalic acid	48
Isophthalic acid	10
Sebacic acid	2
Trimethylol propane	2
Fascat 4201—catalyst (monobuthyltin oxide (MBTO))	0.09

carboxylic acid [11]. Table 13.1 demonstrates a polyester resin composition used in thermosetting polyurethane powder coating systems.

Polyester resins are synthesized in one reactor and the process consists of two steps. During the first step, all the acidic components and part of neopentyl glycol are added to the reactor. Polycondensation of neopentyl glycol and acids produces resin with terminal carboxyl groups, which during the second stage reacts with the remaining amount of glycol, yielding the final product with the required parameters. By replacing glycol with raw materials obtained from renewable sources, e.g., isosorbide [12] or products of PET recycling [13] it is possible to obtain coatings with properties which are as good as those in coatings produced by crosslinking traditional polyester resins. Addition of hyperbranched polyester resins containing −OH groups enhances flow properties of the coating during the curing process [14].

Acrylic resins are synthesized by copolymerization of methacrylates and acrylates containing alkyl and hydroxyalkyl groups in the presence of radical initiators such as 2,2′-azobisizobutyronitryl [5].

Blocked diisocyanates and polyisocyanates (PIC) are used as crosslinking agents. In order to prevent uncontrolled reactions of isocyanate groups in PICs, during the powder coating manufacturing process and while storing the finished coating systems, with the compounds containing reactive hydrogen atoms, e.g., hydroxyl groups of the resin, it is necessary to block these before using. We can distinguish two types of PIC blocking: external and internal.

External blocking is based on a reaction between isocyanate group and a specially introduced blocking agent, containing a reactive atom of hydrogen (e.g., −OH group, −NH group or $-CH_2-$ acidic group), which after deblocking at a higher temperature is released into the atmosphere (Eq. 13.1).

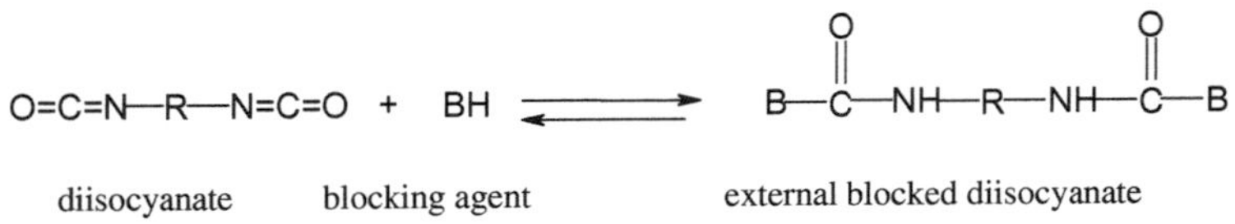

(13.1)

If the applied blocking agent is a hydroxyl group containing compound, the reaction produces a urethane group; on the other hand the use of a compound containing –NH group results in producing urea group [2]. Compounds used as blocking agents include alcohols, ε-caprolactam, 2-butanone oxime, 3,5-dimethylpyrazole, 1,2,4-triazole, and acetylacetone. During the curing process of systems containing the externally blocked PIC, the blocking compound is released into the atmosphere. Due to this, powder systems of this type cannot be applied for forming thick coatings (over 60 μm) because of the bubbles which may be left in the coating. A small quantity of the blocking agent always stays permanently trapped in the coating and this may be the cause of yellowing of the coating.

Diethyl malonate (DEM) is an interesting blocking agent [15]. The process of crosslinking in this case is not a result of reaction between deblocked isocyanate groups contained in PIC and hydroxyl groups in resin, but it is the effect of the transesterification reaction of malonate ethyl groups with hydroxyl groups. Hence, DEM is not released into the atmosphere during the curing process, but it is built into the structure of the formed PUR.

The factors of critical importance for the usefulness of PIC as a crosslinker include: glass transition temperature (T_g=55–90°C), melting point (T_m=80–110°C) and isocyanate groups content (C_{NCO}=10–15%) [11,16]. PICs are obtained mainly to enhance functionality of the isocyanate component. They are most frequently produced through reaction of diisocyanates with small-molecule compounds, containing two, three, or four hydroxyl groups (e.g., ethylene glycol, trimethylolpropane, triethanolamine [17], or pentaerythritol), or, less frequently, amine groups e.g., diamines [18] or as a result of trimerization [19], biuretization [20], as well as reaction of diisocyanates with urethanes yielding allophanates [21].

Powder systems of greatest commercial importance are ε-caprolactam blocked PICs synthesized from isophorone diisocyanate (IPDI) (1) [22]. In addition to isophorone diisocyanate, the compounds used by the industry include: aliphatic 1,6-hexamethylene diisocyanate (HDI) (2), 2,2,4- and 2,4,4-trimethyl-1,6-hexamethylene diisocyanate (TMHDI)

(3, 4), cycloaliphatic 4,4′-dicyclohexylmethylene diisocyanate ($H_{12}MDI$) (5), 1,4-bis(isocyanatomethyl)cyclohexane (6) [23], 1,3,5-tris(isocyanatomethyl) cyclohexane (7), aromatic tolylene-2,4- and 2,6-diisocyanate (TDI) (8) and 3,5-tris(isocyanatomethyl)benzene (9) [24], m- and p-xylylene diisocyanate (XDI) (10), p-tetramethylxylyllene diisocyanate (TMXDI) (11). The use of aromatic diisocyanates (e.g., TDI) in production of coatings, despite the lower prices, is limited due to their lower weathering resistance (application exclusively in coatings designed for interiors). In the following diisocyanates: 3,5-tris(isocyanatomethyl)benzene (9), m- and p-xylylene diisocyanate (XDI) (10) and p-tetramethylxylylene diisocyanate (TMXDI) (11) the isocyanate groups are not linked to the benzene ring directly but via methylene groups. As a result they have intermediary properties between those of aliphatic and aromatic diisocyanates. Due to the presence of quaternary isocyanate group, TMXDI is found with the lowest toxicity in comparison with all the previously mentioned diisocyanates.

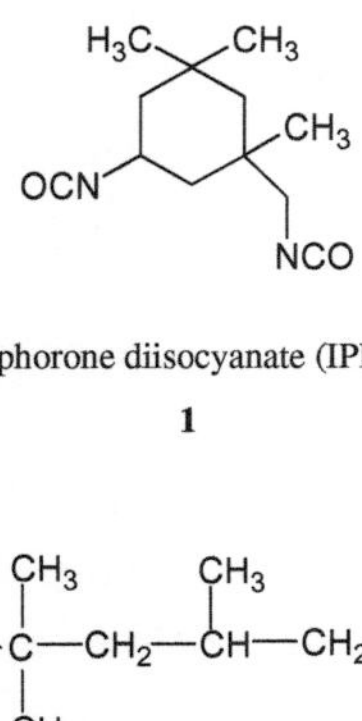

isophorone diisocyanate (IPDI)

1

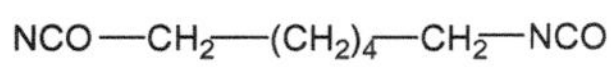

1,6-hexamethylene diisocyanate (HDI)

2

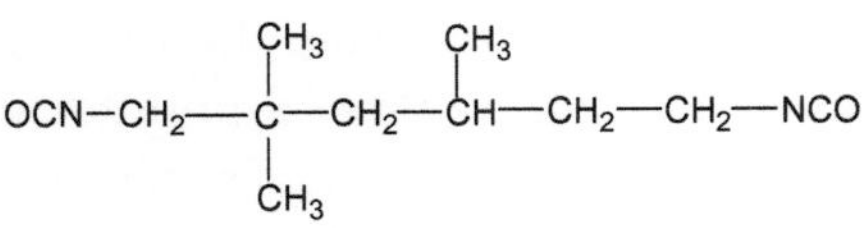

2,2,4-trimethyl-1,6-hexamethylene diisocyanate (TMHDI)

3

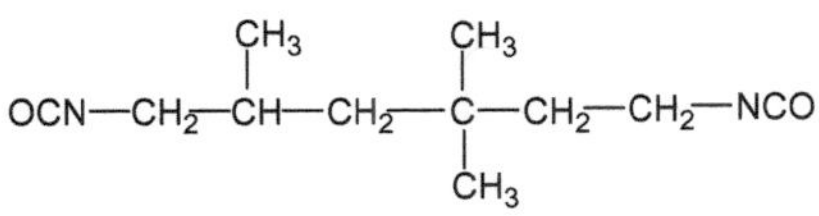

2,4,4-trimethyl-1,6-hexamethylene diisocyanate (TMHDI)

4

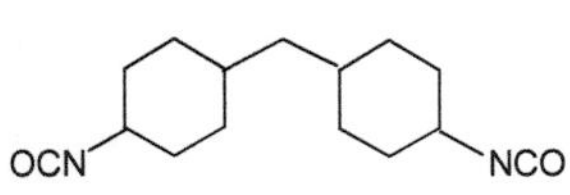

4,4'-dicyclohexylmethylene diisocyanate ($H_{12}MDI$)

5

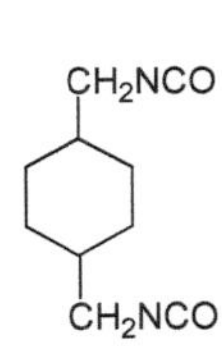

1,4-bis(isocyanatomethyl)cyclohexane

6

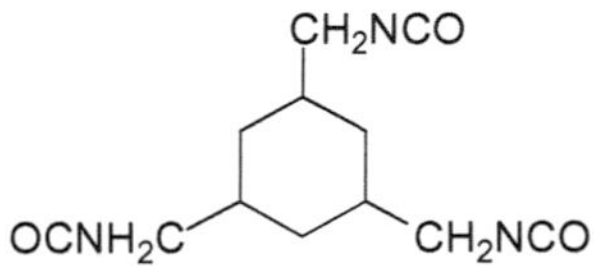

1,3,5-tris(isocyanatomethyl) cyclohexane

7

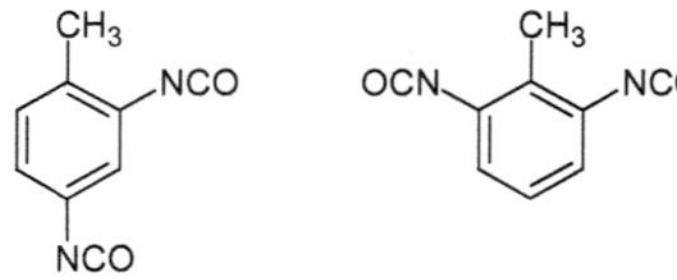

tolylene-2,4- and 2,6-diisocyanate (TDI)

8

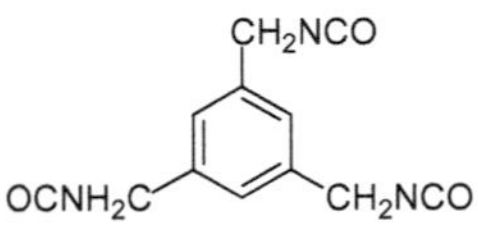

3,5-tris(isocyanatomethyl)benzene

9

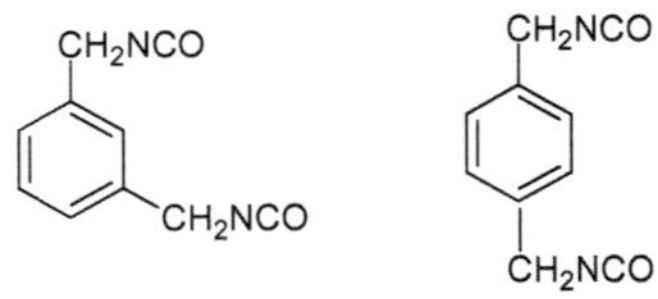

m- and p-xylylene diisocyanate (XDI)

10

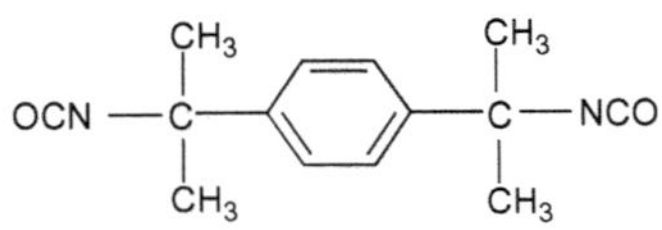

p-tetramethylxylylene diisocyanate (TMXDI)

11

Internal blocking involves formation of uretdione (Formula 13.2), allophanate and biuret groups. This is the best method to avoid unwanted emissions of organic compounds into the atmosphere during the process of deblocking IC groups and coat curing. The curing process in systems which contain internally blocked PICs is completely friendly for the environment.

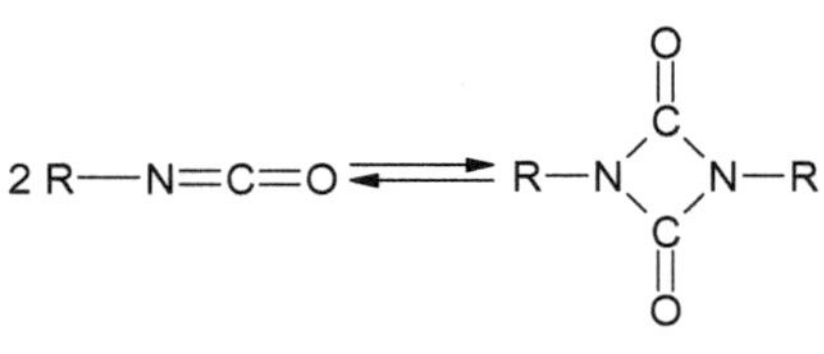

(13.2)

isocyanate uretdione ring

To obtain uretdione bonds it is necessary to use adequate catalysts and reaction conditions which are difficult to control. The obtained four-membered ring is thermolabile and at increased temperature it decomposes back into the initial compounds [5].

The process of dimerization formed internally blocked PICs proceeds in two steps. The first step involves dimerization of diisocyanate, in the presence of tris(dimethylamino)phosphine, and during the second step the dimerization product reacts with diols and/or monoalcohols or monoamines, e.g., butane-1,4-diol and/or 2-ethylhexanol and/or dibutylamine [25,26]. As an example the formula of such polyisocyanate can be presented as follows (12):

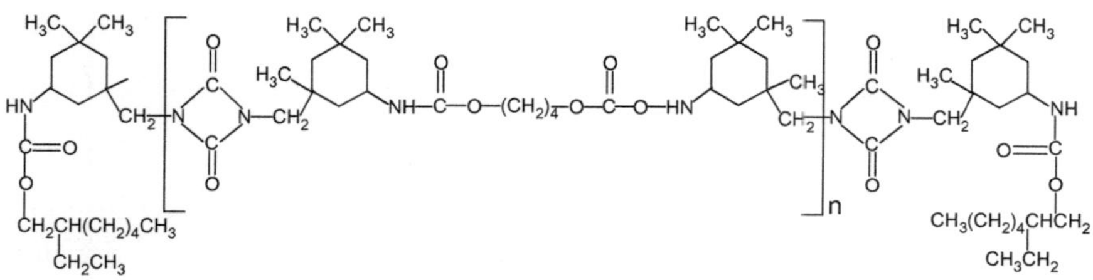

internal blocked polyisocyanate by means of uretdione rings

12

So far, only IPDI has been dimerized with satisfying yield (99% of the dimer in relation to isocyanurate structures). For this reason IPDI is most frequently used to produce internally blocked polyisocyanate. During dimerization of other diisocyanates (HDI, TMDI, or $H_{12}MDI$), commonly used in powder systems, due to the same reactivity of both isocyanate groups, the products always include significant quantities (25–50%) of compounds containing isocyanurate and oxadiazinetrione rings [27]. At present researchers focus on finding highly selective dimerization catalysts, or on optimizing reaction conditions in order to form uretdione rings and to reduce the quantity of unreacted diisocyanates.

Due to the fact that functionality of polyisocyanates which contain uretdione groups is lower than 2, they are less reactive during the process of curing in the presence of the widely used organotin catalysts (dibutyltin(IV) dilaurate, tin(II) octoate) in comparison to externally blocked PICs, and coatings based on them have lower resistance to solvents and other chemicals. An increase in the speed of crosslinking reaction in systems containing uretdione compounds occurs in the presence of tin(II) acetylacetonate [28] or amidine derivatives, e.g., 1,2-dimethyltetrahydropyrimidine [29].

Allophanate (Eq. 13.3) and biuret (Eq. 13.4) linkages are formed more easily and effectively than uretdione rings [30]. Terminal IC groups in oligomer chains obtained in this way are blocked with compounds containing labile hydrogen atoms. PICs blocked in this way share some qualities with externally blocked PICs, yet they contain lower quantity of the blocking agent [31,32].

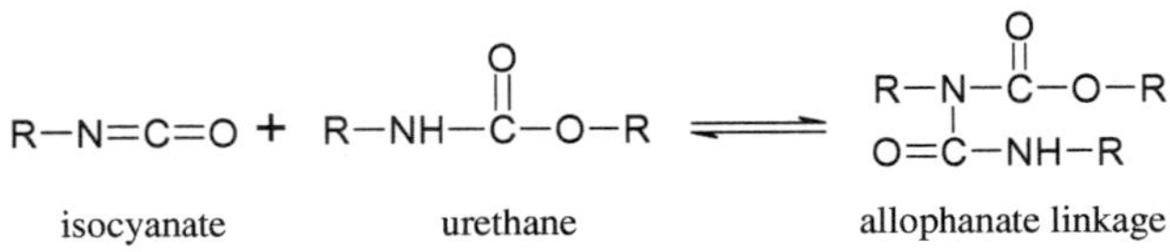

(13.3)

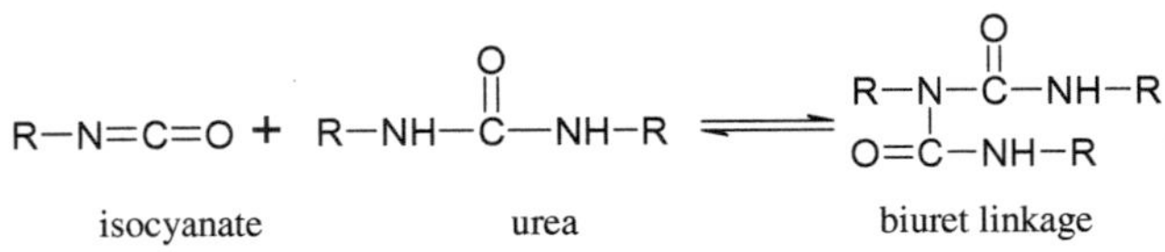

(13.4)

These reactions are reversible at lower temperatures (150°C) in comparison to dimers (180°C), and the obtained PICs may have functionality higher than two, hence they present an advantage in comparison to the difficult and long dimerization process [20,21,33].

In order to prepare powder coatings, resin is blended with blocked polyisocyanate in such amounts which will ensure the molar ratio of hydroxyl and isocyanate groups in the range from 1:1 to 1:1.05. Crosslinking occurs as a result of the reaction between the deblocked isocyanate groups and hydroxyl groups of resin, in accordance with addition-elimination or elimination-addition mechanism, at a temperature higher than polyisocyanate deblocking temperature (Eq. 13.5).

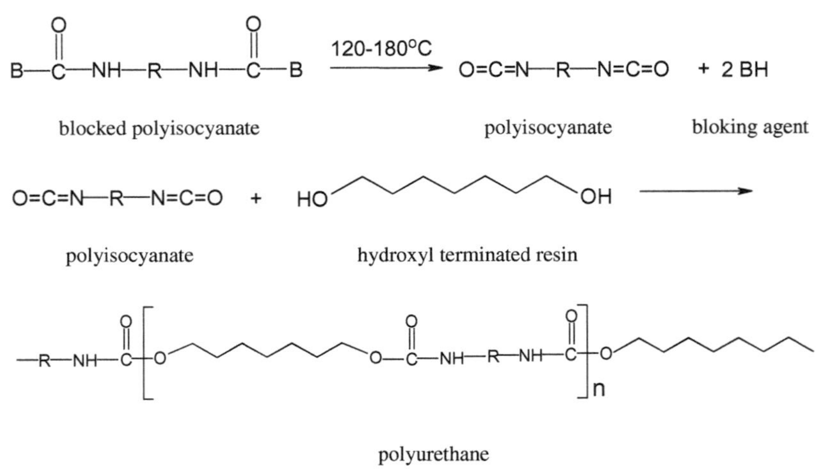

(13.5)

Systems characterized by excellent compatibility of the components are known as high gloss and transparent coatings [27]. High gloss coatings were characterized in gloss measured at the angle of 60° over 90 GU in contrast to gloss coatings (gloss in the range 70–90 GU), semi-gloss (gloss in the range 45–70 GU), semi-matt (gloss in the range 25–45 GU), matt (gloss in the range 10–25 GU) and low matt coatings (gloss below 10). Matt coatings, which account for over 40% of the manufactured powder products, are also interesting from an application viewpoint [34]. Compositions of this type are obtained by selecting components with chemical structure containing hydrophilic and hydrophobic segments, which leads to micro heterogeneity of the final products [35]. Low gloss PUR coatings are obtained from the resin crosslinked with a PIC containing imine groups (formed during reaction of diisocyanate and amine) or by addition of matting agents, crystallizing resins and other powder systems, e.g., epoxy compositions [35,36]. PIC, patented by Bayer, recommended for matt systems, was obtained from 2,2-bis(hydroxymethyl)propionic acid, IPDI and ε-caprolactam, at the molar ratio of 1:2:2 (13) [37]. The carboxyl group can be used for reaction with epoxy resin.

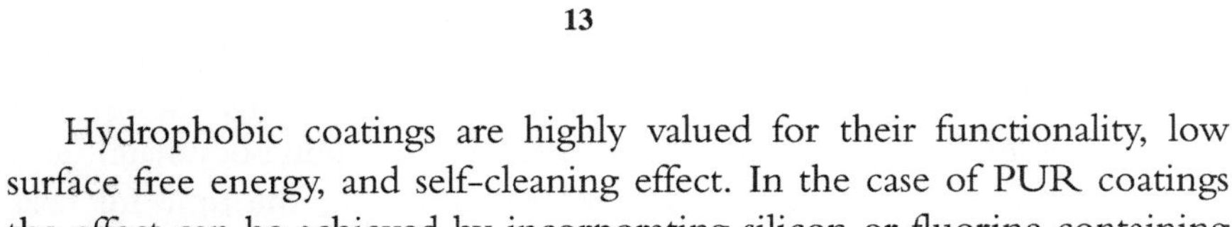

externally blocked polyisocyanate containing carboxyl group

13

Hydrophobic coatings are highly valued for their functionality, low surface free energy, and self-cleaning effect. In the case of PUR coatings the effect can be achieved by incorporating silicon or fluorine containing segments into the PUR structure [38]. The crosslinking of polyester resin with blocked PIC, which was synthesized by the reaction of perfluorinated alcohol with *N,N*-carbonylbiscaprolactam, leads to the formation of hydrophobic coating with water contact angle of 125° at 2% content of fluorine, while in nonmodified coatings the contact angle was 87° [39].

Polyurethane-based powder coating systems are widely used in the automotive industry. Coatings that are typically applied to car body consists of anticorrosive coating most frequently deposited electrophoretically

(*E-coat*) (~25 μm), ensuring both anticorrosive protection and excellent adhesion to the substrate; primer (~30 μm) designed to even out the surface and to prevent chipping; top coat consisting of pigmented basecoat (~20 μm) responsible for the color, and clear coat (~40–50 μm) designed as a protective and decorative layer (*top coat=base coat+clear coat*). The curing of top coat is usually carried out during a single process [40].

Polyurethane-based powder systems are used in car bodies as original coatings consisting of basecoat and clear coat, and less frequently as primers [36]. The top coat must be decorative and provide protection, and hence it is manufactured from raw materials ensuring the best resistance to external factors. The applied diisocyanates include aliphatic HDI, cycloaliphatic IPDI, $H_{12}MDI$ as well as aromatic-aliphatic TMXDI. PIC blocking agents widely used in this type of coatings include oximes, e.g., the most frequently applied 2-butanone oxime (MEKO) as well as derivatives of pyrazole and triazole, because of the low deblocking temperature. The applied types of resin include polyester resins and acrylic resins, highly resistant to yellowing, as well as their mixtures, and additions of small quantities of melamine-formaldehyde resins.

The use of powder rather than solvent-based systems in the coil coating process designed for roof coverings or for automotive industry resulted in a decrease of speed of technology lines from approx. 100 to 15–30 m min^{-1}. Development of advanced methods of applying powder systems by dipping in electrostatic fluidized bed (e.g., MSC Powder Cloud technique) as well as with the use of electromagnetic brush technology (EMB) utilizing magnetic drum attracting positively charged powder particles deposited on a negative carrier made it possible to use powder systems, e.g., in coil coating or can coating [5]. During the curing process the temperature is high (e.g., 280°C during the curing of a coating containing 2-butanone oxime blocked HDI polyisocyanate), and the time of interaction with the coating is short (e.g., the time for coat curing at the rate of 160 m min^{-1}. amounts to 15 s). Coating systems of this type must be resistant to discoloration at high temperatures, if the line stops during the painting process. The temperature and time of curing are selected specifically to match the applied PIC and blocking agent.

Radiation curable powder systems were introduced in order to separate curing from the melting, flow, and coalescence process during the film formation. The UV cured systems contain unsaturated polyurethane binder syntesized from carboxyl groups contained polyester resin, TDI, glycidyl methacrylate, and 2-hydroxyethyl methacrylate-IPDI adduct

Table 13.2 Comparison of selected properties of thermoplastic and thermosetting polyurethane powder coating systems

Thermoplastic systems	Thermosetting systems
The film forms as a result of powder melting	The film forms as a result of powder melting and crosslinking
Higher melting point (approx. 180–300°C)	Significantly lower melting point (80–130°C)
High molecular weight of polymer, high melt viscosity	Lower molecular weight, lower melt viscosity
Difficulties while mixing with pigment	Easier pigmenting; possibility to add greater quantity of pigments
Poor resistance to solvents	No solubility due to the crosslinked structure
Excellent weathering resistance	Excellent weathering resistance
Excellent mechanical resistance	Excellent mechanical resistance
Difficulty in forming thin films	Possibility to form thin films
Fine appearance	Fine appearance
Possibility to use for high-speed coating applications because no chemical reactions take place during the curing process	Lower speed of coating application due to the crosslinking reaction

[41]. The EB curable coating binders were based on a high molecular weight thermoplastic adipic acid-butanediol-TDI polyurethane and a low molecular weight crosslinker. The crosslinkers used in the EB systems were a diacrylate of an epoxy-bisphenol A adduct or a triacrylate of a TDI-trimethylolpropane adduct [42]. The radiation curable powder coatings are first melted by short wave IR radiation without heating the substrate and then after the film formation, cured by UV or EB [5]. The radiation curing process is significantly shorter (by a few seconds) than temperature curing, which leads to reduction of energy costs.

Selected properties of thermoplastic and thermosetting polyurethane powder coating systems have been compared in Table 13.2.

13.3 ADDITIVES USED IN POWDER CLEAR COATINGS

In order to facilitate the processes of applying the powder to the substrate as well as the film formation and to provide the obtained films with required properties, besides the basic ingredients, i.e., resin and crosslinker, the following additives are used: catalysts, flow control additives, degassing additives, tribo-charging additives, depending on the intended application (pigments, fillers, corrosion inhibitors, biocides, flame retardants, antioxidants, as well as matting, antistatic, and structural agents).

13.3.1 Catalysts

Crosslinking reaction catalysts are used to decrease the curing time. The type of catalyst is selected to match the type of resin and crosslinking agent. They are frequently added, in liquid form, to molten resins during synthesis, or deposited on carriers, such as silica. Tin octoate (II) known by the commercial name WorleeAdd ST-70 is most commonly used as the catalyst.

13.3.2 Flow Control Additives

Additives improving flow properties of coatings are used to minimize the roughness of the cured coating and to prevent defects which may occur on the film, e.g., craters and 'orange peel' effect. Melt flow rate is significantly influenced by surface tension and melt viscosity.

Optimum flow properties are obtained in powder systems when:

- the surface tension of powder melt is high, yet lower than the surface tension of the substrate; in this case the substrate is easily wetted by the melt
- the surface tension of powder melt is homogenic at the interface with the air

Additives used to improve flow properties include acrylic and silicone-acrylic homopolymers, and copolymers as well as polysiloxanes. Such additives are manufactured under various trade names, e.g., BYK 360, Resiflow PV 88. As a result of the incompatibility of additives and coating system ingredients, the migration to the surface of the melt occurs. During this process the polar segments of the chain are oriented towards the melt and the nonpolar segments towards the surface. The nonpolar segments originating from the additives, accumulated on the surface of the melt, equalize the surface tension at the surface and in the melt. This results in smoothing out the obtained film.

13.3.3 Degassing Additives

The process of coat curing may be accompanied with a release of bubbles of moisture and air originating from the spaces between powder particles, gases emitted from the substrate, or as byproducts of the crosslinking reaction. Because the crosslinking reaction is associated with a significant increase in viscosity, the bubbles may fail to be released before the end of the curing process. As a result the bubbles are trapped in the film, or they may produce other defects such as pinholes and craters. Degassing additives promotes the effective release of bubbles from the coating during the curing process, preventing any resulting defects. Benzoin is the most commonly used degassing agent [43]. Coatings that contain benzoin are not resistant to yellowing. Degassing agents which do not cause the yellowing phenomenon include: stearic acid, hydrogenated derivatives of castor oil, polyethylene, polypropylene and polyamide waxes, 1,4-cyclohexanedimethanol dibenzoate and acrylate resins, known by their commercial names ARALDIT PT 810 and WorleeAdd 902 [5].

13.3.4 Additives Improving the Charging of Powder

It is possible to increase the charging capacity of paint particles during the deposition to the substrate with the use of electrokinetic method (TRIBO) by adding nitrogen compounds, e.g., sterically hindered amines (HALS) or amino alcohols [5]. The frequently used sterically hindered amines, known by their commercial name (e.g., Tinivin 144) are also light stabilizers. Other recommended compounds include: quaternary ammonium salts, benzotriazole, melamine, oligomeric sterically hindered tertiary amines, phosphor-, arsenic- and antimony-compounds, as well as polymeric ammonium salts with structural formula 14 or 15 [44].

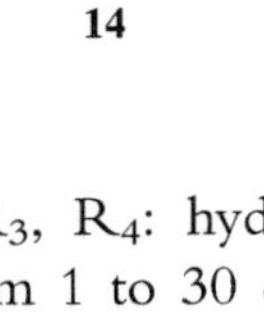

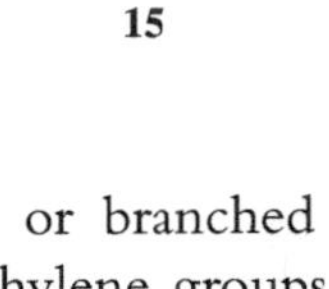

the structure of ammonium salts used as powder charging additives

14 **15**

where:

R_1, R_2, R_3, R_4: hydrogen atoms, linear or branched alkyl groups containing from 1 to 30 carbon atoms, oxyethylene groups, monocyclic or polycyclic cycloaliphatic radical containing 5 to 12 carbon atoms, or monocyclic or polycyclic aromatic radical

R_5: atom of hydrogen, fluorine, bromine or amino group

A^-: anion of organic compound e.g., aliphatic or aromatic carboxylic or sulfonic acid

CONCLUSION

At present, developments in the technologies of polyurethane powder clear coatings focus mainly on the following issues: production of new crosslinking systems to be applied at the lowest possible curing temperature, which will result in energy savings and make it possible to apply this type of clear coating systems to materials with lower thermal resistance. e.g., surfaces of plastics or wood, or prefabricated wood products; synthesis of crosslinking agents with high contents of reactive functional groups and at the same time with minimized quantity of the blocking agent; improvements in application properties of cured coatings obtained from powder coatings, mainly towards increased hydrophobicity and weather resistance, and production of coatings with interesting visual effects. The above development trends focus on both the basic problems associated with increasing material resources for the technologies of polyurethane powder clear coatings and on strictly practical issues taking into account market demand for these products and connected with promoting sustainable chemical technologies.

REFERENCES

[1] Directives of the EU: 199/13/EEC, 2004/42/EC, 2010/79/EU.
[2] A. Prociak, G. Rokicki, J. Ryszkowska, Materiały Poliuretanowe, PWN, Warszawa, 2014.
[3] J. Burke, Enlarging the cure window of powder coatings, Paint Coat. Ind. Mag. 10 (2004).
[4] D.K. Chattopadhyay, K.V.S.N. Raju, Structural engineering of polyurethane coatings for high performance applications, Prog. Polym. Sci. 32 (2007) 352.
[5] E. Spyrou, Powder Coat. Chem. Technol., Vincentz Network Gmbh, Hannover, 2012.
[6] A.A. Tracton, Coatings Technology Handbook, Taylor & Francess Group, New York, 2006.
[7] Y. Tsudo, Patent WO 2010098027, 2010.
[8] Q. Zhang, Patent CN 103740181, 2014.
[9] J. Yang, Patent CN 101245201, 2008.
[10] European Patent EP 0045996, 1980.
[11] USA Patent 2002095019, 2002.
[12] J. Van Haveren, E.A. Oostveen, F. Miccichè, B.A.J. Noordover, C.E. Koning, R.A. T.M. Van Benthem, et al., Resins and additives for powder coatings and alkyd paints, based on renewable resources, J. Coat. Technol. Res. 4 (2007) 177.
[13] A. Minesso, Renewable or recycled raw materials for polyester powder coatings, Focus Powder Coat. 7 (2011) 3.
[14] P.N. Mehta, Hyperbranched polymers: unique design tool for coatings, Surf. Coat. Int. Part B Coat. Trans. 89 (2006) 333.

[15] D. Greszta-Franz, U. Itter, Crosslinking behavior of one component polyurethane coatings based on malonate blocked polyisocyanates: the effect of the malonate structure and optimization of the NCO:OH ratio, Farby i Lakiery 6 (2012) 3.
[16] A.T. Chen, R.T. Wojcik, Polyurethane coatings for metal and plastic substrates, 26. Dimeric isocyanates in polyurethane powder coatings, Met. Finish 98 (2000) 143.
[17] German Patent DE 3 004 876, 1980.
[18] German Patent DE 3 143 060, 1981.
[19] German Patent DE 19729242, 1999.
[20] Polish Patent, translation of European Patent PL/EP 1524284, 2009.
[21] USA Patent 4 177 342, 1979.
[22] P. Gillis de Lange, Powder Coatings Chemistry and technology, Vincentz Network Gmbh, Hannover, 2004.
[23] USA Patent 4 375 539, 1981.
[24] German Patent DE 3 128 743, 1980.
[25] European Patent EP 0045996, 1980.
[26] F. Schmitt, A. Wenning, J.-V. Weiss, Prog. Org. Coat. 34 (1998) 227.
[27] A. Wenning, J.-V. Weiß, W. Grenda, Polyisocyanates today and tomorrow, Eur. Coat. J 4 (1998) 244.
[28] European Patent EP 1137689, 1998.
[29] European Patent EP 0803524, 1996.
[30] A. Lapprand, F. Boisson, F. Delolme, F. Méchin, J.-P. Pascault, Reactivity of isocyanates with urethanes: conditions for allophanate formation, Polym. Degrad. Stab 90 (2005) 363–373.
[31] B. Pilch-Pitera, Polyurethane powder coatings crosslinked with allophanate structures containing polyisocyanates, J. Appl. Polym. Sci. 116 (2010) 3613.
[32] B. Pilch-Pitera, Polyurethane powder coatings containing polysiloxane, Prog. Org. Coat. 76 (2014) 33.
[33] M. Gedan-Smolka, L. Häußler, D. Fisher, Thermal deblocking of masked low molecular isocyanates I. Aliphatic isocyanates, Thermochim. Acta 351 (2000) 95.
[34] P. Thometzek, U. Freudenberg, H.-U. Meier-Westhues, K. Yonek, Weather-stable low-gloss powder coatings, J. Coat. Technol. 72 (2000) 75.
[35] C. King, J.-V. Weiss, W. Grenda, One-shot, low gloss, polyurethane powder coatings, Focus Powder Coat 12 (2004) 4.
[36] D.A. Wicks, Z.W. Wicks, Blocked isocyanates III: Part B: uses and applications of blocked isocyanates, Prog. Org. Coat 41 (2001) 1.
[37] J.W. Rawlins, S. Feng, C. Sullivan, P. Thometzek, New crosslinkers for polyurethane powder coating, Paint Coat. Ind. Mag 18 (2002) 38.
[38] B. Pilch-Pitera, Polyurethane powder lacquers modified with fluorine compounds, Polimery 57 (2012) 830.
[39] L. Van Ravenstein, W. Ming, R.D. Van de Grampel, R. Van der Linde, G. De With, T. Loontjens, et al., Low surface energy polymeric films from novel fluorinated blocked isocyanates, Macromolecules 37 (2004) 408.
[40] M. Horgnies, E. Darque-Ceretti, R. Combarieu, Influence of additives segregation on surface composition of automotive coatings: effects of an ethanol cleaning, Prog. Org. Coat. 47 (2003) 154.
[41] M. Ishitani, S. Asami, T. Kajino, Patent JP 2002206070, 2002.
[42] R.S. Tu, D.L. Anglin, Performance comparison of electron beam and radiation curable powder coatings at an elevated temperature, in Conference Proceedings of RadTech Europe, 10th, 1986, pp. 6/35–6/46.
[43] S. Jahromi, Meechanism of bubble dissolution in reactive coatings: the role of benzoin, Macromolecules 35 (2002) 5883–5890.
[44] P. Malec, Diploma thesis, Rzeszow University of Technology, 2007.

CHAPTER 14

Recycling of Polyurethanes

Janusz Datta and Marcin Włoch
Gdańsk University of Technology, Gdańsk, Poland

Contents

14.1 INTRODUCTION

Concerns regarding the increasing amount of plastics waste has resulted in the development of recycling for these products. *Mechanical recycling methods*, which can be used for all thermoplastics waste, are focused on changing physical form of waste (e.g. by crushing or grinding) to form which is suitable for further processing. *Chemical recycling methods*, which are applied

Polyurethane Polymers: Blends and Interpenetrating Polymer Networks
DOI: http://dx.doi.org/10.1016/B978-0-12-804039-3.00014-2

mostly for poly(ethylene terephthalate) and polyurethanes, are focused on recovering monomers or olygomers, which can be used to synthesize new polymers. Chemical decomposition of polymers obtained by polycondensation, involve chemical reactions such as hydrolysis, glycolysis or aminolysis. The last, but also important method of recycling, is connected with *recovery of energy*, e.g. combustion or incineration. This methods are commonly applied to all polymers, but from an environmental point of view, this method should be used for waste plastics, which cannot be recycled by other methods, or when other methods are not costeffective. It is important to take into consideration the emission of gases during the thermal decomposition of plastics. The further development of the recycling methods should be concerned with the costeffectiveness of the process and their environmentally friendly nature [1–4].

In the case of polyurethane waste, all mentioned methods were applied on a laboratory scale, and some of them have industrial applications. A summary of all possible recycling methods is presented in Fig. 14.1. Polyurethane waste can be recycled by mechanical methods (i.e., grinding and applying as a filler or pressing with a bonding agent) and chemical methods (mainly by e.g., glycolysis, hydrolysis, or aminolysis). There is also possibility to the recover energy from polyurethanes waste (by incineration, gasification, and pyrolysis).

In 2014, the total worldwide production of polyurethanes equaled nearly 16 million tons (US imperial tons), and the largest sectors were flexible foams, rigid foams, coatings and elastomers [5]. Polyurethanes, due to their segmented and domain structure, exhibit a good

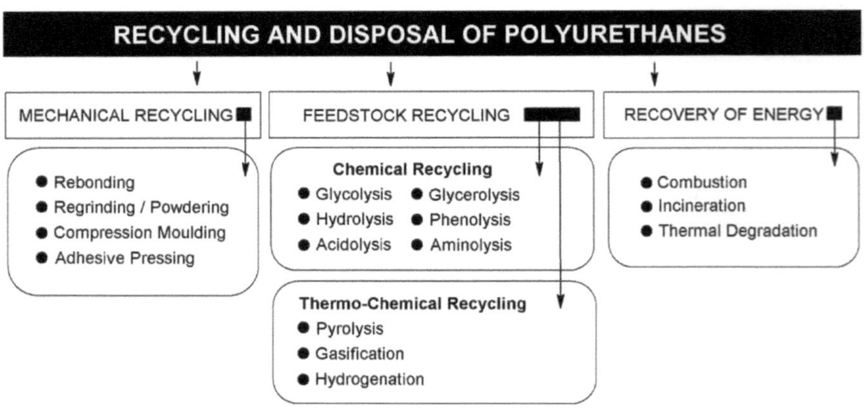

Figure 14.1 Overview of options for polyurethanes waste recycling.

combination of properties (e.g., mechanical properties), which can be created by a suitable component choice for the synthesis. From this reason, polyurethanes are used to solve "problems" in industrial applications. Usually, different synthetic intermediate products are used to obtain polyurethanes and a lot of polyols, isocyantes, and chain extenders that can be used to the synthesis of polyurethanes in the form of elastomers, foams, adhesives, or coatings. The trends in the development of polyurethanes include application of the bio-based substances (e.g., polyols based on vegetable oils [6–8], glycerol [9,10], or lignin [11,12]) to PUs synthesis or preparing of PUs composites (e.g., natural fibers [13–15] or micro- and nano-crystalline cellulose [16,17]).The low price of the natural raw materials resulting from low production costs and renewable nature makes bio-based substances a good alternative to the petrochemical compounds. There are known examples of bio-based and environment-friendly polyurethanes and its composites [6,7,11–17]. Another trend now observed in the scientific world bio-trends is polyurethane recycling, for example the glycerolysis, which involves using a vegetable-based glycerol obtained as a by-product during bio-diesel fuel production.

14.2 MECHANICAL RECYCLING

Mechanical recycling methods of the polyurethanes are connected with crushing of waste, resulting in suitable feedstock (in the terms of physical form) for further processing, e.g., mixing with adhesives or compression molding. The most important benefits of the mechanical recycling is the low cost and the simplicity of the realization, but on the other hand there are some disadvantages, for example, with poor mechanical performance of recovery products that limit applications of the obtained materials or products.

14.2.1 Rebonding

The rebonding process of polyurethane waste is one of the most used recycling methods. The idea of this process is presented in Fig. 14.2. At first polyurethane foam waste is shattered into fine flakes, which are coated, in the next step, with adhesive (e.g., 4,4′-diphenylmethane diisocyanates (MDI) or poly(phenylenemethylene isocyanate) (PAPI)) in the amount about 5%–10% by mass. Next, a water stream (with a high temperature) melts the polyurethane, and after that the obtained system is compressed. This process has simple technology and low cost. From the

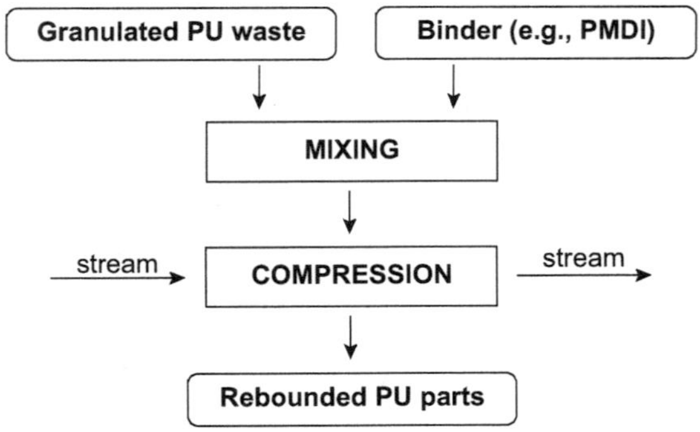

Figure 14.2 Rebonding process of granulated polyurethane waste [18].

environmental point of view, realization of the process includes reusing waste connected with reducing polyurethane foam waste volume [19,20]. The most important disadvantage is the insufficient mechanical performance of regenerated foam product that results and are used as car linings, carpet underlay, sport mats or furniture elements. An example of this process was reported by Nafziger et al., [18] in the case of polyurethane foam waste.

14.2.2 Compression Moulding, Injection Moulding and Extrusion

Compression molding of the polyurethanes waste includes softening, self-bonding under heat and pressing, without adhesives (as in the case of bonding process). The process requires suitable temperature and pressure of melting of the thermoplastic polyurethane. Polyurethanes have a segmented structure, resulting from the presence of hard (isocyanate-based constituents) and soft segments (polyol-based constituents). Soft segments possess a thermoplastic nature in the range of temperature from 150°C to 220°C, related to the polyols used to synthesize the polyurethane. Other factor connected with the melting point of the thermoplastic polyurethanes is the degree of the crosslinking of the polyurethane waste, and it should be low as much as possible. Waste can directly bond together in melted state. The products obtained by this method exhibit poor mechanical properties (e.g., low elongation at break) and performance of the surface (e.g., color of final product), so these products can be applied in items such as fenders [4].

Polyurethanes obtained by reactive injection moulding (RIM) and reinforced reactive injection moulding (RRIM) methods can be grounded into fine particles and compressed to obtain material for the automotive applications [17].

Extrusion, like injection molding, is a combination of mixing, heating and pressing, and can change polyurethane to form thermoplastic material (resulting from the shortening of the polyurethane chains). Heneczkowski and Galina [21] describe a method for recycling of RIM polyurethanes foams by controlled degradation during twin-screw extrusion. The method proposed resulted in obtaining thermoplastic polyurethane elastomers with similar mechanical strength to the original material.

14.2.3 Regrinding, Powdering, and Using as a Filler

Before the polyurethane waste will be used as a filler, it should be reground and powdered, what can be undertaken by shredding, milling (e.g., by using two-roll mill, impact disks), cutting (e.g., by using precision knife), grinding (e.g., by using cryogenic method) or extrusion [3]. An example of powdering process is presented in Fig. 14.3.

Ground polyurethane waste, in the form of powder (with particle size 50–200 μm or higher), can be used as filler in new polyurethane products. The final amount of highly ground polyurethane filler in new material can be equal ca. 20 wt%. The surface of such filler can interact with polyurethane matrix, due to presence of similar chemical groups, and this is connected with the lack of negative effect on the mechanical properties of final products [22]. Obtained by this method, the products can be applied in ways such as energy absorption and sound insulation foams, due to the polyurethane waste filler acting as a damping agent. Ground polyurethane foam can be also applied as a filler for concrete [23,24], natural rubber composites [25], polyester resin [26] and as a component for preparing polymer blends with, e.g., polypropylene [27]. For example, Lin et al. [28] produced polyurethane material by grinding of rigid polyurethane foam waste into powder and mixing it with polyetherol and isocyanate.

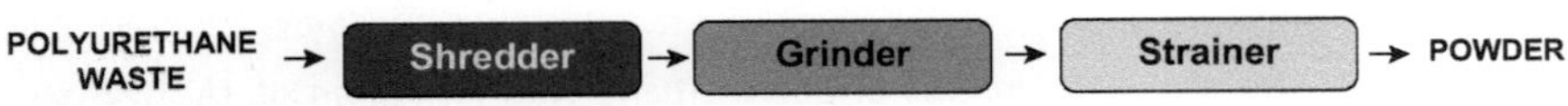

Figure 14.3 Example of powdering process realization.

14.3 CHEMICAL RECYCLING

The polymerization of polyurethanes is partially reversible in the suitable conditions. Functional groups such as urethane, ester, ether, and amine bond can be gradually depolymerized using organic compounds with an active hydrogen atom that can attack the polar groups presented in the main polyurethane chain. Suitable chemical reagents, reaction conditions (temperature, pressure, and atmosphere) and catalysts result in obtaining olygomers or monomers. Moreover, the purification and the distillation of obtained products permit to obtain pure monomers, i.e., polyols and amines, which can be applied in the synthesis of polyurethanes, polyureas, polyamides or polyesters. Different degradation agents mean that the physico-chemical properties and functionality of obtained products are different. The main processes, which are connected with the chemical methods of recycling of the polyurethanes recycling include: glycolysis, hydrolysis, aminolysis, and phosphorolysis.

14.3.1 Glycolysis

Glycolysis is the chemical process that can be performed on a commercial industrial scale, and involves the chemical decomposition of urethane groups by active hydrogen atoms present in the structure of glycols (e.g., ethylene glycol). The mentioned reaction is presented in Scheme 14.1. Glycols can be used as a single reagent or in the mix with short-chain amines and alkanolamines. A few countries of the world (e.g., USA, China, Germany, Poland, Austria, France, Italy) are carried out strong development of this recycling method [2,19,29,30], resulting in the recovery of virgin polyols or the production of glycolysates (olygomers

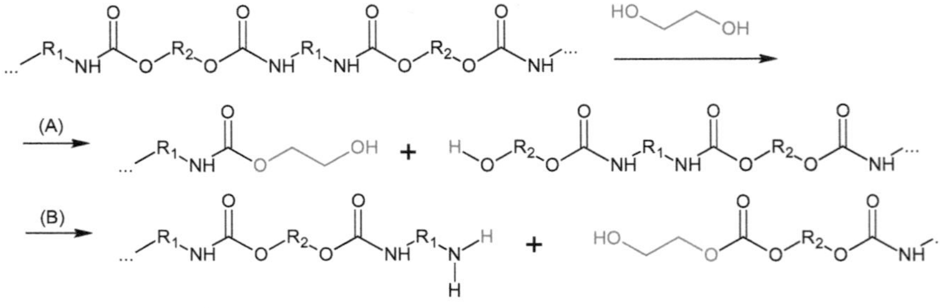

Scheme 14.1 The glycolysis process of polyurethane with formation of: (A) hydroxyl groups or (B) amine and hydroxyl group.

with urethane groups in the structure), and both components can be used in the synthesis of polyurethanes.

The quality of the glycolysates and yield of the reaction depends on many factors:

- type and concentration of a catalyst (e.g., acetate inorganic salts, amines, stannous catalyst)
- type (elastomers, rigid, and flexible foams) and size of waste
- molar/mass ratio of polyurethane waste to depolymerizing agent
- suitable temperature and time of chemical decomposition
- suitable degree of agitation.

14.3.1.1 Glycolysis Agents and Reaction Conditions

Principally, the glycolysis process is conducted in the presence of a glycol at the temperature related to the glycol boiling point, i.e., 180°C–240°C, allowing the process under atmospheric pressure. Glycolysis in an atmosphere of inert gas (e.g., nitrogen) is also possible. The reaction time depends on five major factors: (1) the temperature at which glycolysis is performed, (2) molar/mass ratio of waste to glycol, (3) the presence and amount of the catalyst, (4) the degree of size reduction of polyurethane waste, and (5) the speed with the recycled wasted material is dosed into a chemical reactor [30,31].

During the process two major reactions take place: (1) the transesterification of urethane and urea bonds; (2) breakup of the crosslinking allophanate and biuret groups. Glycolysis of polyurethanes can be realized with the aim of recovery the original polyols, or the oligomers compounds, called glycolystes, that have similar structure to polyols, but usually a higher molecular weight and consist of urethane bonds in the chain of the oligomer. Glycolysates in terms of chemical structure are low-molecular weight mono- and dicarbamates, amines, and low-molecular weight urea compounds as well as urea oligomers [28,29].

It should be noticed that during the glycolysis of waste poly(ester-urethane)s the additional esterification of ester bonds in soft segments (polyol-based moieties) takes place. Moreover, it is also possible to form primary amine groups during the process (Scheme 14.1). The presence of the amines in glycolysates limits the direct application, when glycolysates must be reused in the synthesis of polyurethanes. It results from the higher reactivity of amine groups than hydroxyl groups (in the case of reaction with isocyanate groups), which results in faster crosslinking in the final polyurethane. Moreover during the reaction of

$-NH_2$ and $-NCO$ groups, urea bonds are formed. The methods for purifying glycolysates include extraction, distillation, and purification of amines. Amines can be also removed by the reaction with cyclic carbonates (Scheme 14.2). The most convenient are ethylene and propylene carbonates, which formed urethane bonds and hydroxyl groups [32,33] during the reaction.

Glycolysates may be homogenous products or layered mixture, depending on the used glycol and catalyst, reagents ratio, and reaction conditions. When the glycolysates are nonhomogenous products, the top layer consists of the original polyol (used to produce the polyurethane) and the compounds of a similar structure, whereas the lower layer contains low-molecular weight carbamates, amines, and the unreacted glycol (especially in the situation when an excess mass of glycol was added during the reaction). In some cases the third layer is also formed and contains urea oligomers.

Homogenous glycolysates are mostly synthesized from rigid foams, whereas glycolysates from elastomers are of the layered type, due to the molecular weight of a polyol used in the production of polyurethane. When the molecular weight is quite high, recovered polyol is usually separated from the other products of glycolysis. Glycolysis agents influence on the time and the temperature of the process and the ability of the reaction mixture to separate into distinct phases as well as each phase composition. Glycolysis can be conducted without a catalyst, but the reaction time in such case is significantly longer. Realization of the process without a catalyst requires higher temperature during the process, which may cause thermal degradation of the waste.

Catalysts used in the process of glycolysis can be divided into four groups:

- acetates—e.g., sodium acetate, potassium acetate, lithium acetate
- hydroxides of alkaline metals such as sodium, potassium, and lithium
- metallo-organic compounds—e.g., dibutyltin dilaurate, stannous octoate
- amines—e.g., diethanolamine

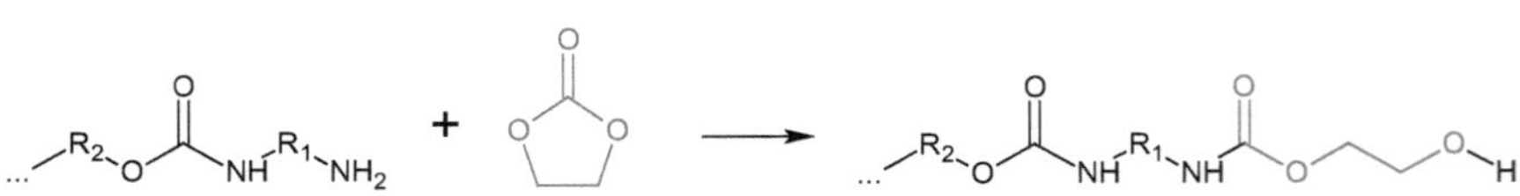

Scheme 14.2 Transformation of amine-terminated glycolsysates into semiproducts terminated with hydroxyl groups by using ethylene carbonate.

In accordance to the literature, there are known two possible ways of glycolysis realization (connected with the mass ratio of polyurethane waste to glycol), i.e.:

- glycolysis with the excess of polyurethane—investigated mainly by Datta et al. [34–39]
- glycolysis with the excess of glycols—investigated mainly by Rodriguez et al. [40–44]

A lot of studies in the area of polyurethane waste glycolysis, were described in the literature. Examples, divided in the terms of type and form of polyurethane waste, type of glycol used and the reaction conditions, are summarized in Table 14.1.

14.3.1.2 Glycolysis Realized With the Mass Excess of Polyurethane

Glycolysates obtained from polyurethanes waste are mainly homogenous, and do not required purification and distillation of glycol excess (in comparison to the method with using mass excess of glycols) is not required. Obtained semiproducts can be directly used in the synthesis of polyurethanes (as a part of the polyols mixture). In terms of chemical structure, glycolysates are oligomers and low-molecular mass molecules, which can contain urethane groups in a chain. The length of the oligomer chains can be controlled by suitable mass amount of glycolysis agent—when the mass of applied glycol rises the molecular weight of oligomers decreases. This type of glycolysis is intensively studied e.g. by Datta et al. [36,38], and some of these results will be discussed.

Datta [38] studied the influence of different glycols on the thermal stability and the chemical structure of the obtained glycolysates. The intermediates were obtained by the glycolysis process of polyurethane foams waste during the reaction with different glycols, i.e., 1,3-propanediol, 1,4-butanediol, 1,5-pentanediol, and 1,6-hexanediol. The constant mass ratio of glycol to polyurethane was used and was equal 10:1. Datta and Kacprzyk [36] also conducted glycolysis of polyurethane with using 1,3-propanediol at different mass ration of polyurethane to glycol, i.e., 2:1, 4:1, 6:1, 8:1, and 10:1. The FTIR spectra of the obtained glycolysates are presented in Fig. 14.4 and Fig. 14.5, which correspond to Table 14.2 with assignments of wavenumbers to chemical bonds. It can be seen that the obtained FTIR spectra indicates the presence of a similar chemical structure in the obtained glycolysates irrespectively of the glycolysis agent used.

Table 14.1 Examples of polyurethane glycolysis in accordance to the literature data

Polyurethane waste (PU)	Glycolysic agent (GA)	Mass ratio PU: GA	Catalyst	Temperature (°C)	Refs.
Flexible foam poly(ether-urethane) based on TDI	Diethylene glycol	1:1.5	Potassium octoate, calcium octoate, stannous octoate	189	[45]
Flexible foam from recycled cars	Diethylene glycol	1:0.5 1:1 1:1.15 1:2 1:2.5	Potassium acetate	215–225	[40]
Flexible foam poly(ether-uretane) based on TDI	Diethylene glycol	1:1.5	Diethanolamine, titanium(IV) butoxide, potassium octoate, calcium octoate	189	[41]
Flexible foam poly(ether-urethane) based on TDI	Diethylene glycol	1:1.5	Stannous octoate	190	[42]
Flexible foam poly(ether-urethane) based on TDI	Diethylene glycol	1:1.5 1:1.125 1:0.9	Stannous octoate	179 184 189	[43]
Flexible foam poly(ether-urethane) based on TDI	Diethylene glycol	1:1 1:1.3 1:1.5 1:2	Potassium octoate	190	[44]
Flexible foam poly(ether-urethane) based on TDI	Ethylene glycol Diethylene glycol Propylene glycol Dipropylene glycol	1:1.5	Diethanolamine	189	[46]
FLEXIBLE FOAM poly(ether-urethane) based on PMDI	diethylene glycol dipropylene glycol	1:1 1:1	Diethanolamine	220	[34]

Flexible foam	Ethylene glycol	1:2 1:1 2:1 5:1 8:1 10:1 20:1	Potassium acetate	195	[35]
Flexible foam poly(ether-urethane)	1,4-butanediol	1:2 1:1 2:1 5:1 8:1 10:1	Potassium acetate	210–230	[36]
Flexible foam poly(ether-urethane)	Poly(ethylene glycol) PEG300	1:2 2:1	Potassium acetate	210–220	[34]
Flexible foam	1,6-heksanediol	1:1 2:1 4:1 6:1 8:1	Potassium acetate	230–245	[37]
Flexible foam	1,3-propanediol	2:1 4:1 6:1 8:1 10:1	Sodium hydroxide	170–190	[38]
FLEXIBLE FOAM	1,3-propanediol 1,5-pentanediol	6:1 8:1	Potassium acetate	190–245	[39]

(*Continued*)

Table 14.1 (Continued)

Polyurethane waste (PU)	Glycolysic agent (GA)	Mass ratio PU: GA	Catalyst	Temperature (°C)	Refs.
	1,6-heksanediol	10:1			
Flexible foam	1,3-propanediol 1,4-butanediol 1,5-pentanediol	10:1	Potassium acetate	190–250	[47]
Rigid foam	Diethylene glycol	1:1 1:2 1:3	Potassium acetate	220–226	[48]
Rigid foam (wood based) poly (ether-urethane) based on PMDI	Ethylene glycol Propylene glycol Diethylene glycol	0.5:1 1:1 1.5:1	Diethanolamine	160 150 190	[49]
Integral skin foam poly(ether-uretane) based on MDI	Diethylene glycol	1:1	Sodium hydroxide and diethanolamine	200–205	[50]
Elastomer poly(ether-urethane) based on MDI	Ethylene glycol/ diethylene glycol Ethylene glycol/ ethanolamine	1:2	Lithium acetate	170	[51]
Elatomer	Ethylene glycol	1:2 1:1 2:1 5:1	Potassium acetate	170–180	[52]
Fibers poly(ether-uretane-urea) based on MDI and propylene diamine	Diethylene glycol	1:3	Dibutyltin dilaurate	210	[53]

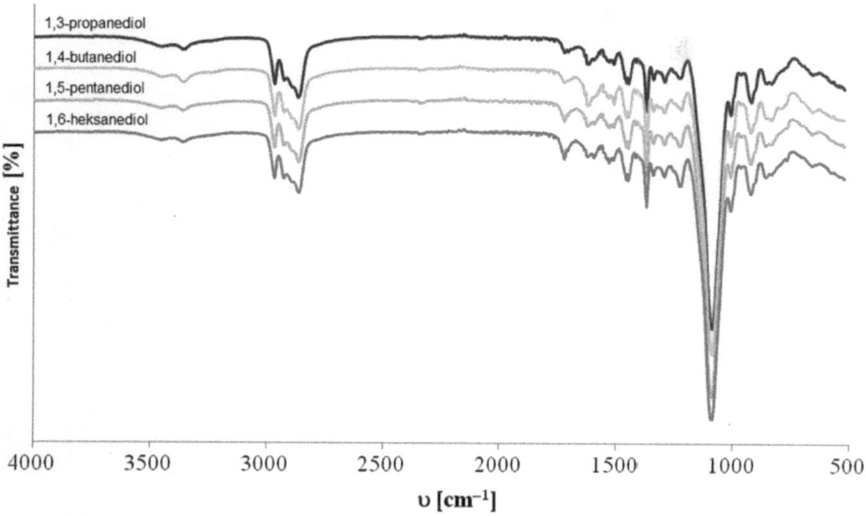

Figure 14.4 FTIR spectra of glycolysates obtained for the mass ratio of polyurethane: glycol of 10:1 and various glycols.

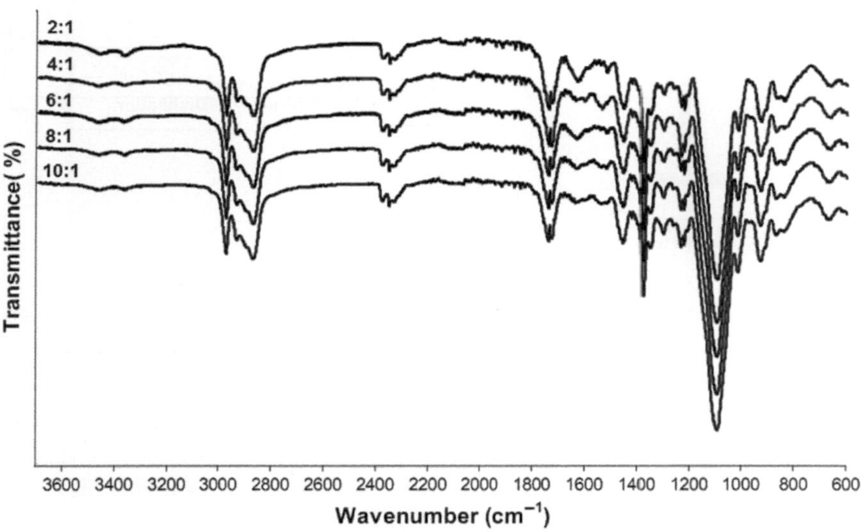

Figure 14.5 FTIR spectra of glycolysates obtained by using 1,3-propanediol at various mass ratio of polyurethane to glycol.

14.3.1.3 Glycolysis Realized With the Mass Excess of Glycols

Glycolysis with the mass excess of glycol is a method connected mainly with the recovery of polyols, which after purification can be used in the synthesis of polyurethanes. This process is called "split-phase" due to the

Table 14.2 FTIR analysis of glycolysates obtained with using different glycols at constant mass ratio of glycol to polyurethane

Wavenumber	Corresponding chemical bond
3343 cm^{-1}	Amine groups (presented in urethane groups)
3440 cm^{-1}	Hydroxyl groups
3000–2800 cm^{-1}	Methyl groups
1222 cm^{-1}	Ester group (presented in urethane bond)
1730 cm^{-1}	Carbonyl group (presented in urethane bond)
1550–1350 cm^{-1}	Methyl groups
1085 cm^{-1}	Ester groups

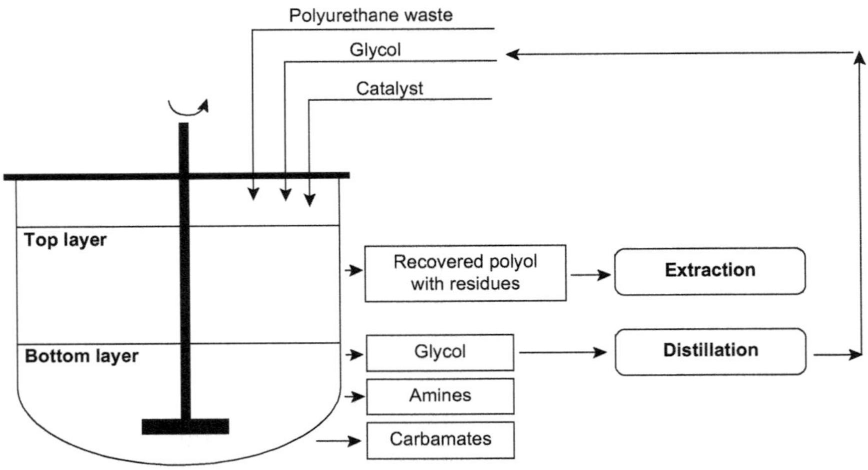

Figure 14.6 "Split-phase" glycolysis of polyurethane waste [50].

formation of two phases during the process, resulting from using an excess of glycolysis agent. This method is still developed and investigated e.g., by Rodriguez et al. [54]. The idea of process is shown in Fig. 14.6. After completion of the reaction, the recovered polyols require purification, and distillation of the applied glycol, which was used in mass excess to the mass of polyurethane waste. The bottom phase is mainly formed by the excess of used glycol and aromatic amines or carbamates, which have high toxicity and are environmentally harmful. The upper phase is mainly composed of recovered polyol, which require purification.

14.3.1.4 Applications of Glycolysis Products

In the literature are lot of examples of successful applying glycolysates as part of the polyols component in the synthesis of polyurethanes.

Investigations are focused on the influence of glycolysates on the final (e.g., mechanical or thermal) properties of polyurethanes synthesized with using them [34–37,46]. Datta and Rohn [35] synthesized polyurethanes, by prepolymer method, from 4,4′-diphenylmethane diisocyanate, glycolysates, and chain extenders, i.e., 1,2-ethanediol, 1,4-butanediol, or 1,6-hexanediol. The thermogravimetric analysis shows that in the temperature about 320°C, the 5% weight loss of obtained polyurethanes was observed. Tensile properties, abrasion resistance and density of obtained materials were also characterized. Datta and Pasternak [46] prepared polyurethane elastomers, by prepolymer method, with using 4,4′-diphenylmethane diisocyanate (MDI) and a polyols mixture containing glycolysis product and polyetherol, and 1,4-butanediol as a chain extender. The effect of the composition of polyols mixture on the polyurethanes phase transitions in the melting area was characterized. The thermal stability, tensile properties, hardness and elasticity were also determined. Datta and Pniewska [34] obtained polyurethanes from MDI, oligo(ethylene-butylene adipate)diol, and chain extenders, i.e., glycolysates or mixture of glycolysates with BDO or PEG300. It was found that the use of glycolysis products in the PURs syntheses improved their thermo-mechanical strength (i.e., storage modulus) at temperatures below zero as well as elongation at break (during static tensile test) and hardness. Datta and Haponiuk [35] showed that polymers produced from PMDI and glycolysates were stable at the temperatures up to about 300°C, which is similar to the thermal stability of polyurethanes synthesized from the commercially available components. This was also confirmed by Datta and Rohn [46], and Datta and Kacprzyk [34] in the case of polyurethanes synthesized from glycolysates and MDI.

Glycolysates can be applied in the production of parts by the RIM, at a mass ratio of up to 50%, without a significant worsening of the mechanical properties of final products [30,33]. The application of glycolysates in the synthesis of rigid foams and adhesives for joining different construction materials has been reported [30]. The use of glycolysates in the production of rigid polyurethane foams (applied as insulation with higher thermal stability), composites (applied in the construction and automotive industry as partitions, panels, and covers) or elastomers (applied as a footwear) is possible.

Current state-of-art research in the area of synthesis and properties of glycolysates or polyurethanes synthesized with using them, allows a closed lifecycle of polyurethane, where the streams from production or waste polyurethane elements are collected and subjected to glycolysis. The obtained semiproduct (in the form of glycolysates or recovered polyols), is used as a part of the polyols system for production of new polyurethane elements. This closed lifecycle of polyurethanes is shown in Fig. 14.7. Some companies successfully implemented the technology of chemical recycling (by glycolysis) using their own streams from production or products with defects.

14.3.2 Hydrolysis

The hydrolysis of polyurethanes waste involves reaction with water or steam under suitable conditions, i.e., high temperature and pressure. The reaction is presented in Scheme 14.3. This process was studied by the Ford Motor Company, which applied a wide range of temperatures (e.g., 300°C–400°C) and pressures (15–50 atm) of water steam. Other scientists applied less extreme conditions resulting in higher level of polyol recovery. For example, Campbell et al. [55] studied the hydrolysis of polyurethane foam under atmospheric pressure at temperatures from 232°C to 316°C, and obtained polyols (in the content of 20 wt%) to synthesize new polyurethanes foam. Gerlock et al. [56] investigated

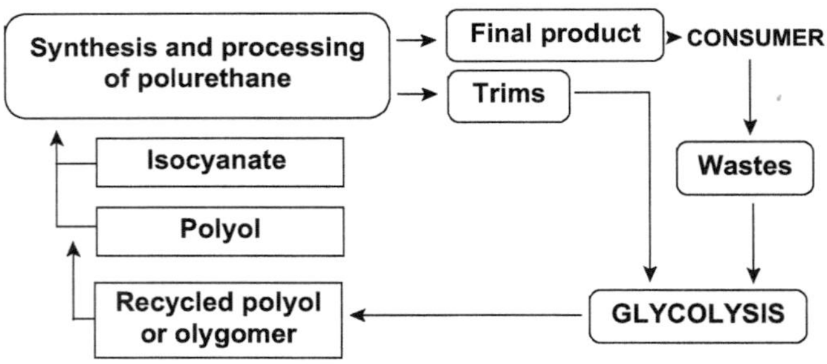

Figure 14.7 Closed lifecycle of polyurethanes with the glycolysis process as chemical recycling method for polyurethane waste.

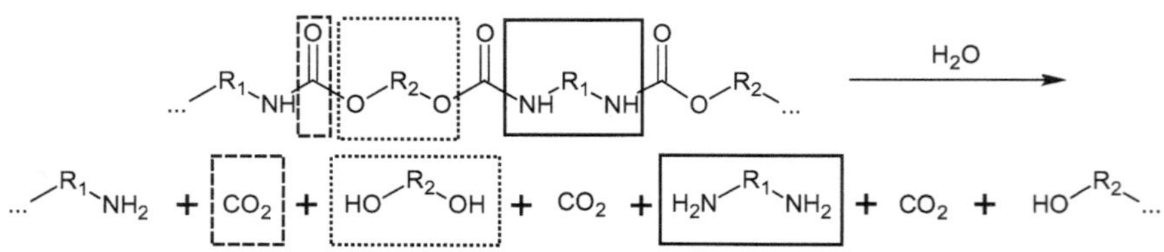

Scheme 14.3 Hydrolysis process of polyurethane.

hydrolysis of poly(ether-urethane) foam (based on TDI) at temperatures from 190°C to 230°C resulting in high-quality polyols, toluenediamines, and also carbon dioxide. The diamine can be removed by distillation or extraction methods, and the polyol is the residue of this processes. The hydrolysis process can be catalyzed by bases (e.g., NaOH, Scheme 14.4) or acids (e.g., HCl, Scheme 14.5), but Matuszak et al. [57] found that using an alkaline catalyst is more effective. Shi et al. [58] also proved that the presence of a catalyst and increasing the temperature caused higher effectiveness of the process.

Dai et al. [59] compared the effectiveness of toluenediamine (TDA) and sodium hydroxide as catalysts in the hydrolysis of the polyurethane waste at a temperature range of 150°C–350°C. Application of TDA effectively reduced the foam reaction activation energy, reaction temperature, and improved the toluenediamine recovery from polyurethane waste. The most important disadvantages of using TDA as a catalyst is difficult purification of polyol and a high cost of process realization (e.g., purification of products, cost of energy consumed by the plant).

Some interesting examples of the chemical recycling of polyurethanes were presented by Nikje et al. [60,61]. The decomposition of the polyurethane were realized using mixtures of glycerol and water or glycerol, sorbitol, and water (a process called hydro-glycolysis) at 160°C–180°C under atmospheric pressure and using microwave irradiation. The resultant semiproducts of glycolysis were used to the synthesize polyurethane foams. The hydro-glycolysis is a combination of hydrolysis and glycolysis

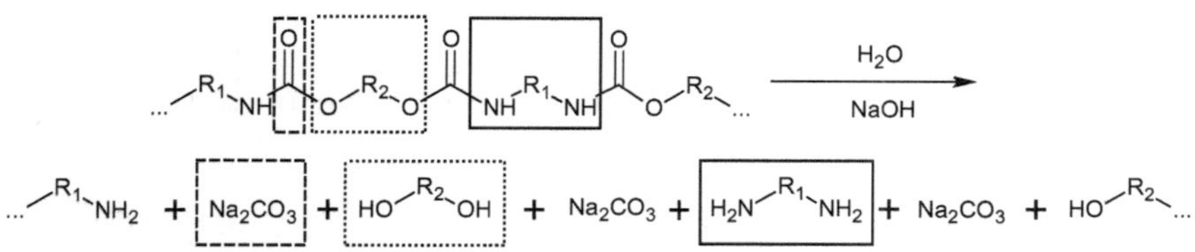

Scheme 14.4 Hydrolysis process of polyurethane catalyzed by base.

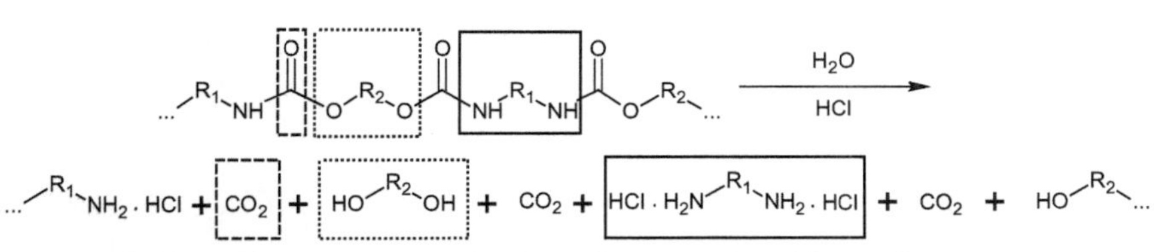

Scheme 14.5 Hydrolysis process of polyurethane catalyzed by acid.

process with softer conditions than simple hydrolysis. In accordance to the literature, there is also known hydrolysis of polyurethane waste with using sub- and supercritical water [62].

The main disadvantages of hydrolysis are high operating conditions (pressure and temperature) connected with high energy consumption by the plant and its application is only the decomposition of polyurethane foams.

14.3.3 Acidolysis

Acidolysis is a chemical decomposition of the urethane groups realized by organic or inorganic acids. In the case of inorganic acid (e.g., hydrochloric acid, Scheme 14.5) three products are formed: amine salt, polyol, and carbon dioxide [63].

Organic acids (especially dicarboxylic acids) that are used for chemical recycling of polyurethanes react with the urethane group that results from the formation of an amide group is the effect of the reaction (Scheme 14.6). These organic acids can be divided into two groups:

- unsaturated dicarboxylic acids (e.g., maleic acid or fumaric acid)—as a result of reaction, the oligomers with unsaturated carbon–carbon bond, amide or acylurea groups are formed—low pH of reaction products eliminate these products from application as a semiproducts in the synthesis of PU, but due to the presence of an unsaturated bond, these products can be applied in the production adhesives by radical polymerization,
- saturated dicarboxylic acids (e.g., adipic acid)—reaction can be carried out at low temperatures (ca. 60°C) in a short time, and oligomers with urea, amide, amine, or hydroxyl groups can be obtained from this reaction.

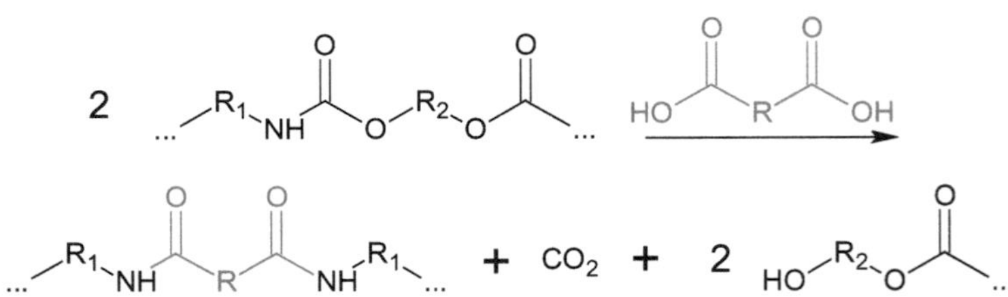

Scheme 14.6 Acidolysis process of polyurethane realized by using dicarboxylic acid.

14.3.4 Phosphorolysis

Phosphorolysis of polyurethanes is realized using esters of phosphonic or phosphoric acid. Phosphorolysis was initially studied by Troev et al. [42–44,46], whose used the following to chemically decompose PU: dimethyl phosphanate, diethyl phosphanate, triethyl phosphate, tris(2-chloroethyl) phosphanate, and tris(1-methyl-2-chloroethyl) phosphanate. The process was realized in the temperatures in the range of 140°C–180°C with an excess of phosphonate esters. The example of the phosphorolysis reaction is presented in Scheme 14.7. The final degradation products were liquid and soluble in organic solvents, e.g., chloroform, tetrahydrofuran, methylene chloride, dimethyl formamide, or dimethyl sulfoxide. The obtained products (with or without chemical treatment) can be applied as components for the synthesis of polyurethanes with flame retardant properties, due to the presence of phosphonate constituents in oligomers chain. See also Table 14.3.

14.3.5 Aminolysis

Polyurethane waste can also be decomposed by using aliphatic diamines or polyamines (with primary and secondary amine groups). Decomposition resulted in formation mainly amino and hydroxyl groups. High activity of amine groups allow the reaction to be carried out low temperatures.

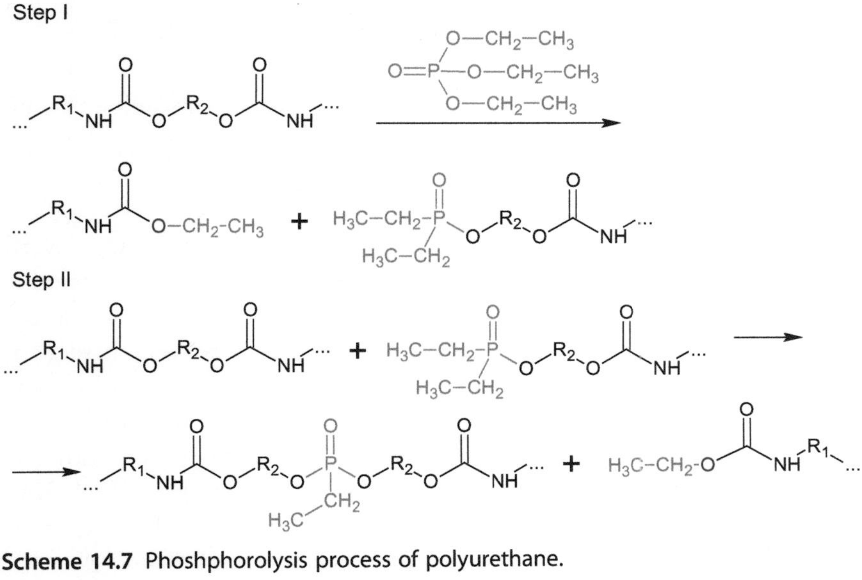

Scheme 14.7 Phoshphorolysis process of polyurethane.

Table 14.3 Examples of polyurethane phosphorolysis in accordance to the literature

Polyurethane waste (PU)	Phosphanate ester (PEs)	Mass ratio PU:PEs	Temperature (°C)	Refs.
Flexible foam Poly (ester-urethane) based on TDI	Dimethyl phosphanate, diethyl phosphanate	1:3	160	[64]
Flexible foam Poly (ether-urethane) based on TDI	Dimethyl phosphanate	1:3	160	[65]
Microcellular elastomer Poly(ester-urethane) based on MDI	Dimethyl phosphanate	1:2	142	[66]
Microcellular elastomer Poly(ester-urethane) based on MDI	Triethyl phosphanate, tris (2-chloroethyl) phosphanate	1:3	180	[67]
Microcellular elastomer Poly(ester-urethane) based on MDI	Diethyl phosphanate, tris(1-methyl-2-chloroethyl) phosphanate	1:3	170	[68]

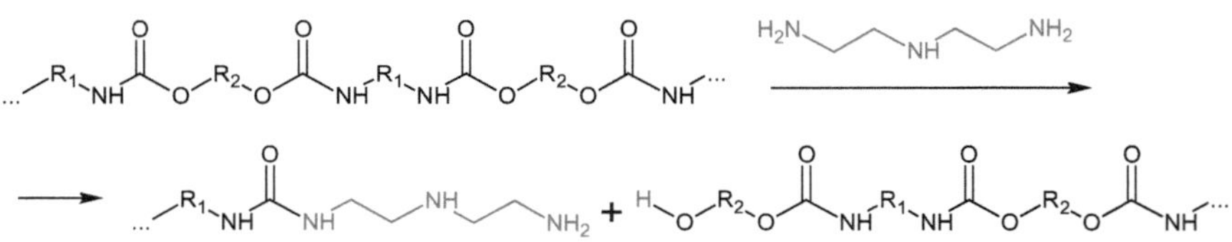

Scheme 14.8 Aminolysis of polyurethane by reaction with diethylenetriamine.

Xue et al. [69] decomposed the rigid polyurethane foams using several polyamines, such as diethylenetriamine (DETA), triethylenetetramine (TETA), fourethylenepentamine (TEPA). The reaction was carried out without using a catalyst, and products were applied as epoxy resin hardeners. Chuayjuljit et al. [70] decomposed rigid polyurethane foam also by using diethylenetriamine at 180°C, during 70 min, but using sodium hydroxide as a catalyst. Applying NaOH resulted in side-reactions connected with formation of carbon dioxide, similar to that presented in Scheme 14.4. Kim et al. [71] also decomposed rigid polyurethane foam by using diethylenetriamine at 180°C, 4 h, without using a catalyst (Scheme 14.8).

Alkanolamines (e.g., diethanolamine) react slower with urethane linkage than diamines or polymaines, due to lower basicity of nitrogen present in the structure of alkanolamine. This fact resulted in higher temperature of the reaction, longer time to realize it, and the possibility of acceleration by using catalyst, e.g., sodium hydroxide. Kanaya et al. [72] reported decomposition of rigid polyurethane foam (based on MDI) by alkanolamines at 150°C without using a catalyst, and the obtained products were completely separated into two layers. The upper liquid layer was a polyether polyol, and the lower liquid layer was 4,4′-methylene diphenylamine (MDA) and alkanolamine derivatives. Kanaya et al. [37] reported that reaction of urethane bond with alkanolamine occurred probably by alcoholysis, not aminolysis, mechanism. See also (Scheme 14.9).

14.3.6 Ammonolysis

Ammonolysis of the polyurethane waste includes chemical decomposition at an elevated temperature using ammonia or ammonium hydroxide. Ammonia exhibited higher nucleophilicity than water (applied in hydrolysis) or glycol (applied in glycolysis). As a result of ammonolysis polyols, amines, and unsubstituted urea are obtained. Polyols produced by this process can be reused for the synthesis of polyurethanes e.g. rigid polyurethane foams. It is favorable to apply large excesses of NH_3 to urethane groups present in the structure of the material, ensuring faster chemical decomposition into the desirable products.

In recent years a growing interest has arisen in the ammonolysis process is observed, wherein polyurethane is reacted with ammonia under supercritical conditions. Lentz and Mormann [73] carried out ammonolysis of

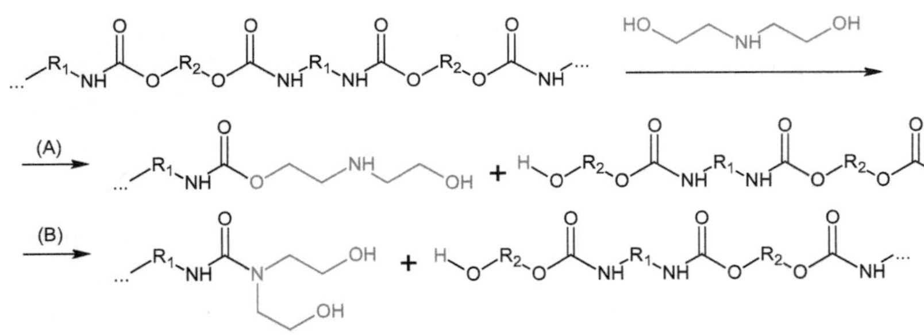

Scheme 14.9 Decomposition of polyurethane during the reaction with diethanolamine by: (A) glycolysis and (B) aminolysis mechanism.

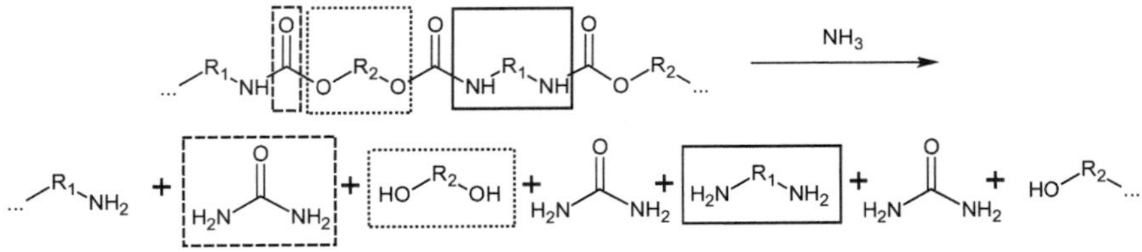

Scheme 14.10 Ammonolysis of polyurethane.

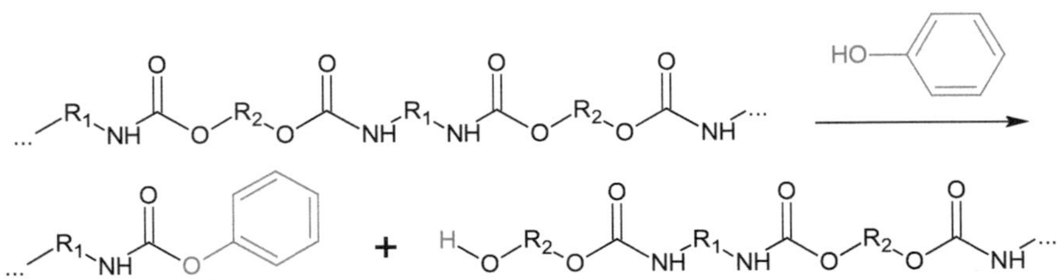

Scheme 14.11 Phenolysis of polyurethane.

flexible polyurethane foam and rigid elastomers, with the mass ratio PU:NH_3 of 1:1. The process was realized at 139°C under a pressure of 140 atm. for 2 h, and the conditions of the process resulted in complete chemical decomposition of polyurethanes into polyols, amines (corresponding to isocyanates), and unsubstituted urea (Scheme 14.10).

14.3.7 Phenolysis

Phenolysis is a type of chemical decomposition of polyurethane waste realized using phenol at a temperature of 50°C (melting temperature of phenol is ca. 40°C) in the presence of acidic and basic catalysts. Synthetic polyurethane leather can be chemically decomposed by heating with phenol in the presence of hydrochloric acid as catalyst. The obtained products can be applied to the synthesis of phenol-formaldehyde resins with improved elasticity (in comparison to resins obtained only with using phenol and formaldehyde) and good mechanical and dielectric properties (see Scheme 14.11).

14.3.8 Glycerolysis

Glycerolysis is a relatively new chemical decomposition method for polyurethane waste. Glycerol, a waste by-product of bio-diesel, can be used as a feedstock for this process. The process is realized at temperatures ranging from 220°C to 240°C under atmospheric pressure, but now the process is carried out on a laboratory scale, differing from glycolysis, which is

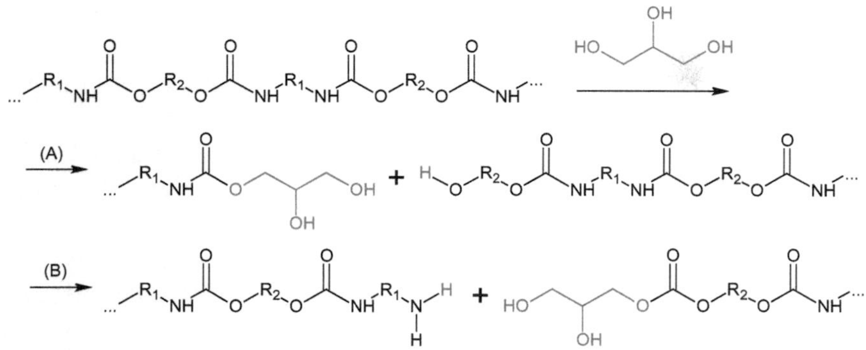

Scheme 14.12 Polyurethane glycerolysis process, with possible routes includes formation of (A) hydroxyl and (B) amine groups.

realized on the large scale in industry. The chemical decomposition of polyurethane by using glycerol is presented in Scheme 14.12.

The disadvantage of glycerol obtained from bio-fuel production, is contamination by glycerol methyl esters, water, and others. The level of impurities present in waste glycerol depends on the bio-diesel production method, and the construction of plant for realization of process. Very often the glycerol from bio-diesel production has a brown color and contains solid fraction. The use of contaminated glycerol results in formation of amines and their derivatives during glycerolysis, which can reduce the effectiveness of the process. It is difficult to obtain a homogeneous product with repeatable physical and chemical properties, so one of the future scientific works should be focusing on developing efficient purification of glycerol from bio-diesel production.

14.4 THERMO-CHEMICAL RECYCLING

An overview of thermo-chemical recycling methods of polyurethanes for the recovery of chemicals, fuels, energy and others is characterized in this section. A special emphasis is on the advantages and disadvantages of the presented methods with consideration of their environmental effects.

14.4.1 Pyrolysis and Thermal Decomposition

Pyrolysis of polyurethane foams and elastomers have been discussed in the literature. The main goal of the investigations is focused on the PUs thermal decomposition mechanisms with special emphasis on the depolymerization of urethane linkage and the range of products formed during the process. Moreover the thermal stability and flame retardancy of polyurethanes are also intensively investigated [74].

Bilbao et al. [75] investigated the thermal decomposition of poly (ether-urethane) foam in different atmospheres, and showed that in a nitrogen atmosphere, significant decomposition is obtained in the range of temperatures from 200°C to 400°C, but in the case of air atmosphere it is 200°C–300°C. Spirckel et al. [76] investigated the thermal degradation and the fire performance of the phosphanate polyurethanes, which were synthesized using phosphanate diols. During the thermal degradation, phosphonate constituents decompose at lower temperatures than the polyurethane matrix, but the presence of this compound favours the formation of a protective thermal layer at the material surface.

Dyer et al. [77–79] studied the thermal decomposition (see Scheme 14.13) of model urethanes (s carbamates) and indicated three ways of the urethane linkage cleavage at low temperatures (200°C–300°C) that there were three ways of the urethane linkage cleavage:

- dissociation to isocyanate and alcohol
- dissociation to primary amine, olefin, and carbon dioxide:
- elimination of carbon dioxide, leading to formation of a secondary amine

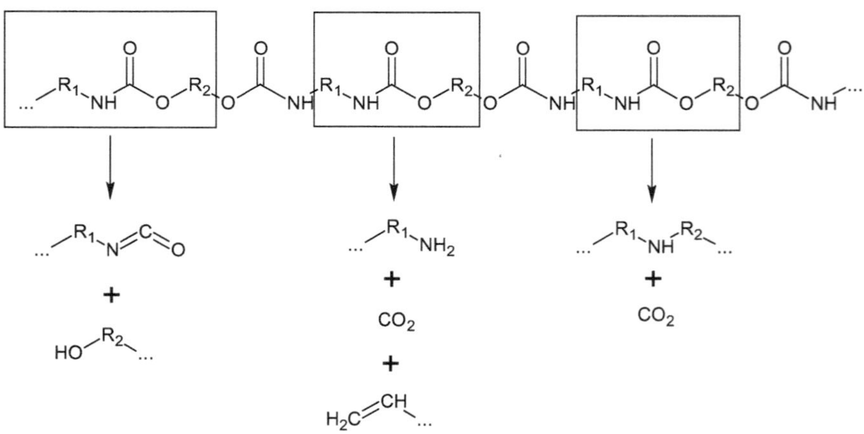

Scheme 14.13 Thermal decomposition ways of urethane bond, proposed by Dyer et al. [69–71].

Saunders and Backus [80] found that dissociation of urethane group to isocyanate and alcohol can occur at low temperatures (150°C–200°C), but with a lower rate than at higher temperatures. When the alcohol can be dehydrated, decomposition to primary amine and olefin occurred. Lattimer et al. [81] studied the degradation of a segmented polyurethane in the lower temperatures than 300°C, and indicated that decomposition occurred in two steps:

- dissociation of the urethane linkage resulting in formation of isocyanato and hydroxyl end groups
- ester exchange resulting in the formation of cyclic (polyester or urethane-containing polyester) oligomers. When the temperature is higher than 300°C dehydration occurs and products terminated with unsaturated groups are formed.

It was also found that decomposition of polyurethane in nitrogen atmosphere at 500°C resulted in the formation of a mixture of gases e.g., toluene, benzene, pentadiene, ethane, ethylene, propylene, and butadiene; when the temperature is 900°C the gases are benzene, ethane, ethylene, and methane but also ammonia (NH_3), pentene, hydrogen cyanide (HCN), aniline (aminobenzene), benzonitrile, and naphthalene are formed. Grittner et al. [82] investigated the pyrolysis of semi-rigid poly (ether-urethane) foam at two different temperatures—700°C and 800°C—and reported formation of gaseous products: methane, ethylene, and benzene. Font et al. [83] used TG thermogravimetric analysis to study the thermal decomposition of a poly(ester-urethane) at temperatures range of 500°C–900°C using a pyrolysis reactor connected with a secondary reactor. It was also found that decomposition of polyurethane in nitrogen atmosphere resulted in formation of several different gases, e.g., carbon monoxide, carbon dioxide, benzene, toluene ethane-ethylene, and methane but also ammonia, pentene, hydrogen cyanide, aniline, benzonitrile, and naphthalene are formed. Esperanza et al. [84] investigated the thermal decomposition of polyurethane-based varnish waste, and results showed that the main products at 850°C are e.g., CO, CO_2, benzene, naphthalene, methane, and ethylene. Voorhees et al [85,86]. investigated the thermolysis mechanisms of polyurethanes (based on poly(phenylene-methylene isocyanate) and propoxylated trimethylolpropane in an inert atmosphere.

The pyrolysis of polyurethane can be combined with utilization of metallic compounds. Moroi et al. [87,88] investigated the influence of metallic salts ($MnCl_2$, $CoCl_2$, $FeCl_3$, $CrCl_3$, $CuCl_2$, NaCl) on the thermal decomposition of poly(ester-urethane). The results showed that chromium and copper salts acted as the catalysts, which enhanced the thermal degradation of polyurethanes at lower temperatures but stabilized the intermediate pyrolysis products. The other investigated metal salts enhanced the thermal degradation of polyurethanes without stabilization of the intermediate semiproducts. Terakado et al. [89] investigated the influence of other metallic salts ($ZnCl_2$, $FeCl_2$, $FeCl_3$, $CuCl_2$,) on the thermal decomposition of polyurethane flexible (based on TDI) and rigid (based on MDI) foams. Metal chlorides can be introduced into polyurethane waste by mixing the powder polyurethane with metal chlorides.

The thermal degradation of PU foams and elastomers has been studied extensively in the scientific world. Presented results include data connected with different types of polyurethanes. The most important factors, which are responsible for a large variety of results, are types of: isocyanates (e.g., MDI, PMDI, TDI, HDI, HMDI, IPDI, and others), polyols (e.g., difunctional or polyfunctional polyetherols and polyestrols) and chain extenders (e.g., 1,2-etanediol, 1,4-butanediol, diethanolamine) used to produce the polyurethanes. Different constituents in the polymer chain structure have different thermal dissociation temperatures, resulting in the diversification of the thermal stability of polyurethanes, obtained using different components. Thermal degradation of the polyurethanes starts in the temperature range 250°C–270°C, from components used to the synthesize PUs and conditions of the analysis (e.g., atmospheric or nitrogen atmosphere).

The mechanism of polyurethanes thermal degradation, described by Cullis and Hirschler [90], includes three possible routes for the dissociation of urethane bonds, i.e., dissociation into isocyanate and alcohol, into carbamic acid and olefin and dissociation of the carbamic acid into CO_2 and secondary amine. These reactions were found to occur at the temperature range from 200°C to 300°C. In the case of flexible polyurethanes foam based on toluene diisocyanates (TDI), Ravey and Pearce [91] found that diisocyanate can be recovered, when it is rapidly removed from the reaction zone. The longer of TDI in the reaction zone resulted in formation of diaminotoluene (DAT).

14.4.2 Gasification

Gasification of waste is an exothermic process resulting in production of heat, ashes, and synthetic gas (which contains large amounts of H_2

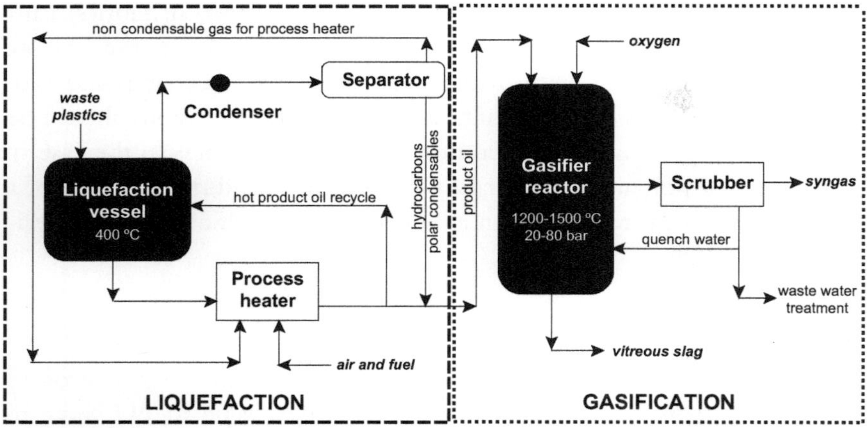

Figure 14.8 Sample realization of polymer waste liquefaction and gasification [92].

and CO). The gasification is carried out in oxygen atmosphere at very high temperatures (1200°C–1500°C) and under high-pressure (20–80 bar), as is presented in Fig. 14.8. The very high conversion, equal to 98%–99%, of fuel is reached after few seconds. The CO and H_2 produced can be used to produce isocyanate components and polyetherols, which are used in the synthesis of polyurethanes [84].

14.4.3 Hydrogenation

Hydrogenation is a process that combines the heat and high-pressure hydrogen treatment of waste, to obtain gaseous and liquid products [93]. These obtained products can be used as fuel and chemical feedstock for other processes. Hydrogenation can be considered as a compromise between pyrolysis (effect of heat) and gasification (effect of high-pressure hydrogen). Hydrogenation includes pyrolysis and an additional step, which is the combination of the heat, pressure and hydrogen treatment. In comparison to pyrolysis, hydrogenation results in produce of more pure gases and oils. Two important issues should be considered, during the hydrogenation of polyurethanes waste, the purity of derived gases and oils, and the costs final products production [94].

14.5 RECOVERY OF ENERGY

Energy recovery (combustion or incineration) is often considered as the only suitable disposal method of polyurethane waste, when the waste material (e.g., laminates, high-filled composites, leathers, or fabrics) cannot be recycled by mechanical and chemical methods. In the case of polyurethane materials with flame retardants, the recycling and also energy recovering is very difficult, so nowadays investigations should be focused on improving the combustion process effectiveness in the case of flame retarded polymeric materials. On the other hand, incineration of a polyurethane foams results also in a high volume reduction of this materials in the landfill.

14.5.1 Combustion

The combustion of plastics waste is directly related to the energy recovery from polyurethane waste. The comparable energy value of PU waste to fossil fuel resulted in a reduction of plastics waste amount in landfill. The Polyurethanes Recycle and Recovery Council (PURRC) realized investigations focused on adding flexible (up to 20% by weight) polyurethane foam waste to municipal solid waste (MSW), result in improvements of energy production and generating emissions below operating limits [95]. The European Diisocyanate and Polyol Producers Association (ISOPA) promote producing energy from plastics waste by using of carefully controlled incineration [96]. The European PU Industry promote recovery of energy from polyurethane waste. Some European countries have modern MSW combustors connected with flue gas cleaning (e.g., Denmark and Germany). Realization of this process can result in a supply of electricity and heat energy for the European society [97–100].

14.5.2 Incineration

Incineration is one-step process connected with heat recovery through the burning of polyurethane waste materials. In the comparison to the combustion, the incineration process is an incomplete combustion. Unfortunately this process can also produce poisonous gases which can seriously pollute the atmosphere, so from the environmental point of view the use of this method should be gradually reduced. Incineration plays an important role in the recycling methods of polyurethane waste, especially due to fast and high reduction level of polyurethane waste volume.

The calorific value of polyurethanes is similar to coal, and is equal 25–30 MJ kg^{-1}. During the incineration of polyurethane waste volume is reduced ca. 99%. Polyurethane foam waste can be ground and pressed to form of pellets, and used as a alternative fuel for coal in recovery of energy, and applied in the power industry. The addition of waste polyurethane elastomer to municipal solid waste can obviously improves its fuel calorific value. A similar solution was proposed by Kim et al. [101], who developed a novel composite fuel obtained by mixing of polyurethane waste (10 wt%) with coal (90 wt%) under high-pressure.

As mentioned earlier, incineration is an incomplete combustion resulting in the production of poisonous gases. This problem was considered by Rogaume et al. [102] during investigation concerning optimization of combustion process conditions to reduce NO and CO emissions. The authors realized the combustion of polyurethane foams from waste automobile seats under grate incineration exhaust gas recirculation conditions, at temperature ranges from 850°C to 1000°C under atmospheric pressure. The applied solution consisted of delivery of primary air below the grid and secondary air above the bed. Moreover, applying a recirculation of flue gases to the primary air, in this solution, resulted in complete burning-out of traces of hydrocarbons, hydrogen cyanide, and ammonia.

14.6 LANDFILL

The first methods of polyurethane waste management include landfill disposal and combustion. The high volume of polyurethane foam waste is one of the most important problems in landfill disposal. On the other hand, the decomposition of polyurethane waste in the natural conditions are very long and difficult. Decomposition of polyurethane in a garbage landfill in leakage water and by soil microorganisms were studied by Filip et al. [103], and results showed that polyurethanes exhibit high level of resistance against abiotic and biotic conditions. Biodegradation of polyurethane waste are described in the literature [104,105]. Some European countries (e.g., Netherlands, Sweden, Denmark, Switzerland, and Germany) restrictplastic waste disposal in a landfill. This situation results in development of mechanical and chemical recycling methods of plastics waste, e.g., polyurethanes. Nowadays, in European Union, is popular project "Zero Plastics to Landfill" which was proposed by PlasticsEurope (Association of Plastics Manufacturers). The key recommendations of this projects include: stopping of the landfilling of recyclable and other recoverable postconsumer waste, pointing that efficient energy-from-waste should be part of Europe's energy strategy, and supporting innovation in plastics recycling technologies [106].

14.7 FUTURE

The area of the polyurethanes waste recycling is still improving. The difficulties, which are meet during recycling, result from diversity of forms (elastomers or foams) and types of substrates, which are used to the synthesize polyurethanes. Different types of isocyanate components (di- or polyfunctional aliphatic, cycloaliphatic, aromatic isocyanates) and polyols components (di- or polyfunctional polyetherols or polyestrols), resulted in the synthesis of polyurethanes with the diversified structure and properties for the different applications. A lot of possible recycling methods of polyurethanes waste have been developed, but only some of them have industrial applications e.g., compression moulding (mechanical recycling), glycolysis (chemical recycling) or adding a fossil fuel as a component to improve energy value (recovery of energy).

The glycolysis of polyurethane foams plays a very important role in the recycling methods, due to wide knowledge about the process, which is still expanded by a lot of researchers. As an example, glycolysis was successful implemented in the industry, e.g., as a technology for recovering a feedstock for the production (from company's own production streams or products with defects). Moreover, some companies are focused on production of polyols from, e.g., polyurethane or poly(ethylene terephthalate) waste. On the other hand, it is also necessary to determine the influence of the recycling products (added as a part of whole system) on the properties of the final products.

The development of the recycling technology must be connected with two important issues, i.e., costeffectiveness and the environmentally friendly nature of the processes. The selective collection of waste should improve the reuse of plastics. When combustion or incineration is one of the possible methods, the gases produced during the processes should be rendered as much as possible, e.g., by returning flue gases to the reactor. It is also important to predict the applications of the chemical recycling products. Research should be focused on the optimizating of the reaction conditions in terms of process costs. Expensive recycling plants are possible only in big companies, but the process must result in production of the high-quality components, which are cheaper than new petrochemical components applied in the production. Moreover, recycling technology can be also connected with current bio-trends present in the area of the polyurethane synthesis, where the bio-based substances are used as components for the synthesis (e.g., chemically modified vegetable oils used as a polyols). As a result of this situation, the vegetable-based glycerol (by-product form the bio-diesel production) was proposed as a compound for chemical recycling of polyurethane waste.

REFERENCES

[1] S.M. Al-Salem, P. Lettieri, J. Baeyens, Recycling and recovery routes of plastic solid waste (PSW): a review, Waste Manage. 29 (10) (2009) 2625–2643.

[2] D. Simon, A.M. Borreguero, A. de Lucas, C. Gutierrez, J.F. Rodriguez, *Sustainable polyurethanes: chemical.* Recycling to get it. Environ. Chem. Pollut. Wastes 32 (2015) 229–260.

[3] K.M. Zia, H.N. Bhatti, I.A. Bhatti, Methods for polyurethane and polyurethane composites, recycling and recovery: a review, React. Funct. Polym. 67 (8) (2007) 675–692.

[4] W. Yang, Q. Dong, S. Liu, H. Xie, L. Liu, J. Li, Recycling and disposal methods for polyurethane foam wastes, Proc. Environ. Sci. 16 (2012) 167–175.

[5] *Polyurethane (PU) Market Analysis By Product (Rigid Foam, Flexible Foam, Coatings, Adhesives & Sealants, Elastomers), By End-use (Furniture and Interiors, Construction, Electronics & Appliances, Automotive, Footwear, Packaging) And Segment Forecasts To 2020*—Grand View Research. <www.grandviewresearch.com/industry-analysis/polyurethane-pu-market>.

[6] P. Rojek, A. Prociak, Effect of different rapeseed-oil-based polyols on mechanical properties of flexible polyurethane foams, J. Appl. Polym. Sci. 125 (4) (2012) 2936–2945.

[7] J. Datta, E. Głowińska, Effect of hydroxylated soybean oil and bio-based propanediol on the structure and thermal properties of synthesized bio-polyurethanes, Ind. Crops Prod. 61 (2014) 84–91.

[8] S. Caillol, M. Desroches, G. Boutevin, C. Loubat, R. Auvergne, B. Boutevin, Synthesis of new polyester polyols from epoxidized vegetable oils and biobased acids, Eur. J. Lipid Sci. Technol. 114 (2012) 1447–1459.

[9] S. Hu, X. Luo, Y. Li, Production of polyols and waterborne polyurethane dispersions from biodiesel-derived crude glycerol, J. Appl. Polym. Sci. 132 (2015) 41425.

[10] X. Luo, S. Hu, X. Zhang, Y. Li, Thermochemical conversion of crude glycerol to biopolyols for the production of polyurethane foams, Bioresour. Technol. 139 (2013) 323–329.

[11] S. Hu, Y. Li, Two-step sequential liquefaction of lignocellulosic biomass by crude glycerol for the production of polyols and polyurethane foams, Bioresour. Technol. 161 (2014) 410–415.

[12] C.A. Cateto, M.F. Barreiro, C. Ottati, M. Lopretti, A.E. Rodrigues, M.N. Belgacem, Lignin-based rigid polyurethane foams with improved biodegradation, J. Cell. Plast. Jan. 50 (2014) 81–95.

[13] J. Datta, P. Kopczyńska, Effect of kenaf fibre modification on morphology and mechanical properties of thermoplastic polyurethane materials, Ind. Crops Prod. 74 (2015) 566–576.

[14] M. Kurańska, A. Prociak, Porous polyurethane composites with natural fibres, Compos. Sci. Technol. 72 (2) (2012) 299–304.

[15] E. Głowińska, J. Kamerke, M. Włoch, Mechanical properties and morphology of biopolyurethane composites reinforced with modified sisal fibers, Przemysł Chem. 94 (5) (2015) 688–693.

[16] L. Rueda, A. Saralegui, B. Fernández d'Arlas, Q. Zhou, L.A. Berglund, M.A. Corcuera, et al., Cellulose nanocrystals/polyurethane nanocomposites. Study from the viewpoint of microphase separated structure, Carbohydr. Polym. 92 (1) (2013) 751–757.
[17] E. Głowińska, J. Datta, Structure, morphology and mechanical behaviour of novel bio-based polyurethane composites with microcrystalline cellulose, Cellulose (2015). Available from: http://dx.doi.org/10.1007/s10570-015-0685-0.
[18] J.L. Nafziger, S.B. Lowenkron, C.E. Koehler, B.N. Stevens: Flexible polyurethane rebond foam having improved tear resistance and method for the preparation thereof, US Patent US 5312888 (1994).
[19] The European Isocyanates Producers' Association (ISOPA) – www.isopa.org.
[20] Efficient Solid Waste Management, InTOUCH 4:1 (1991) – Polyurethane Foam Association – www.pfa.org.
[21] M. Heneczkowski, H. Galina, Material recycling of RIM flexible polyurethane foams wastes, Polimery 47 (7–8) (2002) 523–527.
[22] J.R. Wang, D.J. Chen, The chemical and physical recycling methods for polyurethane wastes, China Elastomerics 13 (6) (2003) 61–65.
[23] A.B. Fraj, M. Kismi, P. Mounanga, Valorization of coarse rigid polyurethane foam waste in lightweight aggregate concrete, Constr. Build. Mater. 24 (6) (2010) 1069–1077.
[24] P. Mounanga, W. Gbongbon, P. Poullain, P. Turcry, Proportioning and characterization of lightweight concrete mixtures made with rigid polyurethane foam wastes, Cem. Concr. Compos. 30 (9) (2008) 806–814.
[25] J. Datta, M. Chojnacka, Effect of polyurethane foam recyclate on the mechanical properties of rubber vulcanizates plasticized with glycolysates, Przemysł Chem. 90 (1) (2011) 1758–1761.
[26] A.J. Hulme, T.C. Goodhead, Cost effective reprocessing of polyurethane by hot compression moulding, J. Mater. Process. Technol. 139 (1–3) (2003) 322–326.
[27] J.S. Park, Y.M. Lim, Y.C. Nho, Preparation of high density polyethylene/waste polyurethane blends compatibilized with polyethylene-graft-maleic anhydride by radiation, Materials 8 (2015) 1626–1635.
[28] Y.F. Lin, Waste rigid polyurethane foam plastics recycling method, Patent CN101096426 (2008).
[29] *Eco Solutions* RAMPF – www.rampf-gruppe.de.
[30] J. Datta, M. Rohn, Glycolysis of polyurethane wastes. Part II. Purification and use of glycolysis products, Polimery 9 (2007) 627–633.
[31] J. Datta, M. Rohn, *Glycolysis of polyurethane wastes*. Part I. Glycolysis agents and catalysts, Polimery 52 (7–8) (2007) 579–582.
[32] M. Modesti, F. Simioni, Chemical recycling of reinforced polyurethane from the automotive industry, Polym. Eng. Sci. 36 (1996) 2173–2178.
[33] B. Naber, W. Schmiade, M. Gassan, Production of rigid polyurethane foams. US Patent 6087409 (2000).
[34] H. Beneš, J. Rösner, P. Holler, H. Synková, J. Kotek, Z. Horák, Glycolysis of flexible polyurethane foam in recycling of car seats, Polym. Adv. Technol. 18 (2007) 149–156.
[35] J. Datta, S. Pasternak, Oligourethane glycols obtained in glycolysis of polyurethane foam as semi-finished products for cast urethane elastomers preparation, Polimery 50 (5) (2005) 352–357.
[36] J. Datta, K. Pniewska, Syntheses and properties of polyurethanes got from glycolysis products obtained from waste polyurethane foams, Polimery 53 (1) (2008) 27–32.
[37] J. Datta, M. Rohn, Thermal properties of polyurethanes synthesized using waste polyurethane foam glycolysates, J. Therm. Anal. Calorim. 88 (2007) 437–440.

[38] J. Datta, M. Kacprzyk, Thermal analysis and static strength of polyurethanes obtained from glycolysates, J. Therm. Anal. Calorim. 93 (3) (2008) 753–757.
[39] J. Datta, J.T. Haponiuk, Influence of glycols on the glycolysis process and the structure and properties of polyurethane elastomers, J. Elast. Plast. 43 (2011) 529–541.
[40] C.H. Wu, C.Y. Chang, C.M. Cheng, H.C. Huang, Glycolysis *of waste flexible polyurethane foam*, Polym. Degrad. Stab. 80 (1) (2003) 103–111.
[41] C. Molero, A. de Lucas, J.F. Rodriguez, Recovery of polyols from flexible polyurethane foam by "split-phase" glycolysis with new catalysts, Polym. Degrad. Stab. 91 (2006) 894–901.
[42] D. Simon, A.M. Borreguero, A. de Lucas, J.F. Rodríguez, Glycolysis of flexible polyurethane wastes containing polymeric polyols, Polym. Degrad. Stab. 109 (2014) 115–121.
[43] D. Simon, M.T. García, A. de Lucas, A.M. Borreguero, J.F. Rodríguez, Glycolysis of flexible polyurethane wastes using stannous octoate as the catalyst: Study on the influence of reaction parameters, Polym. Degrad. Stab. 98 (1) (2013) 144–149.
[44] C. Molero, A. de Lucas, J.F. Rodríguez, Recovery of polyols from flexible polyurethane foam by "split-phase" glycolysis: Study on the influence of reaction parameters, Polym. Degrad. Stab. 93 (2) (2008) 353–361.
[45] C. Molero, A. de Lucas, F. Romero, J.F. Rodriguez, Glycolysis of flexible polyurethane wastes using stannous octoate as the catalyst, J. Mater. Cycles Waste Manage. 11 (2) (2009) 130–132.
[46] C. Molero, A. de Lucas, J.F. Rodríguez, Recovery of polyols from flexible polyurethane foam by "split-phase" glycolysis: glycol influence, Polym. Degrad. Stab. 91 (2) (2006) 221–228.
[47] J. Datta, Effect of glycols used as glycolysis agents on chemical structure and thermal stability of the produced glycolysates, J. Therm. Anal. Calorim. 109 (2012) 517–520.
[48] C.H. Wu, C.Y. Chang, J.K. Li, Glycolysis of rigid polyurethane from waste refrigerators, Polym. Degrad. Stab. 75 (3) (2002) 413–421.
[49] C.Y. Yu, W.J. Lee, Characteristics of glycolysis products of polyurethane foams made with polyhydric alcohol liquefied *Cryptomeria japonica* wood, Polym. Degrad. Stab. 101 (2014) 60–64.
[50] M.M.A. Nikje, M. Haghshenas, A.B. Garmarudi, Preparation and application of glycolysed polyurethane integral skin foams recyclate from automotive wastes, Polym. Bull. 56 (2-3) (2006) 257–265.
[51] X. Wang, H. Chen, C. Chen, H. Li, Chemical degradation of thermoplastic polyurethane for recycling polyether polyol, Fibers Polym. 12 (7) (2011) 857–863.
[52] J. Datta, M. Janicka, Synthesis and properties of polyurethanes made of secondary raw materials, Przemysł Chem. 86 (7) (2007) 624–626.
[53] J. Wang, D. Chen, Q. Zhang, Analysis of glycolysis products of polyurethane fiber waste with diethylene glycol, Fibers Polym. 8 (1) (2007) 13–18.
[54] D. Simón, A.M. Borreguero, A. de Lucas, C. Molero, J.F. Rodríguez, Novel polyol initiator from polyurethane recycling residue, J. Mater. Cycles Waste Manage. 16 (3) (2014) 525–532.
[55] G.A. Campbell, W.C. Meluch, Polyurethane foam recycling – superheated steam hydrolysis, Environ. Sci. Technol. 10 (2) (1976) 182–185.
[56] J.L. Gerlock, J. Braslaw, L.R. Mahoney, F.C. Ferris, Reaction of polyurethane foam with dry steam: kinetics and mechanism of reactions, J. Polym. Sci. Part A Polym. Chem. 18 (2) (1980) 541–557.
[57] M.L. Matuszak, K.C. Frisch, S.L. Reegen, Hydrolysis of linear polyurethanes and model monocarbamates, J. Polym. Sci. Part A Polym. Chem. 11 (7) (1973) 1683–1690.

[58] Y. Shi, X. Zhan, Q. Zhang, F. Chen, Interfacial hydrolysis of isocyanate in monomer miniemulsion, Chem. React. Eng. Technol. 25 (1) (2009) 88–92.

[59] Z. Dai, B. Hatano, J. Kadokawa, H. Tagaya, Effect of diaminotoluene on the decomposition of polyurethane foam waste in superheated water, Polym. Degrad. Stab. 76 (2) (2002) 179–184.

[60] M.M.A. Nikje, M. Nikrah, F.H.A. Mohammadi, Microwave-assisted polyurethane bond cleavage via hydroglycolysis process at atmospheric pressure, J. Cell. Plast. 44 (5) (2008) 367–380.

[61] M.M.A. Nikje, F.H.A. Mohammadi, Sorbitol/glycerin/water ternary system as a novel glycolysis agent for flexible polyurethane foam in the chemical recycling using microwave radiation, Polimery 54 (7-8) (2009) 541–545.

[62] M. Goto, Chemical recycling of plastics using sub- and supercritical fluids, J. Supercrit. Fluids 47 (3) (2009) 500–507.

[63] G. Behrendt, B.W. Naber, The recycling of polyurethanes (review), J. Univer. Chem. Technol. Metall. 44 (1) (2009) 3–23.

[64] K. Troev, A. Tsekova, R. Tsevi, Chemical degradation of polyurethanes: degradation of flexible polyester polyurethane foam by phosphonic acid dialkyl esters, J. Appl. Polym. Sci. 78 (14) (2000) 2565–2573.

[65] K. Troev, A. Tsekova, R. Tsevi, Chemical degradation of polyurethanes II. Degradationof flexible polyether foam by dimethyl phosphonate, Polym. Degrad. Stab. 67 (2000) 397–405.

[66] K. Troev, V.I. Atanasov, R. Tsevi, G. Grancharov, A. Tsekova, *Chemical degradation of polyurethanes*. Degradation of microporous polyurethane elastomer by dimethyl phosphonate, Polym. Degrad. Stab. 67 (2000) 159–165.

[67] K. Troev, V.I. Atanasov, R. Tsevi, Chemical degradation of polyurethanes II. Degradation of microporous polyurethane elastomer by phosphoric acid esters, J. Appl. Polym. Sci. 76 (2000) 886–893.

[68] K. Troev, G. Grancharov, R. Tsevi, Chemical degradation of polyurethanes III. Degradation of microporous polyurethane elastomer by diethyl phosphonate and tris (1-methyl-2-chloroethyl) phosphate, Polym. Degrad. Stab. 70 (2000) 43–48.

[69] S. Xue, M. Omoto, T. Hidai, Y. Imai, Preparation of epoxy hardeners from waste rigid polyurethane foam and their applications, J. Appl. Polym. Sci. 56 (2) (1995) 127–134.

[70] S. Chuayjuljit, C. Norakankorn, V. Pimpan, Chemical recycling of rigid polyurethane foam scrap via base catalyzed aminolysis, J. Metals Mater. Miner. 12 (1) (2002) 19–22.

[71] B.C. Kim, D.S. Lee, S.W. Hyun, Curing behavior of epoxy resins using aminolysis products of waste polyurethanes as hardeners, J. Ind. Eng. Chem. 7 (6) (2001) 449–453.

[72] K. Kanaya, S. Takahashi, Decomposition of polyurethane foams by alkanolamines, J. Appl. Polym. Sci. 51 (4) (1994) 675–682.

[73] H. Lentz, W. Mormann, Chemical recycling of polyurethanes and separation of the components by supercritical ammonia, Makromol. Chem. Macromol. Symp. 57 (1992) 305–310.

[74] D.K. Chattopadhyay, D.C. Webster, Thermal stability and flame retardancy ofpolyurethanes, Progr. Polym. Sci. 34 (2009) 1068–1133.

[75] R. Bilbao, J.F. Mastral, J. Ceamanos, M.E. Aldea, Kinetics of the thermal decomposition of polyurethane foams in nitrogen and air atmospheres, J. Anal. Appl. Pyrol. 37 (1) (1996) 69–82.

[76] M. Spirckel, N. Regnier, B. Mortaigne, B. Youssef, C. Bunel, Thermal degradation and fire performance of newphosphonate polyurethanes, Polym. Degrad. Stab. 78 (2) (2002) 211–218.

[77] E. Dyer, G.E. Newborn, Thermal degradation of carbamates of methylenebis-(4-phenyl isocyanate), J. Am. Chem. Soc. 80 (1958) 5495–5498.
[78] E. Dyer, G.C. Wright, Thermal degradation of alkyl N-phenylcarbamates, J. Am. Chem. Soc. 81 (1959) 2138–2143.
[79] E. Dyer, R.E. Read, Thermal degradation of O-1-hexadecyl N-1-naphthylcarbamates and related compounds, J. Org. Chem. 26 (1961) 4388–4394.
[80] J.H. Saunders, J.K. Backus, Thermal degradation and flammability of urethan polymers, Rubber Chem. Technol. 39 (2) (1966) 461–480.
[81] R.P. Lattimer, H. Muenster, H. Budzikiewicz, Pyrolysis tandem mass spectrometry (Py-MS/MS) of a segmented polyurethane, J. Anal. Appl. Pyrol. 17 (3) (1990) 237–249.
[82] N. Grittner, W. Kaminsky, G. Obst, Fluid bed pyrolysis of anhydride-hardened epoxy resins and polyether-polyurethane by the Hamburg process, J. Anal. Appl. Pyrol. 25 (1993) 293–299.
[83] R. Font, A. Fullana, J.A. Caballero, J. Candela, A. Garcia, Pyrolysis study of polyurethane, J. Anal. Appl. Pyrol. 58-59 (2001) 63–77.
[84] M.M. Esperanza, A.N. Garcia, R. Font, J.A. Conesa, Pyrolysis of varnish wastes based on a polyurethane, J. Anal. Appl. Pyrol. 52 (1999) 151–166.
[85] K.J. Voorhees, F.D. Hileman, I.N. Einhorn, J.H. Futrell, An investigation of the thermolysis mechanism of model urethanes, J. Polym. Sci. Polym. Chem. Ed. 16 (1) (1978) 213–228.
[86] K.J. Voorhees, R.P. Lattimer, Mechanisms of pyrolysis for a specifically labeled deuterated urethane, J. Polym. Sci. Polym. Chem. Ed. 20 (6) (1982) 1457–1467.
[87] G. Moroi, Influence of ion species on the thermal degradation of polyurethane interaction products with transition metal ions, J. Anal. Appl. Pyrol. 71 (2004) 485–500.
[88] G. Moroi, C. Ciobanu, E. Costea, N. Blba, I. Palamaru, Aspects of polyesterurethanes interaction with metallic ions. I. Influence of sodium cations on thermal behavior of some aliphatic linear polyesters, Thermochim. Acta 291 (1997) 95–99.
[89] O. Terakado, H. Yanase, M. Hirasawa, Pyrolysis treatment of waste polyurethane foam in the presence of metallic compounds, J. Anal. Appl. Pyrol. 108 (2014) 130–135.
[90] C.F. Cullis, M.M. Hirschler, The Combustion of Organic Polymers, Clarendon Press, United Kingdom, 1981.
[91] M. Ravey, E.M. Pearce, Flexible polyurethane foam. I. Thermal decomposition of a polyether-based, water-blown commercial type of flexible polyurethane foam, J. Appl. Polym. Sci. 63 (1) (1997) 47–74.
[92] J. Scheirs, Polymer Recycling: Science, Technology, and Applications, Wiley, United Kongdom, 1999.
[93] A. Tohka, R. Zevenhoven, Processing wastes and waste derived fuels containing brominated flame retardants, Energy Engineering and Environmental Protection Publications – Report TKK-ENY-7, Helsinki University of Technology, 2002.
[94] Polyurethane Recycling, American Chemistry Council - polyurethane.americanchemistry.com.
[95] Energy Recovery from Flexible PU Foams – Fact Sheet – Recycling and Recovering Polyurethanes (2001) The European Isocyanates Producers' Association (ISOPA) - www.isopa.org.
[96] Energy Recovery – Fact Sheet – Recycling and Recovering Polyurethanes, The European Isocyanates Producers' Association (ISOPA) - www.isopa.org; 2001.
[97] Integrated Pollution Prevention and Control – Reference Document on the Best Available Techniques for Waste Incineration (2006) European Commission.

[98] Directive 2000/76/EC of The European Parliament and the Council on the incineration of waste.
[99] Directive 2008/98/EC of The European Parliament and the Council on waste and repealing certain Directives.
[100] W. Schüttenhelm, R. Holste, A. Licata, New trends in flue gas cleaning technologies for European and Asian waste incineration facilities, 8th Annual North American Waste-to-Energy Conference, Nashville, Tennesse, 22-24.05.2000.
[101] H. Kim, W. Minami, T. Li, Combustion characteristics and pollutant control by eco-fuel from polyurethane foam, Energy Fuels 20 (2) (2006) 575–578.
[102] Y. Rogaume, F. Jabouille, M. Auzanneau, J.C. Goudeau, Thermal degradation and incineration of polyurethane: parameters to reduce the NOx emissions, 5th International Conference on Technologies and Combustion for a Clean Environment, Lisbon, Portugal (1999).
[103] Z. Filip, Decomposition of polyurethane in a garbage landfill leakage water and by soil microorganisms, Eur. J. Appl. Microbiol. Biotechnol. 5 (3) (1978) 225–231.
[104] G.T. Howard, Polyurethane biodegrdation, in: S.N. Singh (Ed.), Microbial Degradation of Xenobiotics, Springer, Berlin, 2012.
[105] M. Cregut, M. Bedas, M.J. Durand, G. Thouand, New insights into polyurethane biodegradation and realistic prospects for the development of a sustainable waste recycling process, Biotechnol. Adv. 31 (8) (2013) 1634–1647.
[106] PlasticsEurope - www.plasticseurope.org.

CHAPTER 15

Application of Blends and Polyurethane Interpenetrating Polymer Networks

Sainath Ashok Jadhav[1], Ajay V. Rane[2], Krishnan Kanny[2], Saket S. Mulge[1], Vayyaprontavida K. Abitha[3] and Sabu Thomas[3]
[1]Institute of Chemical Technology, Mumbai, India
[2]Durban University of Technology, Durban, South Africa
[3]Mahatma Gandhi University, Kottayam, Kerala, India

Contents

15.1 APPLICATIONS OF BLENDS OF POLYURETHANES

Development of new polymers through synthetic routes for specific applications has become an expensive and time-consuming process involving several tedious and hazardous reaction steps. Therefore, blending of polymers has become an easier, more popular, promising, and inexpensive process to produce new materials with the desired end use properties, in addition to having greater design flexibility for specific applications. Properties of polymer blends depend on the properties of the individual constituents, the processing conditions, the morphology developed during

Polyurethane Polymers: Blends and Interpenetrating Polymer Networks
DOI: http://dx.doi.org/10.1016/B978-0-12-804039-3.00015-4

processing, and the interaction between the blend constituents and synergism if any. Most polymer blends are immiscible in nature because of their macromolecular structure dissimilarity and wide difference in surface energy. Quite often people confuse compatibility and miscibility of polymer blends. However, Utracki made it clear that miscible blends are those that are thermodynamically compatible and compatible blends are those that are technologically but not thermodynamically compatible. Compatibilizers are often used as additives to improve the compatibility between immiscible polymers, and it improves the morphology and properties of the blends. The process of compatibilization, whether by adding a third component (compatibilizer) or by in situ chemical reaction between blend constituents (reactive blending) has played an important role in the development of polymer blend systems with better physico-mechanical properties. It has been well established that compatibilization leads to a reduction of the interfacial tension that facilitates fine mixing, dispersion, and controls, as well as stabilizes phase morphology due to the higher wetting and enhancement of adhesion between phases in the solid state, which enhances the stress transfer between the components, leading to higher impact strength and technical properties.

15.1.1 Applications of Blends of Polyamide with Polyurethane

The use of renewable bio-based carbon feedstock is has received great interest recently. This is mainly as it offers the intrinsic value of a reduced carbon footprint as well as an improved life cycle analysis (LCA), which is also in agreement with sustainable development. This is one of the major reasons why forthcoming chemicals and materials for daily use are obtained from biomass in increasing quantities, with the help of various chemical and/or biochemical (white biotech) processes [1]. These bio-based products with particular chemical architectures succeed as a good alternative to conventional and fossil-based chemical feedstock. Many fatty acid-based polymers with new macromolecular architectures such as thermoplastic polyamides and polyurethanes (TPUs) are being synthesized [2]. TPUs can be used in numerous fields like construction engineering, automotive industry, consumer or domestic equipment, even in the biomedical field. Recently, it has even been confirmed that PA with biobase content close to 100% can also be synthesized from fatty acids [3]. Similarly [4], PA can be synthesized from dimers obtained by dimerization which follows Diels Alder mechanism from different fatty acids.

These corresponding dimer fatty acid polyamides (DAPA) display interesting properties [5]. For instance, DAPA are more flexible than the conventional PA6-6 [6]. TPU and DAPA can be mixed together in order to increase the properties of TPUs [7] and also to increase the bio-based content. Intermolecular interactions occur between the amide and urethane, a group which has also been explained by different studies on conventional fossil-based TPU/PA blends [8]. This is mainly because PA and TPU possess close hydrophobic/hydrophilic balances.

Compared to original TPU, the mechanical performances (yield stresses and Young's modulus) of all TPU/DAPA blends have been vastly improved and are also associated with others positive effects such as a reduction of the viscosity in the melting sate, a higher bio-based content and a lower density of the final systems. These renewable multiphase systems may find application in different fields such as the automotive or coating industry. TPU/DAPA blends are in good agreement with the concept of sustainable development.

15.1.2 Applications of Blends of Polydimethysiloxane Rubber with Polyurethane

Thermoplastic polyurethanes (TPU) have attracted attention from prominent polymer industries. Their use includes items for the biomedical, automotive, and footwear, as well as coating adhesive industries, as they possess excellent mechanical strength properties, higher durability, excellent fatigue resistance, and higher abrasion resistance as compared to many other synthetic polymers. Polydimethylsiloxane (PDMS), a borderline polymer between organic and inorganic polymers, finds a variety of applications over a wide range of temperatures. They are used across wire, cable, and biomedical applications, due to its excellent low-temperature flexibility, lower surface energy, and very high thermal stability. It also possesses excellent bio-inertness. [9] TPU and PDMS are biomaterials with excellent biocompatibility and biostability.

Blending of polymers combines the desirable properties of the constituent polymers and also improves additional properties, uncommon to the individual blend constituents, because of synergism. This typical blend imparts excellent mechanical strength, fatigue resistance, bio-inertness, and biocompatibility combined with easy processability.

One of the foremost advantage of blending is the formation of a thermoplastic elastomer material. It has properties of an elastomer at room temperature and also allows processing in conventional polymer processing equipments which can include injection molding machine or an extruder. Alterations in the end use property requirements can be made by adjusting the ratio of the blend components. Being a thermoplastic elastomer, material the need for crosslinking of siloxane component can be avoided, which is an added advantage. There is a chance of migration of curing agents to surrounding tissues during long term implantation period of the material which can make them toxic. This can result in rejection of implant from the body.

One of the major hurdles in blending urethanes with siloxanes is the formation of an immiscible blend which leads to phase separation. This can be answered by undertaking the reactive blending or in situ compatibilization technique by using a copolymer capable of making specific interactions with the blend components. Such a blend can be used for a variety of biomedical applications where the biocompatibility, strength, and toughness of TPU and the inertness, biocompatibility, and flexibility of PDMS are required. Some major applications include:

- Soft tissue replacements
- Artificial organs
- Cartilage replacements
- Membranes for guided tissue regeneration
- Scaffolds for tissue engineering

Implantation biomaterials usually need microporous structures. A controlled network of porosity will improve the invasion and proliferation of cells on to the biomaterial surface in a microporous implant [10].

There are also various techniques for the development of microporous structures in implants. In the blend system, the role of stress concentrator is played by a well dispersed minor phase in the matrix. Porosity is introduced by drawing samples. Micro-cracks are initiated at points of high stress concentration. They are at interface between the two phases. Further growth occurs by crazes propagating into the major phase of blends leading to shear yielding. Rates of craze initiation and growth depend upon the rate of applied stress [11].

Hydrophilicity/hydrophobicity of the material play an important role in controlling the cell proliferation and development over the membrane. Wettability information about the surfaces are provided by contact angle measurements which help in assessing these parameters.

Microporosity is not a necessary condition for cells to proliferate in the blend sample surfaces. They show a flattened morphology that provides good adherence to the surface. In a blend sample without any surface microporosities, the cells however exhibit round morphology, showing low cell adherence and proliferation. In order to mimic the topological and microstructure characteristics of the extracellular matrix, a biomaterial surface must possess:

- High degree of porosity
- High surface-to-volume ratio
- High degree of pore interconnection
- Appropriate pore size
- Geometry control [12].

Melt blending of small quantity of PDMS fluid with TPU has a significant positive impact on the coefficient of friction (CoF) and on the mechanical and wear properties of the polyurethanes, largely due to modification of the bulk by PDMS as it improves packing efficiency (antiplasticization) in the polyurethane soft domain.

Mining industry applications demand high performance from these materials with respect to their mechanical performance and wear resistance. Thermoplastic polyurethanes are mainly used because of the phase separation of their hard and soft segments and physical crosslinking through the hard crystalline domains. Further improvement of the properties (PDMS > 3%) is undesirable as in many applications, abrasion is the predominant cause of component failure.

TPU and PDMS rubber are usually not compatible because of differences in their structures, wide dissimilarities in their surface energies, and noninteraction between the constituents. Hence, compatibilization of the blend with the help of a third component is essential. Ethylene-co-methacrylate (EMA) can be used effectively. Around 5% by weight of ethylene-co-methacrylate (EMA) copolymer can be used with a blend of 1:1 (TPU:PDMS). It forms a hydrogen bond with urethane and becomes completely soluble in the system. It reacts with PDMS rubber chemically to form a graft copolymer (EMA-g-PDMS rubber blend system). So we get improved physio-mechanical and dynamic properties in an alloy. There is a huge scope in studying the effect of EMA as a polymeric reactive compatibilizer for controlling the phase morphology of such systems.

15.1.3 Applications of Blends of Micro Fluidic Devices with Polyurethane

PDMS has been widely used in microfluidics for rapid prototyping because:

- It is easy to fabricate
- It bonds strongly to glass and PDMS substrates

- It has good optical transparency
- There are good elastomeric properties.

Proteins being amphiphilic in nature, the intrinsic hydrophobic nature of PDMS causes significant protein adsorption from protein-containing fluids. Blood contains various plasma proteins at different concentrations and with different surface affinities. Fibrinogen one such protein adsorbs significantly more on hydrophobic than on hydrophilic surfaces [13] which potentiates the process of thrombosis including platelet adhesion and clot formation. Several attempts have been made to reduce protein adsorption on PDMS surfaces and micro-channels. Plasma [14], UV irradiation and gas phase modification have been used to make the PDMS surface hydrophilic. These modifications are however temporary and the surfaces revert back to the hydrophobic state. Polyurethanes [15] have been widely used in various blood contact applications such as the artificial heart, intra-aortic balloons pacemaker leads heart valves and hemodialysis membranes. Hydrophobic surfaces being water repellent provoke adverse reactions in blood contact. This can be improved by making urethane surfaces more hydrophilic [16]. In blood contact, plasma proteins that include albumin, hemoglobin, thrombin, fibrinogen (Fg), fibronectin, complement components, whereas immunoglobulins (IgG) gets adsorbed onto polymeric surfaces immediately. Adsorbed fibrinogen results in thrombus formation. Blood compatibility may be improved by controlling interfacial interactions so that nonspecific protein adsorption is prevented. This must happen when selective adsorption of proteins for a desired bioactive function is being promoted [17]. Limiting nonspecific adsorption approach includes modification of the surface with hydrophilic polymers like polyethylene oxide (PEO) and poly(2-methacryloyloxyethyl phosphorylcholine) (poly-MPC). PEO is well known for its protein and cell resistance properties. PEO-rich surfaces can be generated using adsorption, blending, and grafting [18]. Major challenges faced during surface modification include rapid depletion of reagents due to high surface-to-volume ratio that the modification chemistry has to take into account.

Traditional fabrication techniques like vertical dipping, rotating mandrel, rotating plate method are not suitable for replicating the intricate and detailed microscale features that are present in microfluidic devices. Polyurethane based micro fabrication involves injection molding [19], hot embossing, imprinting, plasma etching, and sacrificial material and reaction polymerization. These methods are very costly and the substrates produced are rigid and not transparent. Solvent casting is more suitable as the intermediate mold can be fabricated using photolithography and the fabrication equipment is low-cost. However, opaque substrates produced do not bond well and are difficult to connect with tubing.

Three major bonding methods have been developed to cater for various application specific factors:

- Convenience (dip coating)
- High bonding strength (micro-contact printing)
- Nano-structural features (oxygen plasma treatment).

Micro-contact printing delivers great performance results for most of the microfluidic applications that require micrometer channel sizes. Integration of interconnects during fabrication enhances the tubing processes for flow connection in microfluidic devices providing ability to withstand high pressures. Reduced levels of fibrinogen adsorption (about 80%) are observed on PEO modified PU devices in comparison with native PDMS and native PU. Successful modification of the inner device surfaces is observed along with a more biocompatible surface with potential blood clot formation resistance. Hence, the micro-contact printing technique to smaller dimension channels and testing of clotting behavior in surface modified PU channels can be quite promising possible future avenues that would definitely result to some unique and beneficial applications.

15.1.4 Applications of Blends of Polyacrylonitrile with Polyurethane

Polyacrylonitrile (PAN) is an important industrial material that has excellent physical and chemical properties. They lack segment mobility which results from intensive molecular orientation. Fibers manufactured from pure acrylonitrile pose problems like low dye affinity, low moisture regain, and high static charge. Several attempts have been made to solve such challenges by many scientists. Incorporation of a comonomer into the PAN chain [20] and blending with other polymers is one such attempt. PAN blends with side-by-side, sheath-core, and matrix-fibril morphology are another novel idea.

N,N-dimethylformamide is used to prepare solution blends of a modified PAN with PU ionomers. Usually urethane and ionomer blends show phase separation up to several micrometers. However, almost homogeneous morphology is obtained using cationomer blends. In PU cationomer blends, the tensile strength increases considerably to almost more than double the original strength of poly acrylonitrile. This happens when the blend concentration of PU cation approximately equals that of the sulfonate anion belonging to the modified PAN.

Only if certain specific interactions exist, the fiber forming PAN generally becomes compatible with other polymers [21]. The segmented PU ionomers are composed of soft and hard segments. Effects on the compatibility as well as properties with other polymers are mainly due to soft segment-hard segment phase separation, which mainly is controlled by the type of polyol and molecular weight of the polyol incorporated in it [22].

PU increases the membrane porosity in casting of a flat sheet of the blend. However, it also increases the blend's thermodynamic instability, hydrophilicity and surface roughness.

Polyurethane is one of the most versatile, biocompatible, biodegradable, and viscoelastic polymers which posesses two alternating hard and soft segments [23]. The hard segment is composed of aliphatic or aromatic diisocyanates, diols or diamine chain extenders. The latter segment consists of dihydroxy ordiamine terminated reactive oligomers, like polyethers, poly-esters, poly-butadienes, and poly-acrylates, with a wide molecular weight range [24]. The intermolecular hydrogen bonding between the hard crystalline and soft amorphous parts with the carbamate (−NH−CO−O−) network offers resistance to pH and temperature conditions and makes PU a suitable material for membrane preparation.

PU membranes are used in pervaporation industries for the separation of aliphatic hydrocarbons and water vapor permeability studies due to its flexible permeability and diffusivity. [25] These membranes have found applications in drug delivery systems, their antimicrobial properties and in antifouling studies. [26]. When combined with fillers like zeolite and silver doped fly ash they can be used for treating textile effluents and to remove arsenic along with microorganisms [27]. Polyacrylonitrile is a very useful polymer in membrane industry due to its low fouling nature, despite its brittleness in dry conditions, because of hydrophilicity as compared to other commercial polymers such as PE, PP, PSF, PES, good thermal stability, commercial availability, resistance against organic solvents and improved chemical stability against chlorine, sodium hypochlorite, and sodium hydroxide.

15.1.4.1 Antifouling properties of the membrane

Usually as the membrane become more porous upon increasing the PU concentration in the blend, two phenomena occur:

- Particles enter into the pores causing pore blocking

- As the surface roughness of the membranes increases, the fouling of the membrane increases.

These two effects act in tandem leading to an increase in flux decline. This is due to an increase in the hydrophilicity of the membranes with increasing PU concentration.

The concept can be summarized as follows:

1. Inclusion of PU made the membrane more porous and the permeability of the membrane was increased.
2. PU made the membrane more hydrophilic and the contact angle was reduced.
3. The surface roughness of the membrane was increased with increasing PU concentration in the blend and made them more prone to fouling.
4. Good compatibility exhibits between the PAN and PU membranes.

15.1.5 Applications of Blends of Urea-Formaldehyde Resin with Polyurethane

Urea and PU both participate in the addition and condensation reactions with formaldehyde to generate a blend with improved toughness, by introducing soft PU segment chemically into Urea-Formaldehyde (UF) system. The toughness of UF can be improved without a decrease of strength by introducing PU, resulting from the formation of reaction-induced microphases separated morphology and the strong chemical interactions between the UF matrix and PU. The toughening mechanism of the blend corresponded to the energy-dissipation mechanisms.

Urea-formaldehyde resin is a very important type of amino plastic resin, based on the manifold reaction between two monomers, urea and formaldehyde. Urea-formaldehyde can be used as wood glue and surface coatings and can be used to improve the wet strength of paper and the crease-resistance of textiles. UF resin powders can also be molded and used as containers, electric fittings, bottle caps, lavatory seats, etc. [28].

Being a thermosetting resin, urea-formaldehyde is very brittle and easily fractures during application. Therefore, despite the high strength and low-cost of UF, this drawback largely limits its usage. One approach to improve its toughness is reducing the mole ratio of formaldehyde/urea, however, this leads to a decrease of crosslinking degree and thus results in inferior performance of the resin, relating to its mechanical strength and water resistance [29]. Resin modifiers can also be used to toughen UF. Polyurethane elastomers, because of their unique property of combining high strength, high toughness, together with high damping, are widely used in many fields. The soft segments in PU largely control the toughness and low-temperature properties, while the crystalline part of segment gives high modulus, hardness, and tear strength of the elastomer [30].

Improvement in toughness is achieved by the use of UF/PU blends. NH groups on the polyurethane molecule participate in addition and condensation reactions with formaldehyde. The curing reactions are always one-step reactions for PU and UF. As PU content increases both the reaction activation energy (E) and end temperature of cure reaction increase. The crosslinking degree in blend increases considerably so it shows a lower modulus, but a higher T_g compared with neat UF. An increase in elongation at break and notched impact strength of the blend is observed although the tensile strength still remains constant at a relative high level of PU. Homogeneous micro phase separated morphology is observed for the blends, which improves the toughness of the blend. The toughening mechanism of the blend corresponds to the energy-dissipation mechanism [31].

15.2 APPLICATIONS FOR POLYURETHANE INTERPENETRATING POLYMER NETWORKS

An interpenetrating polymer network (IPN) can be defined as a blend of two or more polymers in which at least one of the polymer has a cross-linked structure [32]. An IPN is formed when two or more different polymer chains get entangled with each other (see Fig. 15.1). They do not form permanent bonds with each other and these two chains cannot be separated without breaking these bonds [33].

IPN technology helps combine properties of two different polymers in a single polymer network.

15.2.1 Acrylics

Polyurethane and acrylic IPN have blended the properties of the individual networks quite well. Some of the properties that the IPN have helped

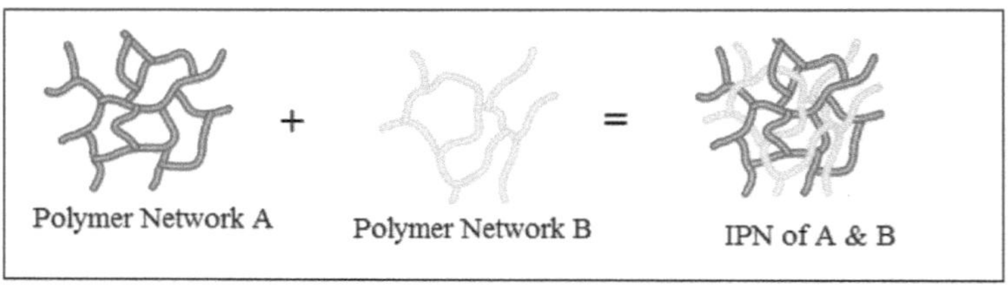

Figure 15.1 Schematic representation of IPN formation.

in improving are impact resistance, increased T_g and also higher Young's Modulus. Let's look at some of the application of these polyurethane-acrylic based IPNs [34].

15.2.1.1 Acrylic Plastisol

Plastisols have a wide range of application in the coatings industry. They are used for dip coated tools, dishwasher, racks and also hard gloves. Gloves are generally cold dip coated with viscosity and composition affecting the thickness and strength of coating. Plastisols show maximum application in textiles as they are generally opaque on dark garments. Adhesive coated films for decorative uses, decals, wall covering, mat casting, and floor coverings can be manufactured [35,36]. PVC has been leading the plastisol market for a long time. Phthalate based plastisols have also been used on a wide scale. Environmental effects of these plastisols are mainly ozone depletion, carcigenocity and endocrine disruptions. An acrylic plastisol can act as replacement to the PVC and phthalate based plastisol. They show workability similar to PVC. They are halogen free and also have good 'pot life' (the period for which two mutually reactive chemicals remain usable when mixed).

Applications: *Toys, Textile Coatings, Protective Coatings, Automotive* [37].

15.2.1.2 PU-Acrylic IPNs for Urethral Application

Biomaterials have been used for centuries for various devices in the urinary tract. Common biomaterials devices used in urinary tract include urethral and supra pubic catheters, urethral and urethral stents, and nephrostomy tubes. The biocompatibility of materials used in urinary tract is guiding parameter for its use. Biomaterials help in achieving this compatibility as to date no synthetic material has shown absolute biocompatibility. The ideal biomaterial should show properties such as biological inertia, chemical stability in urine, resistance to infection and encrustation, excellent long term urinary flow, stability following placement, no significant discomfort to the patient and availability at an affordable price [38,39].

Urethral encrustation is obstruction of urethral devices by calcified debris. The problem of encrustation stems from infection by urease-producing bacteria that colonize the devices forming extensive biofilms [40,41]. PMMA/PU IPN can be used as a biomaterial for urethral devices. This IPN combination gives a slower rate of encrustation. Also, the tensile strength of urethral devices increases using PMMA/PU IPN. Thus, further research may lead to development of acrylic based PU IPNs [42].

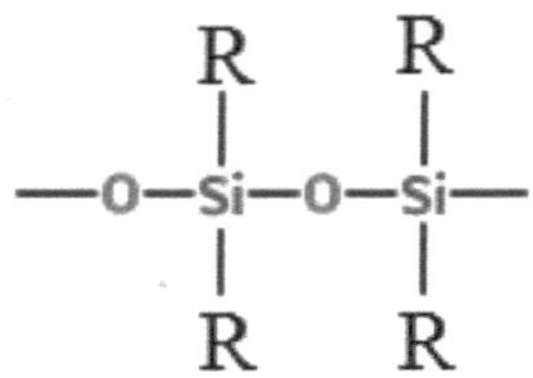

Figure 15.2 Structure of silicone.

15.2.1.3 Drug Delivery

Current research in drugs concerns targeted and constant rates of drug delivery. A PU-acrylic based IPN can be used for controlling the rate of drug delivery. Here, one network will form suspension particles while the other polymer network that contains the drug acts as a condensation polymer [43].

15.2.2 Silicones

Silicones are highly versatile materials (see Fig. 15.2). They can seen as a 'Jack of All Trades'. The silicone resins show properties such as:

- moisture and dirt resistance
- water repellence
- gloss
- blood contact compatibility.

Silicone has very wide application in coatings as binders and modifiers. Silicone resin offers good weathering resistance and durability to the product. It also helps provide water repellent properties to stone and brick. As compared to organic resin, they show a higher temperature resistance making them for useful in industrial applications.

Silicone and urethane copolymers have been synthesized that show good blood contact as well as reasonable mechanical properties. Cardiothane was initially made from a silicone/urethane copolymer. Its composition was 10% silicone. However, its application became limited due to the unfavorable processing parameters and economic issues. Silicone/urethane copolymers are generally prepared by the synthesis of PU and crosslinking with siloxane.

The development of polyurethane and silicone based IPNs has helped overcome problems related to copolymers. In an IPN, both polymers exhibit their individual properties [44,45]:

1. Silicone
    - Can be processed using conventional machines for TPU
    - Lower oxygen permeability

- Better wear resistance
- Reduce blood/polymer interactions

2. Polyurethane
   - It has silicone release properties
   - Improved dielectric properties
   - High temperature performance
   - Improved elastic recovery
   - Improved stress and solvent resistance

Applications of a silicone/polyurethane based IPN have been mainly for biocompatible based applications. An aliphatic PU-based IPN has been used in long term implants while aromatic PU-based IPNs are used in short term implants [46,47].

Pressure sensitive adhesive is an adhesive that adheres to the substrate when applied with certain pressure, e.g., thumb pressure. The main force for adhesion is the polar attraction between the adhesive and the substrate. A silicone polyurethane semi-IPN is incorporated with acrylic PSA to form thermally stable PSA. The resultant PSA formed showed better adhesion as well as water resistance [48].

15.2.3 Polystyrene

Over last century, petroleum products have been used on a large scale. The extensive use of these petroleum products has led to its depletion. Due to this reason, use of renewable sources has become important. The synthesis of polyurethane involves use of petroleum based polyols and cyanides.

Castor oil based Polyurethanes have attracted a large amount of research opportunities. They also show superior properties as compared to traditional PU:

- Better water resistance
- Better insulation
- Better heat and humidity resistance

Recent research has shown that the castor oil based PU and polystyrene IPN have shown better properties [49,50]. Polyurethane imparts abrasion resistance to the polymer structure with castor oil improving the flexibility and hydrolysis resistance of polymers. However, the presence of this fatty acids increases curing time; thus polystyrene is added which improves mechanical properties along with increase in curing time [51]. Polystyrene/PU IPN when grafted with polyethylene oxide has shown blood compatibility. They form a micro domain structure consisting of hydrophilic and hydrophobic block copolymers. This micro domain results antithrombogenicity properties in PS/PU IPN. These polymer networks could be used for preparation of blood contacting as well as organ reconstruction applications. It will help in generating biocompatible materials [52,53].

The application of this IPN can be:
- Marine Industry
- Furniture
- Electronics
- Packaging

15.2.4 Vinyl Ester

Vinyl ester resins are polymers that have properties in between polyester and epoxy resin. They are generally prepared by condensation of epoxy with unsaturated carboxylic acids.

Vinyl esters have good impact strength due to the presence of a double bond that gives flexibility to the chain. They have excellent resistance to water, alkali, and solvents. They have poor resistance to acids. Vinyl esters have better resilience than epoxies. The presence of unsaturation in terminal groups makes the chain reactive and thus, vinyl esters are favorable for composite applications. In glass fiber composites, corrosion resistance can be obtained due to presence of vinyl esters [54,55].

Polyurethane based composites have shown large application as damping materials. Damping materials are used to reduce the vibration and mechanical energy that can damage materials [56]. A PU/VER IPN has shown better damping than PU alone. VER gives extra strength and flexibility required to create damping. Methyl methacrylate (MMA) and styrene are also added to VER to help in curing. One study has found that presence of MMA gives better damping effect than styrene [57].

A polyurethane/VER IPN can also be used in corrosion protection. VER with its higher solvent, alkali, and water can play a significant role. The PU/VER IPN needs to be further investigated for additional applications.

15.2.5 Polyacrylamide

Polyacrylamide is a water soluble polymer prepared from an acrylamide monomer. Polyacrylamides are part of diverse class of water soluble and water swellable polymers. They possess hygroscopic properties which has led to their application in dewatering and related applications. They are used as flocculants, viscosity inducer (thickner), and in many solid–liquid separation processes. They have been used in municipal waste management and industrial waste management, and for deinking in the paper recycling process.

One of the applications in the paper industry is for strengthening paper and acting as a binder. Polyacrylamides have excellent binding properties but the swollen gel that are formed are weak [58,59]. A PU/PAA IPN is used to improve the strength of the polymer gel. The swollen gel used is a combination of PU/PAA IPN and AB crosslinked polymer. Depending upon the combination of IPN and ABCP, the properties of gel are affected. When AB CP and PU/PAA IPN are used individually, the strength is moderate. PU/PAA IPN and AB CP in combination lead to a synergistic effect. The strength of PAA gel is increased by nearly 20%. Paper processing may use it as a strengthening agent and also dewatering aid [60].

REFERENCES

[1] M. Patel, I. Bartle, C. Bastioli, K. Doutlik, J. Ehrenberg, D. Johansson, et al., Towards the integration of renewable raw materials in EU climate policy, Part 1. Agro Food Ind. Hi-Tech 13 (6) (2002) 28–31.

[2] C. Bueno-Ferrer, E. Hablot, F. Perrin-Sarazin, M.C. Garrigós, A. Jiménez, L. Avérous, Structure and morphology of new bio-basedthermoplastic polyurethanes obtained from dimeric fatty acids, Macromol. Mater. Eng. 297 (8) (2012) 777–784.

[3] E. Hablot, B. Donnio, M. Bouquey, L. Avérous, Dimer acid-based thermoplastic bio-polyamides: reaction kinetics, properties and structure, Polymer 51 (25) (2010) 5895–5902.

[4] E. Hablot, A. Tisserand, M. Bouquey, L. Avérous, Accelerated artificialageing of new dimer fatty acid-based polyamides, Polym. Degrad. Stab. 96 (6) (2011) 1097–1103.

[5] Hablot E., Matadi R., Ahzi S., Avérous L. Renewable biocomposites of dimer fatty acid-based polyamides with cellulose fibres: thermal.

[6] X.-D. Fan, Y. Deng, J. Waterhouse, P. Pfromm, Synthesis andcharacterization of polyamide resins from soy-based dimer acids and different amides, J. Appl. Polym. Sci. 68 (2) (1998) 305–314.

[7] W. Li, J. Liu, C. Hao, K. Jiang, D. Xu, D. Wang, Interaction of thermoplasticpolyurethane with polyamide 1212 and its influence on the thermal and mechanical properties of TPU/PA1212 blends, Polym. Eng. Sci. 48 (2) (2008) 249–256.

[8] J.T. Haponiuk, Phase transitions of polyamide 6/polyurethane blendscompatibilized by copolyurethaneamides, J. Therm. Anal. Calorim. 60 (1) (2000) 45–52.

[9] P.K.B. Davis, S.M. Jones, The complications of silasticimplants. Experiencewith 137 consecutive cases, Br. J. Plast. Surg. 24 (1971) 405–411.S. Kole, R. Santra, A.K. Bhowmick, Rubber Chem. Technol. 67 (1994) 119.

[10] L. Hench, E. Ethridge, Biomaterials: An Interfacial Approach, Academic Press, New York, NY, USA, 1982.

[11] M. Xanthos, C. Chandavasu, K.K. Sirkar, C.G. Gogos, Melt processed microporous films fromcompatibilized immiscibleblends with potential as membranes, Polym. Eng. Sci. 42 (4) (2002) 810–825.

[12] B.-S. Kim, I.-K. Park, T. Hoshiba, et al., Design of artificial extracellular matrices for tissue engineering, Progr. Polym. Sci. 36 (2) (2011) 238–268.

[13] T.A. Horbett, P.K. Weathersby, Adsorption of proteins from plasma to a series of hydrophilic-hydrophobic copolymers. I. Analysis with the in situ radioiodination technique, J. Biomed. Mater. Res. 15 (1981) 403–423.

[14] X. Ren, M. Bachman, C. Sims, G.P. Li, N. Allbritton, Electroosmotic properties of microfluidic channels composed ofpoly(dimethylsiloxane), J. Chromatogr., Biomed. Appl 762 (2001) 117–125.

[15] D.J. Lyman, et al., The development and implantation ofa polyurethane hemispherical artificial heart, Trans. Am. Soc. Artif. Internal Organs 17 (1971) 456–463.

[16] A. Takahara, J.I. Tashita, T. Kajiyama, Microphase separatedstructure, surface composition and blood compatibility of segmented poly(urethane ureas) with various soft segment components, Polymer 26 (1985) 987–996.

[17] J.L. Brash, Exploiting the current paradigm of blood-materialinteractions for the rational design of blood-compatible materials, J. Biomater. Sci., Polym. Ed. 11 (2000) 1135–1146.

[18] J. Lee, H. Lee, J. Andrade, Blood compatibility of polyethylene oxide surfaces, Prog. Polym. Sci. 20 (1995) 1043–1079.

[19] Y.H. Kim, D.K. Han, K.D. Park, S.H. Kim, Enhanced blood compatibility of polymers grafted by sulfonated PEO via a negativecilia concept, Biomaterials 24 (2003) 2213–2223.

[20] A. Folch, S. Mezzour, M. Du, Stacks of microfabricatedstructures as scaffolds for cell culture and tissue engineering, Tissue Eng. (2000) 207–214.

[21] D.J. Poyton, Acrylic Fiber, The Production of Man-Made Fiber, Text. Prog 8 (1976) 51.

[22] N.S. Schneider, C.S. Paiksung, Transition behavior and phase segregation in TDI polyurethanes, Polym. Eng. Sci. 17 (1977) 73.

[23] E.A. Collins, J. Bares, F.W. Billmeyer Jr, Polyurethane Preparation, Experiments in Polymer Science, Wiley, New York, 1973 347

[24] I. Yilgor, E. Yilgor, Hydrophilic Polyurethane urea membranes: Influence of soft block composition on the water vapor permeation rates, Polymer 40 (1999) 5575.

[25] L. Lin, Y. Kong, K. Xie, F. Lua, R. Liu, L. Guo, et al., Polyethylene glycol/polyurethane blend membranes for gasoline desulphurization by pervaporation technique, Sep. Purif. Technol. 61 (2008) 293.

[26] M. Mandru, C. Ciobanu, M.E. Ignat, M. Popa, L. Verestiucand, S. Vlad, Sustained release of Nystatin from polyurethane membranes for biomedical applications, Digest. J. Nanomater. Biostruct. 6 (3) (2011) 1227.

[27] M.A. Frommer, R.M. Massalem, Mechanism of membrane formation.VI. Convective flows and larger void formation during membrane precipitation, Ind. Eng. Chem. Prod. Res. Dev. 12 (4) (1993) 328.

[28] A.S. Singha, V.K. Thakur, Synthesis and characterization of short Grewia optivafiber-based polymer composites, Polym. Compos. 31 (3) (2010) 459–470.

[29] E. Roumeli, E. Papadopoulou, Synthesis, characterization and thermal analysis ofurea–formaldehyde/nano SiO_2 resins, Thermochim. Acta 527 (2012) 33–39.

[30] J. Guan, Y. Song, Y. Lin, Progress in study of non-isocyanate polyurethane, Ind. Eng. Chem. Res. 50 (11) (2011) 6517–6527.

[31] J. Wang, L. Ye, Structure and properties of polyvinyl alcohol/polyurethane blends, Compos. Part B Eng. 69 (2015) 389–396.

[32] INTERPENETRATING POLYMER NETWORKS by L. H. Sperling and R. Hu, Polymer Blends Handbook Kluwer Academic Publishers, 2003.

[33] <http://www.polymerics.de/technology/ipn_en.html>.

[34] M.-S. Lin, G.-A. Chiu, J. Polym. Res. 3 (3) (1996) 165–171.

[35] Polyvinyl Chloride and its copolymers in Plastisol Coatings, Coatings Handbook (3rd Edition),59.3.

[36] <http://www.unionink.com/unionink/articles/geninfo.html>.
[37] Acrylic Plastisols, Dianal America Inc.
[38] D.T. Beiko, B.E. Knudsen, J.D. Watterson, P.A. Cadieux, G. Reid, J.D. Denstedt, J. Urol. 171 (2004) 2438–2444.
[39] J. Denstedt, A. Atala (Eds.). Biomaterials and Tissue Engineering in Urology.
[40] <http://medical-dictionary.thefreedictionary.com/encrustation>.
[41] R.J. Broomfield, S.D. Morgan, A. Khan, D.J. Stickler, J. Med. Microbiol. 58 (2009) 1367–1375.
[42] K.F. Mueller, S.J. Heiber, Gradient-IPN-modified hydrogel beads: their synthesis by diffusion-polycondensation and function as controlled drug delivery agents, J. Appl. Polym. Sci. 27 (1982) 4043–4069.
[43] D.S. Jones, M.C. Bonner, S.P. Gorman, M. Akay, P.F. Keane, J. Mater. Sci. Mater. Med. 8 (1997) 713–717.
[44] B.B. Barry, IPN Modified Silicone Thermoplastic. Petrarch System Inc.
[45] T. Easton. Silicones in Coatings. Dow Corning Ltd, Barry (Wales).
[46] M.W. Mazurek. Silicone copolymer networks and interpenetrating polymer networks.
[47] <http://www.wacker.com/cms/en/products/product_groups/silicone_resins.jsp>.
[48] P.-S. Kim, S.-W. Lee, J.-W. Park, C.-H. Park, H.-J. Kim, Synthesis and characterization of thermally stable acrylic PSA using silicone urethane methacrylate with a semi IPN structure, J. Adhes. Sci. Technol. 28 (2013) 15–30.
[49] D.S. Lee', S.C. Kim, Polyurethane interpenetrating polymer networks (IPN's) synthesized under high pressure., Morphology and Tg behavior of polyurethane-polystyrene semi-IPN's and linear blends, Macromolecules 17 (1984) 268–272.
[50] J.V. Dave, S.P. Hasmukh, Synthesis and characterization of interpenetrating polymer networks from trans esterified castor oil based polyurethane and polystyrene. J. Saudi Chem. Soc. (2013).
[51] S. Bai, D.V. Khakhar, V.M. Nadkarni, Mechanical properties of simultaneous interpenetrating polymer networks of castor oil based polyurethane and polystyrene, Polymer. 38 (17) (1997) 4319–4323.
[52] J.H. Kim, M.J. Song, H.W. Roh, Y.C. Shin, S.C. Kim, The in vitro blood compatibility of poly(ethylene oxide)-grafted polyurethane/polystyrene interpenetrating polymer networks, J. Biomater. Sci. Polym. Ed. 11 (2) (2000) 197–216.
[53] Y.C. Shin, D.K. Han, Y.H. Kim, S.C. Kim, Antithrombogenicity of hydrophilic polyurethane-hydrophobic polystyrene IPNs II. In vitro and ex vivo studies, J. Biomater. Sci. Polym. Ed. 6 (3) (1995) 281–295.
[54] F.A. Cassis, R.C. Talbot, Polyester and vinyl ester resin, in: S.T. Peters (Ed.), Handbook of Composites, Chapman & Hall, London, 1998.
[55] <http://www.moldedfiberglass.com/materials/vinyl-ester-resins>.
[56] <http://www.sorbothane.com/what-material-is-best-for-damping.aspx>.
[57] C. Qin, D. Tang, J. Cai, J. Zhang, W. Cai, X. Sun, Study on polyurethane/vinyl ester IPN damping materials. J. Mater. Sci. Technol. 19(1) (2003).
[58] <http://www.zlpam.com/wp-content/uploads/2011/11/About-Polyacrylamide-Fact-Sheet.pdf>.
[59] Acrylamide Polymers, Volume I, Encyclopaedia of Polymer Science and Technology. Copyright John Wiley & Sons, Inc.
[60] S.H. Baek, B.K. Kim, Synthesis of polyacrylamide/polyurethane hydrogels by latex IPN and AB crosslinked polymers, Colloids and Surfaces A: Physicochem. Eng. Aspects 220 (2003) 191–198.

CHAPTER 16

Mechanical and Dynamic Mechanical Properties of Polyurethane Blends and Interpenetrating Polymer Networks

Nikesh Samarth[1], Vinayak Kamble[1], Ajay V. Rane[3], Vayyaprontavida K. Abitha[2], Omdeo Gohatre[4] and Krishnan Kanny[3]
[1]Institute of Chemical Technology, Mumbai, India
[2]Mahatma Gandhi University, Kottayam, Kerala, India
[3]Durban University of Technology, Durban, South Africa
[4]North Maharashtra University, Jalgaon, Maharashtra, India

Contents

16.1 INTRODUCTION

The mechanical properties of polymers are highly dependent on temperature and on the time-scale of any deformation; polymers are viscoelastic

Polyurethane Polymers: Blends and Interpenetrating Polymer Networks
DOI: http://dx.doi.org/10.1016/B978-0-12-804039-3.00016-6

and exhibit some of the properties of both viscous liquids and elastic solids. At low temperatures or high frequencies, a polymer may be glass-like, with a value of Young's modulus in the region 109—1010 Pa, and it will break or yield at strains greater than a few per cent. At high temperatures or low frequencies it may be rubber-like, with a modulus in the region 10^5—10^6 Pa, and it may withstand large extensions of order 100% or more with no permanent deformation.

Fig. 16.1 shows how Young's modulus of a polymer varies with temperature in the simplest case. At still higher temperatures, the polymer may undergo permanent deformation under load and behave like a highly viscous liquid. In an intermediate temperature range, called the glass-transition range, the polymer is neither glassy nor rubber-like; it has an intermediate modulus and has viscoelastic properties. This means that, under constant load, it undergoes creep, i.e., the shape gradually changes with time, whereas at constant strain it undergoes stress—relaxation, i.e., the stress required to maintain the strain at a constant value gradually falls [1].

Dynamic mechanical analysis (DMA) is a technique that does not require a lot of specialized training to use for material characterization. It supplies information about major transitions as well as secondary and tertiary transitions not readily identifiable by other methods. It also allows characterization of bulk properties directly affecting material performance. Depending on whom you talk to, the same technique may be called DMA, forced oscillatory measurements, dynamic mechanical thermal analysis (DMTA), dynamic thermomechanical analysis, or even dynamic rheology. This is a function of the development of early instruments by different specialties (engineering, chemistry, polymer physics) and for different markets.

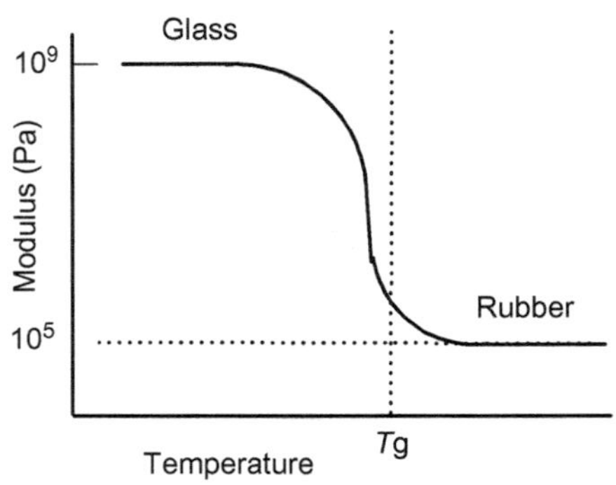

Figure 16.1 Young's modulus versus temperature for a model polymer. T_g is the glass-transition temperature.

The interest in studying curing behavior and curing materials may be even greater. The high sensitivity of the DMA and its ability to measure viscosity quickly make it one of the most valuable tools for studying curing systems. We personally have found it more useful even than DSC, although characterizing a thermoset without having both techniques available would be inefficient at best. Thermoset are materials that change chemically on heating. This can occur in one step or in several, and those multiple steps do not need to be immediately sequential. Thermosetting reactions can be classed into those that involve the loss of a molecule on reacting, the condensation resins, and those that join 'mers' together without changes in the repeat structure, the addition condensation reactions. This classification is based on the reaction mechanism of the polymers and is manifested in very different kinetics.

The DMA's ability to give viscosity and modulus values for each point in a temperature scan allows us to estimate kinetic behavior as a function of viscosity. This has the advantage of telling us how fluid the material is at any given time, so we can determine the best time to apply pressure, what design of tooling to use, and when we can remove the material from the mold [2].

In this chapter we are focusing on the general terms relating to mechanical and dynamic mechanical properties of polyurethane (PU) blends and interpenetrating network polymers (IPNs).

16.2 DEFINITIONS OF MECHANICAL PROPERTIES AND DYNAMIC MECHANICAL PROPERTIES

In this chapter, mechanical and dynamic mechanical properties of a PU blend over a wide range of experimental conditions are discussed. The stress-strain curves of PU blends were elucidated.

Rheology: Rheology is the science that studies the deformation and flow of materials in liquid, melt, or solid form in terms of the material's elasticity and viscosity. This is accomplished by applying a precisely measured strain to the sample to deform it, and accurately measuring the resulting stress developed in the sample. The developed stresses are related to material properties through Hooke's and Newton's laws [3].

Elasticity: Elasticity is the ability of a material to store deformational energy, and can be viewed simply as the capacity of a material to regain its original shape after being deformed.

Viscosity: Viscosity is a measure of a material's resistance to flow.

Viscoelasticity: Materials respond to an applied displacement or force by exhibiting either elastic or viscous behavior, or a combination of these, called viscoelastic behavior. Most polymers are viscoelastic, their mechanical properties showing a marked time- and temperature-dependence.

Stress: Stress is a distribution of forces over an infinitesimal area.

Strain: Strain is a measure of a body's change in shape. The change in strain with time is the shear rate. The change in strain with time is the shear rate.

Hooke's law: Hooke's law defines the mechanical behavior of an ideal solid, relating the applied strain (ε or γ) to the resultant stress (σ or τ) through a factor called the modulus (E or G). Thus, $\sigma = E\varepsilon$ (tension, bending) or $\tau = G\gamma$ (shear). The modulus is a measure of the material's stiffness (i.e., its ability to resist deformation). The linear region in which the modulus does not change when the strain is changed is called the Hookean region.

Newton's Law: Newton developed a relationship similar to Hooke's law for ideal viscous fluids, relating the stress (τ) linearly to the shear rate ($d\gamma/dt$). Thus, $\tau = \eta d\gamma/dt$, where η is the coefficient of viscosity. A fluid is Newtonian if, when sheared, its viscosity does not depend on shear rate.

Theory of Linear viscoelasticity: The foregoing relationships are tied together by an elegant, overall theory of viscoelasticity [1]. So long as the measurements of stress and strain are made in the linear viscoelastic region, any of various viscoelastic functions can be calculated from another, and the various superposition methods can be employed.

Newtonian Versus Non-Newtonian Flow: Water and mineral oil are Newtonian, but not many other materials behave this way. Most fluids are non-Newtonian, their viscosity changing with changes in shear rate. This means the expression $\tau = \eta d\gamma/dt$ is only valid at a particular shear rate. Changing the shear rate changes the viscosity. Six types of non-Newtonian behavior are known, differing by how the viscosity changes with shear rate, and whether or not the flow pattern is time-dependent. The time-independent types are called Bingham, plastic, pseudoplastic, and dilatant; the time-dependent types are called thixotropic and rheopectic.

16.2.1 Mechanical Properties

Polymers mostly behave as viscoelastic materials, which shows their intermediate position between viscous liquids and elastic solids. The elastic solid follows Hooke's law, i.e. stress is proportional to strain. Mechanical properties of solid polymers have been divided into two approaches: (1) microscopic description of the particular facet of the polymer behavior; and (2) molecular description using chemical composition and physical structures. Both approaches are important to understand the mechanical properties and hence this chapter will highlight these areas [4].

16.2.2 Dynamic Mechanical Properties

DMA is a shortened form of dynamic Mechanical Thermal Analysis (DMTA). It is also used term used for Dynamic Mechanical Rheological Testing (DMRT). Most of time DMRT is preferred because this mode of thermal analysis is rooted in the rheological science [3,4].

16.3 EXPERIMENTAL METHODS AND APPARATUS FOR MEASUREMENT OF MECHANICAL AND DYNAMIC MECHANICAL PROPERTIES

16.3.1 Mechanical Interfacial Properties Tester

The mechanical interfacial properties of DGEBA/PU blend measure by Instron tester (Model 125). The fracture toughness parameters, i.e., (KIC) critical stress intensity factor is determined by the state of stress in the vicinity of the tip of crack just for function of specimen geometry, the geometry of crack, and on the basis of load applied of linear elastic fracture mechanics. Soo-Jin Park et al. KIC characterized by a single-edge-notched (SEM) beam fracture toughness test (ASTM E399) in three-point bending flexure [5]. A particular size specimen that had dimensions of $6.0 \times 1.2 \times 0.6\ cm^3$ and by using a diamond saw notches were taken, approximately half of depth of the specimen. A span-to-depth ratio 4:1 and cross-head speed of 1mm/min were taken.

16.3.2 Tensile Test Purpose

The primary use of the testing machine is to build the stress-strain relationship of material [6]. A tensile test determines the strength of the material subjected to a simple stretching operation. Typically, standard dimension test samples are pulled slowly (static loading) and at uniform rate in a testing machine while the *strain* (the elongation of the sample) is defined as: Engineering Strain $= \varepsilon =$ (change in length)/(original length) $=$ L/L0 and the *stress* (the applied force divided by the original cross-sectional area) is defined as: Engineering Stress $= \sigma =$ (applied force)/(original area) $=$ P/A0. The aim of the test is to assess some mechanical characteristics of testing material: its elasticity, ductility, resilience and toughness. There are two kinds of typical diagrams: with distinct yielding limit without distinct yielding limit as shown in Figs. 16.2–16.4.

Below are the useful terms related to mechanical properties:

Elastic Limit (proportional limit): The highest magnitude of stress for which the stress and strain are proportional to each other.

Elastic modulus (Young's modulus): The ratio of stress to strain below the elastic limit.

Elongation: The strain at fracture expressed as a percentage; this is a measure of the ductility of the material.

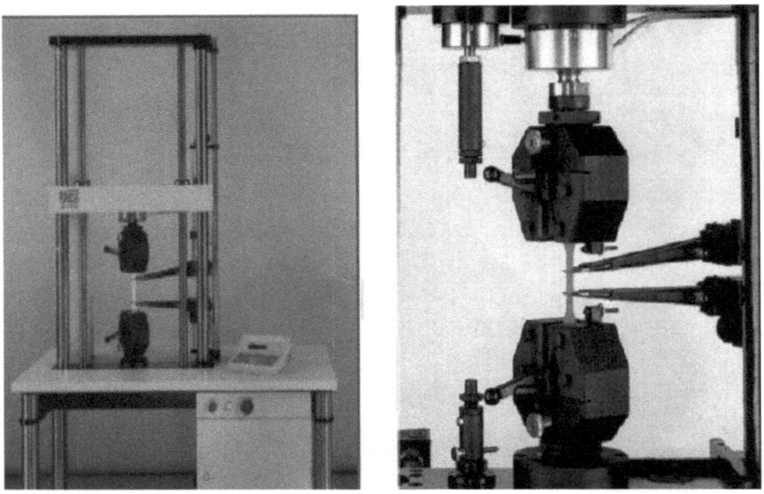

Figure 16.2 Universal testing machine UTS 100K has Self-locking jaws, a tested sample and extensometer.

Figure 16.3 Three stages of the necking.

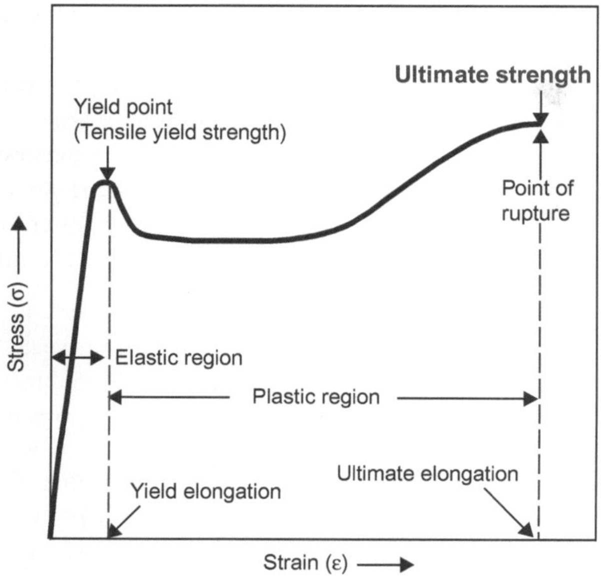

Figure 16.4 Interpretation of the state of stress.

Modulus of resilience: The amount of energy (or work) stored per unit volume at the elastic limit.
Modulus of toughness: The amount of energy stored per unit volume at fracture of the material; this is a measure of the ductility of the material.
Percent Area Reduction: Reduction in area at fracture in necking region with respect to original cross-section area; this is a measure of the ductility of the material.
Strain (engineering): The unit deformation of the material under load. Strain is not normally measured. Deformation is typically measured using extensometers with strain subsequently computed by dividing the measured deformation by the original.
Strain hardening: Portion of the stress–strain curve between the elastic limit and the ultimate stress.
Stress (engineering): Load (force) per unit area; the normal (axial) stress is determined by dividing the load by the original cross-sectional area of the specimen.
Stress-strain curve: An x–y plot of stress vs. strain through the entire range of loading of the specimen until specimen failure.
Ultimate stress: The maximum observed stress that the specimen will withstand.

Yield stress: The stress at which the material begins to 'yield'; for mild steel there is a noticeable increase in deformation with little increase in load. For steel and most metals, a 0.2% offset is used to define the yield stress. A strain value of 0.002 is selected and a line parallel to the elastic portion of the stress–strain curve is constructed. The intersection of this line with the stress–strain curve defines the value of the yield stress.

16.3.3 Dynamic Mechanical Analysis

Dynamic mechanical analysis (DMA) yields information about the mechanical properties of a specimen placed in minor, usually sinusoidal, oscillation as a function of time and temperature by subjecting it to a small, usually sinusoidal, oscillating force. The applied mechanical load, that is, stress, elicits a corresponding strain (deformation) whose amplitude and phase shift can be determined. Fig. 16.5 states that the mode of deformation governs whether the complex modulus is E^*, G^*, K^*, or L^*. The other relationships are shown in Fig. 16.5 for the elastic modulus E, see also ASTM D 4092.

16.4 PRACTICAL ASPECTS OF MECHANICAL AND DYNAMIC MECHANICAL PROPERTIES OF PU BLENDS AND IPNs

16.4.1 Thermoset/Thermoplastic Blend

16.4.1.1 Epoxy/Polyurethane Blend

As per Park et al.'s research [5], the preparation of blend thermoset/thermoplastic, i.e., diglycidylether of bisphenol A (DGEBA) and modified

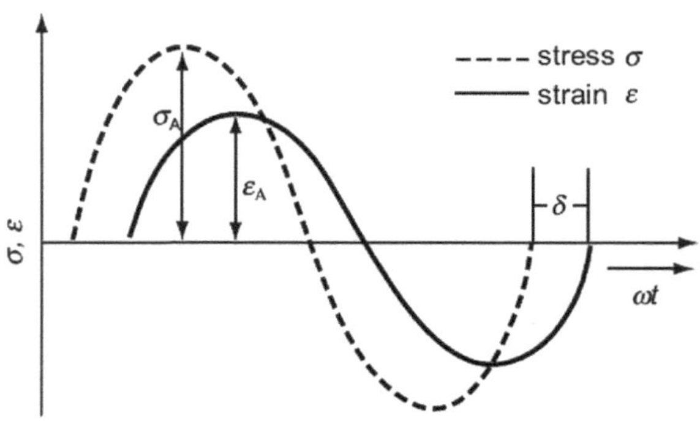

Figure 16.5 Sinusoidal oscillation and response of a linear-viscoelastic material; η phase angle, E = tensile modulus, G = shear modulus, K = bulk compression modulus, L = uniaxial strain modulus.

polyurethane (PU) are characterized with the cure behavior and mechanical interfacial properties and also thermal stability by using TMA and TGA analysis. Epoxy resin is a thermosetting polymer and has several uses. Due to its high modulus and thermal property it is most widely use in fiber composites [7,8].

There is modification in the epoxy system by incorporating the rubber phase into the epoxy network to increase their toughness and it is possible only when micro level rubber particles are used [9,10]. There is some amount of decrease in the modulus and the glass transition temperature (T_g) of the cured epoxy resins by improvement of toughness in most of the elastomer-modified epoxy systems [11,12].

16.4.1.2 PVC/Thermoplastic Polyurethane Blends

Poly(vinyl chloride) (PVC) plays an important role in the plastic industry and is one of the most versatile thermoplastics, but it must be combined with a number of additives before processing [13]. Low-molecular-weight plasticizer is one of the major additives used in PVC compounding. The addition of plasticizers to a PVC formulation decreases many mechanical properties of the PVC product (hardness, tensile strength, modulus); however, low-temperature flexibility, elongation, and the ease of processing are all improved. Polyester-based thermoplastic polyurethane (TPU) is commonly blended with plasticized PVCm resulting in a PVC blend with improved abrasion and fatigue resistance [14–16].

A Twin screw extrusion is one of these. In this way, the rheological properties can be observed and related to the mechanical properties obtained under different conditions of processing [17].

Mechanical and processing properties of blends of poly(vinyl chloride) (PVC) with di-isodecyl phthalate (DIDP) and thermoplastic polyurethane (TPU) were compared with di-2-ethylhexyl phthalate (DOP). The influence of processing conditions on the mechanical properties was studied by changing the content of the additives and using different twin screw speeds. The mechanical properties (tensile strength, modulus, and energy) of PVC/DIDP presented the same behavior as PVC/DOP at all concentrations. The mechanical characteristics of PVC/DOP and PVC/DIDP show that these systems are probably affected by the speed of the twin screw. Polyurethane blends exhibit better structural properties compared to the other plasticizers. The twin screw speed exhibited no influence on the stress–strain property profile of the PVC/TPU blends. The results indicate a minor influence on the decrease of PVC properties when the plasticizer used is a TPU. These facts suggest that the structure of PVC/TPU mixtures could be maintained more efficiently up to 50 phr TPU content. In general terms, the PVC/TPU systems presented higher elastic energy absorption. Therefore, small changes in the TPU content could easily promote the desired properties for selected applications. However, mixtures plasticized with DOP and DIDP showed significant alteration on the product property profile, even at low concentration changes. The results also showed that the stress–strain curves of PVC/TPU blends are not affected by changes in the blend processing conditions. However, it is possible that the toughness data are affected by the screw speed. The use of TPU as a plasticizer can also achieve new horizons considering the large number of possibilities for variation in the polyurethane chain structure.

16.4.1.3 PA(Polyamide)/Thermoplastic Polyurethane Blends

A polyamide (nylon-66, N66) was blended with a series of polyurethanes (PU) based on different polyols or chain extenders [18]. The effects of PU structure and content on the mechanical properties of the blends were investigated. The morphology of the poly blends was examined by scanning electron microscopy. The lzod impact strength of the blends increased with increasing PU content and showed a maximum value at the composition of 15 wt% PU. The effects of moisture content on the mechanical properties of the blends were also investigated. It was found that the water toughened N66 could be further toughened by the addition of PU. A polyester-type PU rather than polyether type PU was recommended for the toughening of N66 because the blends containing the former showed higher tensile strength and Izod impact strength except at very low moisture (0.3 wt%) content. Carboxyl groups were introduced into the molecular chains of PU by using dimethylol propionic acid as chain extender replacing part of 1,4-butanediol. The blends containing PU with carboxyl groups showed a lower Izod impact strength than those without carboxyl groups in the structure of PU.

16.4.1.4 PBT/Thermoplastic Polyurethane Blends

Thermoplastic polyurethane (TPU)/poly (butylene terephthalate) (PBT) blends with different ratios were prepared by extrusion and injection molding. The morphology, dynamic viscoelastic, capillary rheological, thermal, and mechanical properties of the blends w ere studied. Results showed that there was good compatibility between TPU and PBT. The capillary rheological properties showed that the apparent viscosity decreased with the TPU content. DSC analysis [19] indicated that with increasing TPU content the crystallization temperature (T_c), the melting point (T_m) and the percent crystallinity (X_c) decreased. Mechanical properties showed that the addition of TPU could lead to a remarkable increase, about 368.18%, in impact strength with a small reduction in tensile and flexural strength of TPU/PBT blends.

16.4.2 Thermoset/Thermoset Blend

16.4.2.1 Intermolecular Interaction of Silicone Rubber/Polyurethane/Epoxy Blends

The aggregation structure and mechanical properties of liquid silicone rubber, polyurethane (PU), and epoxy (EP) blends were studied [20]. The molecular structure was evaluated by FTIR, and the intermolecular interaction of the three-phase polyblends was measured by dynamic mechanical analysis. The mechanical properties, including the tension, compression, shear, and tear performance, were measured by a material testing system. From the relationship between the intermolecular interaction and mechanical properties, it was found that the aggregation structure of the three-phase blends was influenced by the reaction between silicone and PU, silicone and EP, and PU and EP, which resulted in a change of the crosslinking density and an interpenetrating polymer network structure. Thus, the dominant three-phase mechanical properties are closely related to the aggregation structure (see Fig. 16.6).

According to the analysis of the IR, the absorption peak of the polymer changes with different blending proportions. The chemical structure changes because of the reactions between the silicone, PU, and EP, which influences the crosslinking network of the aggregate structure. The DMA results verify that the three-phase polymer blends of silicone, EP, and PU have a semi-IPN structure. Interactions between silicone/PU, silicone/EP, and PU/EP really exist, which forces the PU and EP peaks to dramatically move toward low temperatures [3]. Due to the reaction between SiOOH and PUONCO in two-phase polymer blends of silicone/PU,

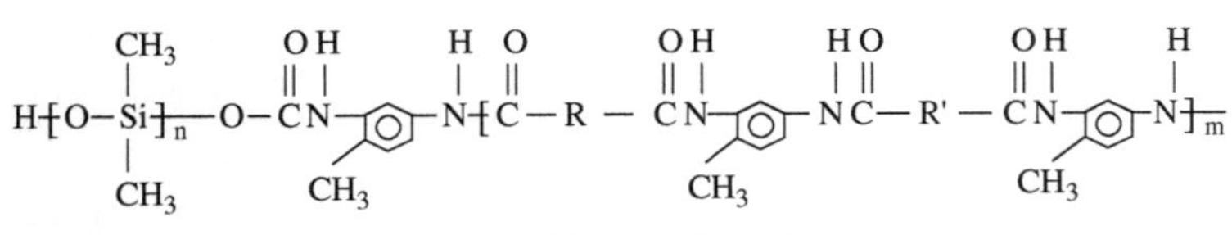

Figure 16.6 The structure of silicon rubber/PU blend.

SiOOH and COOH in two-phase polymer blends of silicone/EP, and PUONCO and COOH in two-phase polymer blends of PU/ EP, the crosslinking distance is enlarged, which significantly decreases the mechanical properties of the three-phase SPE system [4]. In the two-phase polymer blends of silicone/PU, silicone/EP, and PU/EP, the graft-IPN structure makes the mechanical properties of silicone/ PU/EP polymer blends incline toward a compound rule.

Hsien-Tang Chiu et al. studied the aggregation structure and mechanical properties of liquid silicone rubber, polyurethane (PU), and epoxy (EP) blends. From the relationship between the intermolecular interaction and mechanical properties, it was found that the aggregation structure of the three-phase blends was influenced by the reaction between silicone and PU, silicone and EP, and PU and EP, which resulted in a change of the crosslinking density and an interpenetrating polymer network structure. The mechanical properties, including the tension, compression, shear, and tear performance, were measured by a material testing system. Silicone rubber has been widely applied to various industrial products because of the characteristics of silicon oxide, including thermal stability, high mechanical and chemical resistance, weatherproof, ozone and radiation resistance, and so forth [21,22]. The molecular structure of PU is composed of urethane compounds (ONHCOO). Generally speaking, the block copolymer PU comprises rigid diisocyanate, short-chain diol or diamine, and yield polyol. Because it is resistant to chemicals, soft, waterproof, and weatherproof, it can be used for different purposes [23].

Epoxy (EP), a widely applied thermosetting macromolecular material, is extensively used by industry because of its superb mechanical properties, heat and melting resistance, and dimensional stability [24–27]. Because silicone and PU have a flexible network structure and EP has a hard one, the aggregate structure of the mixture can be adjusted effectively through an easy processing approach. Consequently, this study makes use of PU features like flexibility and adhesion properties and EP mechanical properties and rigidity. We use silicone rubber as the main body and mix PU and EP in different proportions to explore the mechanical properties and aggregate structure of a mixed rubber consisting of silicone, PU, and EP (SPE).

The DMA results verify that the three-phase polymer blends of silicone, EP, and PU have a semi-IPN structure. Interactions between silicone/PU, silicone/EP, and PU/EP really exist, which forces the PU and EP peaks to dramatically move toward low temperatures. In the two-phase polymer blends of silicone/PU, silicone/EP, and PU/EP, the graft-IPN structure makes the mechanical properties of silicone/ PU/EP polymer blends incline toward a compound rule.

16.4.3 IPN Blends

16.4.3.1 Surface Modified Graphene in PU (SMG/PU Elastomer)

There was double improvement in the thermal stability and storage modulus of the SMG/PU blend by incorporating 5 wt% of SMG in the matrix [28]. This property adversely depends on the percentage of incorporation SMG, i.e., highly exfoliated graphene layers in the matrix, their size, shape, and the number of graphene layers. The mobility of flexible chains decreases by incorporation of SMG in PU matrix and eventually increases the thermal stability of polymer matrix [29–31].

Darko Ljubic et al. [28] reported easiest way for preparation of PU/ surface modified (SMG) elastomers by using synthesis of soft, single-phase and noncrystalline PU elastomers which is filled with various parameters of SMG (surface modified graphene), and by comparing it with other researcher studies, where TPU and different thermoplastics polymers were studied [32].

16.4.3.2 Fe-octacarboxyl Acid Phthalocyanine/PU Blend

In this paper [33], Fe-octacarboxyl acid phthalocyanine (Fe-OCAP)/ polyurethane (PU) blends were prepared by solution blending. The mechanical and dynamic mechanical properties of the samples were evaluated by tensile tests and dynamic mechanical analysis. The results showed that incorporation of Fe-OCAP significantly improved the tensile strength, elongation at break, and thermal deformation property of PU matrix. The interaction between Fe-OCAP and PU was investigated by infrared analysis. The result indicated that a strong hydrogen bond was formed between Fe-OCAP and hard segments of PU and increased with increasing Fe-OCAP content. Quantitive evaluations of the interaction between Fe-OCAP and PU were conducted by analyzing the physical cross-linking density in samples. The results indicated that the physical crosslinking density of the samples was increased with increasing Fe-OCAP content.

The tensile strength, elongation at break and thermal deformation property of the samples was improved by addition of Fe-OCAP. The hydrogen bond formed between Fe-OCAP and the PU matrix increased the microphase separation degree of PU. Quantitive evaluations of the interaction between Fe-OCAP and PU suggested that the higher physical cross-linking density in the sample was obtained with higher Fe-OCAP content.

16.4.3.3 Block Copolymers/PU blends

Polyurethane grafted onto styrene-styrene sulfonic acid copolymer by stepwise grafting onto PSSSA (sulfonated polystyrene) chains of MDEA (N-methyldiethanoloamine) followed by chain growth on the hydroxyl sites of a wide range of polyurethane structure [34]. Dynamic mechanical tests show the occurrence of a single glass transition temperature for the grafts consisting only of hard segments. If long soft segments are added to the grafts, then these phases separate and yields a new glass transition temperature. The values of the high and low T_g values depend on the length of the soft segments and on the sulfonate ion concentration within the styrene copolymer (see Table 16.1).

16.4.3.4 PLLA [Poly(L-lactide)]/PU Blends

To date, there has been considerable research devoted to improving the impact toughness of PLLA, such as graft and block copolymerization as well as physical blending with a second polymer [[35−45]. In particular, the method of physical blending can be envisaged as a relatively economical and practical manner to modify the physico-mechanical properties of polymers. It has been reported that a variety of flexible polymers can be used to toughen PLLA, including polyethylene (PE), acrylonitrile-butadiene-styrene copolymer (ABS), and poly(ethylene oxide) (PEO). In most cases, the second polymer within the blends is incompatible with PLLA matrix, which would induce the macroscopic phase separation,

Table 16.1 Mechanical properties of PS/PU Blend without and with compatibilizer

Mechanical property	PS/PU	PS/PU/C
	90/10	89.8/10/0.2
Hardness HR (N mm^{-2})	64.35	62.8
Tensile strength (MPa)	24.4	26.1
Elongation (%)	1.71	1.71
Elastic modulus (MPa)	1649	1725
Flexural Stress H = 3.5% (MPa)	29.4	31.9
Flexural Modulus (MPa)	1477	1573
Charpy Notched Impact Strength 1 J (kJ m^{-2})	5.6	6.0
Izod Impact Strength 4 J (kJ m^{-2})	3.9	4.3

resulting in poor toughening effect. At the same time, the rigidity of PLLA blends would decrease generally during the modification of impact toughness. Thermoplastic polyurethane (PU), as an important elastomer, has great potential application in diverse areas such as automobile and medical devices.

The main chain of PU is generally composed of thermodynamically incompatible hard and soft segments. The soft segments are usually polyether or polyester with a high flexibility at room temperature, whereas the hard segments consist of diisocyanate and a low molecular-weight diol (chain extender), which particularly provides the high modulus, hardness and tear strength. There are many polar groups along the backbones of PU, including —NH and —CO groups, which are assumed to have interaction with the —CO groups within PLLA chains and thus enhances the compatibility between PLLA and PU. Therefore, PU has been adopted to mediate the impact resistance as well as other physical properties of PLLA [46—49]. In addition, the intermolecular interaction of PU molecules can be altered through adjusting the hard segment ratio [50]. This may produce great effect on the interaction between PU and PLLA. It arouses our great interest what would happen when PU with different hard segment ratios is added into PLLA? Therefore, in present study, polyester-based PU with four kinds of hard segment ratios was selected to blend with PLLA so as to obtain PLLA/PU blends with increased impact resistance and maintaining biodegradability. The specific objective was to illustrate the influence of the content and hard segment ratio of PU on the phase morphology, crystallization behavior and mechanical properties of PLLA.

Xing et al. studied polymer blends, looking at the domain size and size distribution of dispersed phases as well as the interfacial adhesion play important roles in determining the actual mechanical properties. It is essential to fully understand the phase morphology of PLLA/PU blends and its correlation with mechanical properties. The dynamic viscoelastic curves of tan δ versus temperature of pristine polymers and PLLA/PU blends are shown in Fig. 16.7.

Each pristine polymer shows one obvious glass transition peak, which appears around 60 °C and 30 °C for neat PLLA and PU1, respectively. All the PLLA/PU blends exhibit two distinguished glass transition domains, indicating the immiscibility of PLLA and PU. For PLLA/PU1 blends, the glass transition temperature (T_g) of neat PLLA remains almost constant with the content of PU1, while that of PU1 shifts to lower temperatures.

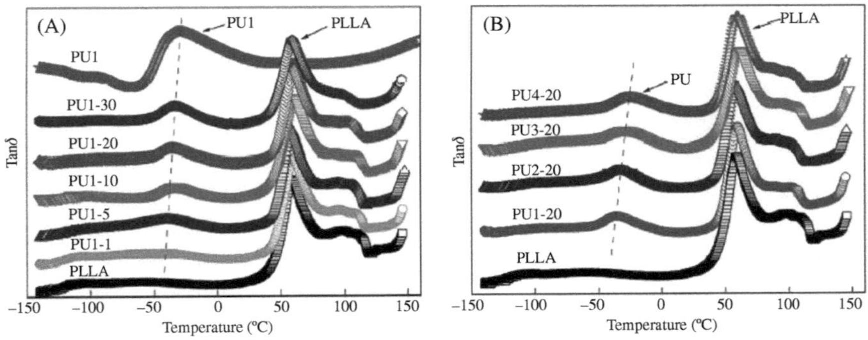

Figure 16.7 Dynamic viscoelastic curves of Tan δ versus temperature for PLLA/PU1 blends (A) and PLLA/PU (80/20) blends (B).

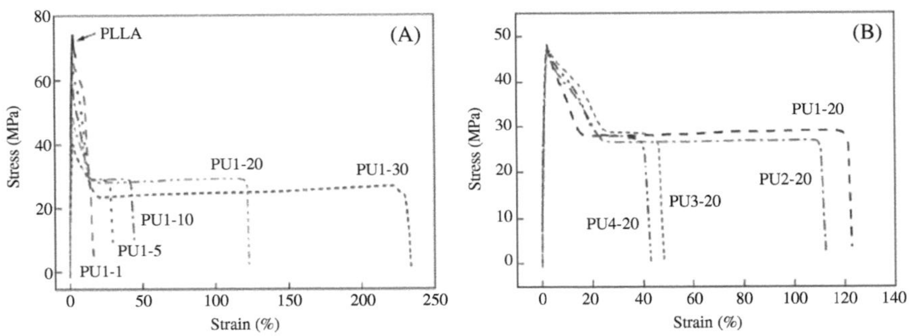

Figure 16.8 Stress–strain curves for PLLA/PU1 blends (A) and PLLA/PU (80/20) blends (B).

16.4.4 Mechanical Properties

The stress–strain curves and Izod impact strength of neat PLLA and PLLA/PU blends are shown in Figs. 16.8 and 16.9. Neat PLLA exhibits the best rigidity with a tensile stress at yield beyond 70 MPa, but the impact strength is only 2.1 kJ m^{-2}, resulting in a brittle fracture with the elongation at break below 5%. The addition of PU can effectively enhance the toughness of PLLA.

In a study by Xing et al. the authors concluded that the effect of blending composition and hard segment ratio of PU on the phase morphology, crystallization behavior and mechanical properties of PLLA should be systematically evaluated. It has been found that the interchain interaction between PLLA and PU was beneficial to the homogeneous distribution of PU, and promoted the mobility of PU molecules with the decrease of glass transition temperature. Additionally, the glass transition

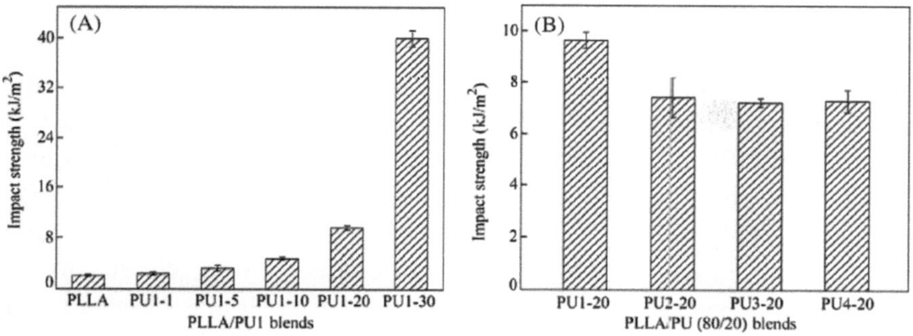

Figure 16.9 Impact strength for PLLA/PU1 blends (A) and PLLA/PU (80/20) blends (B).

temperature of PU increased slightly with the increase of its concentration and hard segment ratio.

16.4.4.1 PC/TPU Blend (Polycarbonate/Thermoplastic PU)

The present study [51] was taken up to investigate the effect of addition of Montmorillonite (MMT) to polycarbonate/thermoplastic polyurethane (PC/TPU) blends. Polycarbonates (PC) known by the trademarked names of Lexan, Makrolon, Makroclear and others, are a particular group of thermoplastic polymers. PC is a tough, transparent, impact-resistant thermoplastic engineering polymer with a very wide range of service temperatures. One of the biggest advantages of polycarbonate is its impact strength. TPU are high-tech materials with a unique combination of very useful properties. They have high abrasion resistance, flexibility and shock absorbing capability. It has certain disadvantages, including high cost, moderate thermal stability and mechanical strength [52]. MMTs are nanoparticles of layered mineral silicates. They consist of stacked, layered silicates about 1 nm thickness including two tetrahedral sheets sandwiching an edge-shared octahedral sheet of either aluminum or magnesium hydroxide [53]. The literature study revealed that the polycarbonate nanocomposites showed rather good dispersion of MMT with a mixture of exfoliated, intercalated and confined morphology. The effect of MMT on the mechanical response of specimens is subjected to tensile and impact loading was investigated results in increased Young's modulus and yield strength. Impact strength decreases significantly as the MMT content increases from 1% to 5% and thermal analysis demonstrated that addition of MMT leads to decreased thermal stability of polycarbonate [54]. From the literature survey it is concluded that MMT is added to either PC or TPU. However no studies on effect of MMT on properties of PC/TPU blends are reported in the literature. In view of this, the study was taken up to investigate the effect by addition of MMT 5 parts per hundred parts of blend. The TPU content in PC/TPU blend was varied from 10% to 30%. PC/TPU blends of similar composition without MMT were also prepared and their mechanical, thermal and dynamic mechanical properties were compared with PC/TPU/MMT compositions. Incorporation of MMT induces brittleness within the PC matrix whereas incorporation of TPU improves flexibility. It was observed that there is an improvement in elongation with an increase in TPU content in case of PC/TPU blends. From the results, it was observed that there is an improvement in tensile properties of PC/TPU blends reinforced with MMT compare to non reinforced blends. The addition of rubbery TPU to PC matrix decreases the flexural properties. This can be attributed to the involvement of the elastomeric nature of the TPU molecular structure. The addition of TPU to PC results a gradual decrease in hardness (PC- 62 Shore D, 70/30—40 Shore D). This is due to the presence of soft elastomeric segments in TPU. It was observed that there is an improvement in hardness values by about 4–16% for PC/TPU blends reinforced with MMT compare to nonreinforced blends. Incorporation of organically modified clay results in improvements in mechanical properties. The primary cause for such improvements is due to the presence of immobilized or partially mobilized polymer phases as a consequence of interaction of polymer chains with organically modified clays and large number of interacting molecules due to the dispersed phase volume ratio characteristics of largely exfoliated nanocomposites [55]. There is a slight increase in abrasion resistance of the PC/TPU blends containing MMT. Similar behavior was observed in surface-modified silica nano particle–reinforced poly (ethylene 2,6-naphthalate) system [56].

REFERENCES

[1] David I. Bower, An Introduction to Polymer Physics, Cambridge University Press, Cambridge, 2002, pp. 162–185.

[2] P. Kevin, Menard, Dynamic Mechanical Analysis, A Practical Introduction, CRC Press, Boca Raton, 1999.

[3] T.A. Luckenbach, DMTA: Dynamic Mechanical Thermal Analysis. Ph.D. Rheometrics, Inc. Piscataway, NJ, 08854, USA.

[4] A.K. Bhowmick, Materials science and engineering, in: Mechanical Properties of Polymers, Encyclopedia of Life Support Systems (EOLSS), 2009, pp. 157–220.

[5] S.J. Park, S.J. Seok, B.G. Min, Thermal and mechanical properties of epoxy/polyurethane blend system initiated by cationic latent thermal catalyst, Solid State Phenom. 119 (2007) 215–218.
[6] J.R. Davis, Tensile Testing, second ed, ASM International, Materials Park, OH, 2004, pp. 1–260.
[7] C.A. May, Epoxy Resins, Marcel Dekker, New York, 1998.
[8] H. Lee, K. Nevile, Handbooks of Epoxy Resins, McGraw-Hill, New York, 1986.
[9] J.Y. Qian, R.A. Pearso, V. Shaffer, M.S. El-Aasser, Polymer Vol. 38 (1997) 21.
[10] Y. Yanaka, Epoxy Resins Chemistry and Technology, Marcel Dekker, New York, 1998.
[11] S.J. Park, M.K. Seo, J.R. Lee, Isothermal cure kinetics of epoxy/phenol-novolac resin blend system initiated by cationic latent thermal catalyst, J. Polym. Sci. 38 (2000) 2945–2956.
[12] J.H. Saunders, K.C. Frisch, Polyurethane Chemistry and Technology Chemistry, Wiley, NewYork, 1962.
[13] V.J.R.R. Pita, E.E.M. Sampaio, E.E.C. Monteiro. Mechanical properties evaluation of PVC/plasticizers and PVC/thermoplastic polyurethane blends from extrusion processing.
[14] Y.V. Ovchinnikov, B.Y. Tetel'Baum, A.J. Maklakov, Effects of aging and temperature on changes in the states of poly(vinylchloride)–plasticizer systems, Vysokomol. Soedin. Ser. A 13 (11) (1971) 2422–2428.
[15] J. Piglowski, W. Laskawski, Compatibility of poly(vinyl chloride)/polyurethane blends, Angew. Makromol. Chem 82 (1979) 157–170.
[16] J. Piglowski, T. Skowronski, B. Masiulanis, Properties of poly(vinyl chloride) modified with thermoplastic polyurethane elastomers, Angew. Makromol. Chem. 85 (1980) 129–136.
[17] C.S. Ha, Y. Kim, W. Lee, W. Cho, Y. Kim, Fracture toughness and properties of plasticized PVC and thermoplastic polyurethane blends, Polymer 39 (20) (1998) 4765–4772.
[18] W.B. Wu, K.H. Hsieh, W.Y. Chiu, Effect of polyurethane structure on mechanical properties of nylon-66/polyurethane blends, J. Polym. Res. 1 (4) (1994) 337–344.
[19] J. Huang, X. Lu, G. Zhang, J. Qu, Study on the rheological, thermal and mechanical properties of thermoplastic polyurethane/poly (butylene terephthalate) blends, Polym. Test. 36 (2014) 69–74.
[20] H.-T. Chiu, S.-H. Chiu, J.-H. Wu, Study on mechanical properties and intermolecularinteraction of silicone rubber/polyurethane/epoxy blends, J. Appl. Polym. Sci. 89 (2003) 959–970.
[21] C. Eaborn, Organosilicon Compounds, Butterworths, London, 1962, p. 100.
[22] W. Noll, Chemistry and Technology of Siloxanes, Academic, New York, 1968, p. 17.
[23] A.K. Sha'aban, N. Patel, I. Yilgor, J.S. Riffle, D.W. Dwight, J.E. McGrath, Polym Prepr 24 (1983) 130.
[24] J.S. Shim, W. Lee, J. Jang, Cure kinetics of high performance epoxy resin systems, J. Polym. Bull. 25 (1991) 661.
[25] J.S. Shim, W. Lee, J. Jang, DSC cure studies of high performance epoxy resins I. The reaction with 4,4′-diaminodiphenyl sulfone (DDS), J. Polym. J. 23 (1991) 903.
[26] J.S. Shim, W. Lee, J. Jang, Cure kinetics of high performance epoxy resin systems, J. Polym. Bull. 25 (1991) 669.
[27] N.B. Samarth, P.A. Mahanwar, Modified vegetable oil based additives as a future polymeric material: a recent trend, Open J. Org. Polym. Mater. 5 (2015) 1–22.
[28] D. Ljubic, M. Srinivasan, R. Szoszkiewicz, I. Javni, Z.S. Petrovic, Surface modified graphene/single-phase polyurethane elastomers with improved thermo-mechanical and dielectric properties, Eur. Polym. J. 70 (2015) 55–65.

[29] H. Kim, A.A. Abdala, C.W. Macosko, Graphene/polymer nanocomposites, Macromolecules 43 (2010) 6515–6530.
[30] R. Verdejo, M.M. Bernal, L.J. Romasanta, M.A. Lopez-Manchado, Graphene filled polymer nanocomposites, J. Mater. Chem. 21 (2011) 3301.
[31] J.R. Potts, D.R. Dreyer, C.W. Bielawski, R.S. Ruoff, Graphene-based polymer nanocomposites, Polymer 52 (2011) 5–25.
[32] K. Appel, R. Thomann, R. Mülhaupt, Polyurethane nanocomposites prepared from solvent-free stable dispersions of functionalized grapheme nanosheets in polyols, Polymer 53 (2012) 4931–4939.
[33] Y. Ti, D. Chen, Mechanical and dynamic mechanical properties of polyurethane/Fe-octacarboxyl acid phthalocyanine blends., Progr. Org. Coat. 76 (2013) 119–124.
[34] F.B. Calleja, Z. Roslaniec (Eds). Block Copolymers. Marcel Dekker, Inc.2000, pp. 215–250.
[35] Q. Xing, R. Li, X. Dong, X. Zhang, L. Zhang, D. Wang, Phase morphology, crystallization behavior and mechanical properties of poly(L-lactide) toughened with biodegradable polyurethane: effect of composition and hard segment ratio, Polym Sci. 33 (2015) 1294–1304.
[36] Y. Sun, C. He, Biodegradable "core–shell" rubber nanoparticles and their toughening of poly(lactides), Macromolecules 46 (24) (2013) 9625–9633.
[37] M.T. Martello, M.A. Hillmyer, Polylactide–poly(6-methyl-ε-caprolactone)–polylactide thermoplastic elastomers, Macromolecules 44 (21) (2011) 8537–8545.
[38] C.L. Wanamaker, M.J. Bluemle, L.M. Pitet, L.E. O'Leary, W.B. Tolman, M.A. Hillmyer, Biomacromolecules 10 (10) (2009) 2904.
[39] A.P.T. Pezzin, G. van Ekenstein, C.A.C. Zavaglia, G. ten Brinke, E.A.R. Deuk, J. Appl. Polym. Sci. 88 (12) (2003) 2744.
[40] K.A. Afrifah, L.M. Matuana, Impact modification of polylactide with a biodegradable ethylene/acrylate copolymer, Macromol. Mater. Eng 295 (9) (2010) 802.
[41] K.S. Anderson, S.H. Lim, M.A. Hillmyer, Toughening of polylactide by melt blending with linear low-density polyethylene, J. Appl. Polym. Sci. 89 (14) (2003) 3757.
[42] Y. Li, H. Shimizu, Improvement in toughness of poly(l-lactide) (PLLA) through reactive blending with acrylonitrile–butadiene–styrene copolymer (ABS): morphology and properties, Eur. Polym. J. 45 (3) (2009) 738.
[43] A.N. Gaikwad, E.R. Wood, T. Ngai, T.P. Lodge, Two calorimetric glass transitions in miscible blends containing poly(ethylene oxide), Macromolecules 41 (7) (2008) 2502.
[44] T. Semba, K. Kitagawa, U.S. Ishiaku, H. Hamada, The effect of crosslinking on the mechanical properties of polylactic acid/polycaprolactone blends, J. Appl. Polym. Sci. 101 (3) (2006) 1816.
[45] R.Y. Wang, S.F. Wang, Y. Zhang, C.Y. Wan, P.M. Ma, Toughening modification of PLLA/PBS blends via in situ compatibilization, Polym. Eng. Sci 49 (1) (2009) 26.
[46] Q. Xing, X. Dong, R.B. Li, H.J. Yang, C.C. Han, D.J. Wang, Morphology and performance control of PLLA-based porous membranes by phase separation, Polymer 54 (21) (2013) 5965.
[47] Y. Li, H. Shimizu, Toughening of polylactide by melt blending with a biodegradable poly(ether)urethane elastomer, Macromol. Biosci 7 (7) (2007) 921.
[48] F. Feng, L. Ye, Morphologies and mechanical properties of polylactide/thermoplastic polyurethane elastomer blends, J. Appl. Polym. Sci. 119 (5) (2011) 2778.
[49] F. Feng, X. Zhao, L. Ye, Structure and properties of ultradrawn polylactide/thermoplastic polyurethane elastomer blends, J. Macromol. Sci. B 50 (8) (2011) 1500.
[50] R.L. McKiernan, A.M. Heintz, S.L. Hsu, E.D.T. Atkins, J. Penelle, S.P. Gido, Macromolecules 35 (10) (2002) 6970.

[51] G.M. Shashidhara, S.H. Kameshwari Devi, D.K. Shreyas, Effect of thermoplastic polyurethane content on properties of PC/TPU blend filled with montmorillonite, Carbon Sci. Technol. 5 (1) (2013) 218–224.
[52] M. Kannan, S.S. Bhagawan, T. Jose, S. Thomas, K. Joseph, . Preparation and characterization of MMT-filled polyurethane/polypropylene blends, Polym. Eng. Sci. 50 (2010) 1878–1886.
[53] Nevalainen Katija, Vasa Villman, Pentti Jarvela, Janne Sundelin, Toivolepistio, Eng. Sci. 49 (2009) 631–640.
[54] A.S. Luyt, M. Missori, P. Fabbri, J.P. Mofokenge, R. Tauurino, T. Zanasi, et al., Polycarbonate reinforced with silica nanoparticles, Polym. Bull. 66 (2011) 991–1004.
[55] Jayavani, Smita Mohanty, Sanjay Nayak, M. Rahaili Parvaiz, Influence of MMTs and Nano-TiO_2 on the mechanical and thermal properties of polycarbonate nanocomposite, Macromol. Res. 19 (2011) 563–572.
[56] Seon Hoon Ahn, Seong Hun Kim, Seung Goo Lee, Surface modified silica nanoparticle–reinforced poly(ethylene 2,6- naphthalate), J. Appl. Polym. Sci. 94 (2004) 812–818.

INDEX

Note: Page numbers followed by "*f*" and "*t*" refer to figures and tables, respectively.

D

E

H

I

O

P

Q

R

S

T

U

FSC
www.fsc.org
MIX
Papier | Fördert
gute Waldnutzung
FSC® C083411

Kolibri 360 GmbH
Ferdinand-Jühlke-Straße 7
99095 Erfurt, Deutschland
produktsicherheit@kolibri360.de